History-making paper model of the then hypothetical C_{60} structure prepared by Professor Richard Smalley in the kitchen of his home on the night of discovery, 1985.

Buckminsterfullerenes

Buckminsterfullerenes

Edited By

W. Edward Billups

and

Marco A. Ciufolini

W. Edward Billups
Department of Chemistry
Rice University
Houston, TX 77251-1892

Marco A. Ciufolini
Department of Chemistry
Rice University
Houston, TX 77251-1892

Library of Congress Cataloging-in-Publication Data

Buckminsterfullerenes/W. Edward Billups
  & Marco A. Ciufolini, editors.
     p. cm.
  Includes index.
  ISBN 1-56081-608-2
    1. Buckminsterfullerene.   2. Fullerenes.   I. Billups, W.
Edward.
  II. Ciufolini, Marco A.
  QD181.C1B83   1993
  546'.681—dc20   93-3028
                                      CIP

Printed in the United States of America
ISBN 1-56081-608-2
ISBN 3-527-89608-2

Printing History:
10 9 8 7 6 5 4 3 2 1

Published jointly by

VCH Publishers, Inc.
220 East 23rd Street
New York, New York 10010-4606

VCH Verlagsgesellschaft mbH
P.O. Box 10 11 61
D-6940 Weinheim
Federal Republic of Germany

VCH Publishers (UK) Ltd.
8 Wellington Court
Cambridge CB1 1HZ
United Kingdom

Foreword

Buckminsterfullerene is a name I never completely liked. It takes too long to write, and way too long to pronounce. Still it has, I feel, an interestingly clumsy sort of beauty. In choosing it for the name of this book, my colleagues Ed Billups and Marco Ciufolini have made another major step along the road of ensuring that Buckminsterfullerene and the far snappier term "buckyballs" will be remembered for many decades to come. So it is, perhaps, appropriate in this foreword to begin with a description of the origins of these new additions to otherwise sober scientific jargon.

The name was born in the dimmest early thinking of how a pure carbon cluster of 60 atoms could eliminate its dangling bonds, well before we knew it was a soccer ball. Harry Kroto and I had been kidding around in long conversations with each other on all sorts of possibilities. Of all these the "closed solution"—the one where the graphite sheet had somehow curled around and closed on itself—seemed by far the least likely. Yet there was no getting around the increasingly clear results that were coming out of the cluster beam experiments.

My students, Jim Heath and Sean O'Brien, had found that carbon aggregates together in an inert atmosphere in a way that makes C_{60}, and to a lesser extent, C_{70}, very specially abundant—a "magic number" in the cluster distribution. Together with my colleagues, Bob Curl and Frank Tittel, and two other students, Qing-Ling Zhang and Yuan Liu, we had been involved for several years in similar studies of semiconductor clusters such as silicon, germanium, and gallium arsenide. The prime motive of these studies was to find out how elements like silicon contrive to minimize their "dangling" bonds on the surface of a small, nanoscopic bare cluster composed of only 10–100 atoms. We had found that some silicon clusters adopt particularly stable structures, but they never become so inert that they will not readily react with yet another silicon atom. But here, in the case of carbon, the specific cluster, C_{60}, was clearly behaving as if it had absolutely no dangling bonds whatever. While other carbon clusters continued to grow to ever larger sizes in the condensing carbon vapor, C_{60} remained. Ultimately, it was the lone survivor. Somehow it had to have arranged to a geometrical form that eliminated even the last vestige of a dangling bond.

In an effort to make clearer what the "closed solution" involved, I asked Harry if he remembered who was the architect who worked with big domes. Didn't the structure of these domes look something like a curved lattice of hexagons? He said it was Buckminster Fuller. Within a few moments we drew a ball on the blackboard and shouted, with rather Monty Pythonesque humor, "IT'S BUCKMINSTER FULLER. . .ENE!," paying fastidious attention, like all good chemists, to attaching the appropriate suffix. We joked around with it for a while, and then promptly dismissed it. It was just too fantastic to be real. There had to be some other, less sensational, explanation. But there wasn't.

Within a week we had found by using not only hexagons but pentagons as well, one could beautifully explain why C_{60} would be unique. Not only did this structure tie up all dangling bonds, but it did so with perfect icosahedral symmetry, with every carbon identical to every other. The strain of curvature of the graphite sheet was therefore smoothly distributed in the most symmetrical possible way. Here was the roundest of all possible round molecules, a molecular soccerball. The pattern of pentagons and hexagons turned out to be exactly that patented by Fuller in his original disclosure of the geodesic dome.

When we quickly wrote up our paper for *Nature,* we typed the title "C_{60}: Buckminsterfullerene," initially just daring each other actually to submit a serious paper to this illustrious journal with such a frivolous title. But we found there was no other name that worked quite so well. Certainly it would serve as a hook to catch readers, and what else could "Buckminsterfullerene" possibly mean other than an aromatic molecule in the form of a geodesic dome? So we gathered our courage and sent the paper in, after adding a paragraph at the end somewhat apologizing for the title and leaving the ultimate name for the molecule to be settled by consensus. Within a few days my students and I had shortened the name to "bucky balls." We felt somewhat scandalized a few months later when this piece of irreverent lab jargon leaked out in phone conversations, and emerged (inevitably) in the open press.

Both the names and the molecule have long since attained a life of their own, and there is no calling them back. In the subsequent 7 years since our paper appeared in *Nature* in 1985, over 1300 papers have appeared in the scientific literature on this subject. The worldwide rate of submission of new manuscripts specifically devoted to bucky and the other fullerenes is roughly one every 13 hours, and continues to increase. Under such circumstances any book such as this is virtually assured of being out of date even as it hits the presses. The authors, however, have done an excellent job of summarizing the current state of knowledge of this new class of nanoscale materials, and have captured much of the sense of excitement that still fills the hearts of "fullereners" worldwide.

R. E. Smalley
Houston
November 30, 1992

Preface

Few areas of chemical research have attracted the attention of the scientific community as have the fullerenes. Buckminsterfullerene itself has been named the molecule of the year by *Science* (December 20, 1991). The array of new chemistry associated with these structurally pleasing molecules includes functionalization both exohedrally and endohedrally. In addition, the fullerenes have been used as radical sponges, and potentially useful electrical and magnetic properties have already been discovered.

This monograph brings together the work of leading investigators who have provided an account of their own work in this rapidly developing field. Under the circumstances, the task of this monograph is designed to provide the reader with as much of an up-to-date account of this emerging field as possible. We had originally considered including a chapter on the newly discovered stretched fullerenes (bucky tubes), but unfortunate constraints of various nature did not allow this to materialize.

This book is organized loosely into four sections. The first one recounts the main events leading to the discovery of the fullerenes as well as several preliminary investigations. A second section on theoretical studies of fulleroid cages is followed by a third part where the superconductivity properties of the new clusters are explored. This is followed by a discussion of the exo- and endohedral complexes in the gas phase. The book concludes with a discussion of some of the chemical properties of these new materials.

We wish to express our gratitude to all who have contributed their efforts to the preparation of this treatise. As in all works involving multiple authorship, a great deal of the success of the project depends upon the cooperation of the

ix

authors as well as the publisher's staff. This treatise is no exception. The staff of VCH provided assistance in many different ways. We are especially thankful to the authors for their timely submission of manuscripts and for updating them during the past few weeks. Special recognition goes to Professor Helmut Schwarz, who not only submitted his chapter well ahead of the deadline but also updated it as the target date approached. Helmut, we appreciate your forebearance.

W. E. Billups
M. A. Ciufolini
Houston, Texas
November 1992

Contents

Contributors

Chapter 1. Mass Spectrometric Studies of the Fullerenes   1
R. F. Curl

Chapter 2. Fullerene Studies at Sussex   21
H. W. Kroto, K. Prassides, A. J. Stace, R. Taylor, and D. R. M. Walton

Chapter 3. The Higher Fullerenes   59
Carlo Thilgen, François Diederich, and Robert L. Whetten

Chapter 4. Fullerene Structures   83
T. G. Schmalz and D. J. Klein

Chapter 5. Ab Initio Theoretical Predictions of Fullerenes   103
Gustavo E. Scuseria

Chapter 6. Predicting Properties of Fullerenes and their Derivatives   125
C. T. White, J. W. Mintmire, R. C. Mowrey, D. W. Brenner, D. H. Robertson, J. A. Harrison, and B. I. Dunlap

Chapter 7. Electronic Structure of the Fullerenes:
            Carbon Allotropes of Intermediate
            Hybridization   185
            R. C. Haddon and K. Raghavachari

Chapter 8. Theory of Electronic and Superconducting
            Properties of Fullerenes   197
            Marvin L. Cohen and Vincent H. Crespi

Chapter 9. Electronic Structure of the Alkali-
            Intercalated Fullerides, Endohedral
            Fullerenes, and Metal-Adsorbed Fullerenes
            217
            Steven C. Erwin

Chapter 10. Exo- and Endohedral Fullerene Complexes
            in the Gas Phase   257
            *Helmut Schwarz, Thomas Weiske, Diethard K.
            Böhme, and Jan Hrušák*

Chapter 11. Fullerene Electrochemistry: Detection,
            Generation, and Study of Fulleronium and
            Fulleride Ions in Solution   285
            *Lon J. Wilson, Scott Flanagan, L. P. F. Chibante,
            and J. M. Alford*

Chapter 12. Improved Preparation, Chemical Reactivity,
            and Functionalization of C_{60} and C_{70}
            Fullerenes   301
            *G. K. Surya Prakash, Imre Bucsi, Robert
            Aniszfeld, and George A. Olah*

Chapter 13. Chemistry of Fullerenes   317
            Fred Wudl

Index   335

Contributors

J. M. ALFORD. Department of Chemistry, Rice University, Houston, TX 77251-1892

ROBERT ANISZFELD. Department of Chemistry, University of Southern California, University Park, Los Angeles, CA 90089-1661

DIETHARD K. BÖHME. Department of Chemistry, York University, North York, Ontario, Canada M3J 1P3

D. W. BRENNER. Theoretical Chemistry Section, Naval Research Laboratory, Washington, DC 20375-5000

IMRE BUCSI. Department of Chemistry, University of Southern California, University Park, Los Angeles, CA 90089-1661

L. P. F. CHIBANTE. Department of Chemistry, Rice University, Houston, TX 77251-1892

MARVIN L. COHEN. Department of Physics, University of California, Berkeley, Berkeley, CA 94720

VINCENT H. CRESPI. Department of Physics, University of California, Berkeley, Berkeley, CA 94720

ROBERT F. CURL.   Department of Chemistry, Rice University, Houston, TX 77251-1892

FRANÇOIS DIEDERICH.   Laboratorium für Organische Chemie, ETH-Zentrum, Universitätsstrasse 16, CH-8092 Zürich, Switzerland

BRETT I. DUNLAP.   Theoretical Chemistry Section, Naval Research Laboratory, Washington, DC 20375-5000

STEVEN C. ERWIN.   Department of Physics, University of Pennsylvania, Philadelphia, PA 19104-6272

SCOTT FLANAGAN.   Department of Chemistry, Rice University, Houston, TX 77251-1892

ROBERT C. HADDON.   AT&T Bell Labs, Murray Hill, NJ 07974-0636

J. A. HARRISON.   Theoretical Chemistry Section, Naval Research Laboratory, Washington, DC 20375-5000

JAN HRUŠÁK.   Institute of Macromolecular Chemistry, Czechoslovak Academy of Sciences, Heyrovsky Square 2, CS-16202 Prague, Czechoslovakia

DOUGLAS J. KLEIN.   Department of Marine Sciences, Texas A&M University at Galveston, Galveston, TX 77553-1675

HAROLD W. KROTO.   School of Chemistry & Molecular Sciences, University of Sussex, Brighton BN1 9QJ, UK

J. W. MINTMIRE.   Theoretical Chemistry Section, Naval Research Laboratory, Washington, DC 20375-5000

R. C. MOWREY.   Theoretical Chemistry Section, Naval Research Laboratory, Washington, DC 20375-5000

GEORGE A. OLAH.   Department of Chemistry, University of Southern California, University Park, Los Angeles, CA 90089-1661

G. K. SURYA PRAKASH.   Department of Chemistry, University of Southern California, University Park, Los Angeles, CA 90089-1661

K. PRASSIDES.   School of Chemistry & Molecular Sciences, University of Sussex, Brighton BN1 9QJ, UK

K. RAGHAVACHARI.   AT&T Bell Labs, Murray Hill, NJ 07974-0636

D. H. ROBERTSON. Theoretical Chemistry Section, Naval Research Laboratory, Washington, DC 20375-5000

THOMAS G. SCHMALZ. Department of Marine Sciences, Texas A&M University at Galveston, Galveston, TX 77553-1675

HELMUT SCHWARZ. Institut für Organische Chemie, Technische Universität Berlin, W-1000 Berlin 12, Germany

GUSTAVO E. SCUSERIA. Department of Chemistry, Rice University, Houston, TX 77251-1892

A. J. STACE. School of Chemistry & Molecular Sciences, University of Sussex, Brighton BN1 9QJ, UK

R. TAYLOR. School of Chemistry & Molecular Sciences, University of Sussex, Brighton BN1 9QJ, UK

CARLO THILGEN. Laboratorium für Organische Chemie, ETH-Zentrum, Universitätsstrasse 16, CH-8092 Zürich, Switzerland

D. R. M. WALTON. School of Chemistry & Molecular Sciences, University of Sussex, Brighton BN1 9QJ, UK

THOMAS WEISKE. Institut für Organische Chemie, Technische Universität Berlin, W-1000 Berlin 12, Germany

ROBERT L. WHETTEN. Department of Chemistry & Biochemistry, University of California at Los Angeles, Los Angeles, CA 90024-1569

CARTER T. WHITE. Theoretical Chemistry Section, Naval Research Laboratory, Washington, DC 20375-5000

LON J. WILSON. Department of Chemistry, Rice University, Houston, TX 77251-1892

FRED WUDL. Institute for Polymers and Organic Solids, University of California Santa Barbara, Santa Barbara, CA 93106

1

Mass Spectrometric Studies of the Fullerenes

R. F. Curl

1.1 Cluster Beam Mass Spectrometry

In the early 1980s, my colleague Richard Smalley and his co-workers began developing methods for mass spectrometric investigation of cold supersonic molecular beams of clusters of refractory materials [1]. The approach was based upon the fact that any material, no matter how refractory, can be volatilized at the focus of a laser pulse. Figure 1.1 illustrates one of the many similar cluster beam sources that have been designed and utilized for these experiments. In operation, the pulsed valve is opened, producing an intense gas pulse about 500 μsec long. The vaporization laser is fired in the middle of the gas pulse, vaporizing the refractory material and superheating the volatilized material into a plasma. The plasma plume expands into the carrier gas and is cooled by expansion and mixing with the cooler carrier, and small clusters of the refractory form. After clustering the gas pulse expands into a large vacuum chamber, producing a supersonic jet containing clusters cooled to a few degrees kelvin. The jet is then skimmed into a molecular beam and interrogated downstream by various mass spectrometries. Depending upon the nature of the refractory, the carrier gas backing pressure, the duration of the gas pulse, and the time point chosen for vaporization, clusters of from 1 to 100 atoms are produced. In the special case of carbon, very large clusters containing several hundred atoms have been observed.

These clusters have been probed by a variety of mass-spectrometry-based techniques. Generally both positive and negative cluster ions that are residual ions from vaporization plasma are present in the cluster beam [2]. The mass

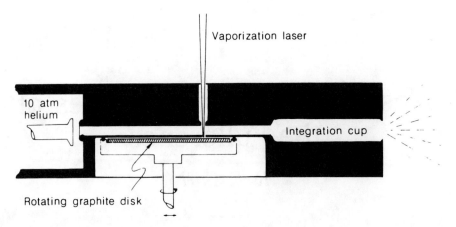

Figure 1.1 Typical cluster beam source. The helium carrier gas pressure behind the pulsed valve is about 10 atm. The approximately 10 nsec long 532 nm wavelength vaporization laser pulse is obtained by doubling the output of a Nd:YAG Q-switched laser and has a pulse energy of 10 to 30 mJ. It is focused to a spot of about 1 mm diameter. (Reprinted by permission for *Nature* **318**, pp. 162–63; Copyright (©) MacMillan Magazines Limited.)

distributions of these residual ions can be observed by extraction with pulsed electric field into the drift tube of a time-of-flight mass spectrometer. Alternatively, the cluster beam can be introduced into an ion cyclotron resonance trap [3]. The neutral cluster distribution may be observed by photoionization of the cluster beam by an ultraviolet laser pulse followed by extraction of the resulting positive ions with a dc electric field into the drift tube of a time-of-flight mass spectrometer [4]. Ions of a particular mass may be selected by one time-of-flight mass spectrometer, then laser photofragmented, and the fragment distribution obtained by mass analysis in a second time-of-flight mass spectrometer [5]. Chemical reactions taking place on the surface of the clusters may be observed by either pulse injecting reagents into an extension tube on the end of the cluster generation source [6], or reactions of a selected cluster ion may be observed by introducing neutral reagents into the ion cyclotron resonance cell [7]. Photoelectron spectroscopy on cold residual anions can be carried out [8, 9] by pulse extracting them from the beam and selecting anions of a particular mass with a time-of-flight mass spectrometer, then subjecting the mass-selected anions to a uv laser pulse. The energies of the electrons thus photodetached are determined by a time-of-flight technique.

The purpose of the present chapter is to review the results obtained when these powerful techniques for studying clusters were applied to clusters of carbon atoms. A review of this subject was published several years ago [10]. However, since this previous review, with the isolation of macroscopic samples of C_{60} and C_{70} by Krätschmer et al. [11], the entire field of fullerene chemistry has undergone a revolutionary change. Recently, cluster beam research not possible when carbon clusters had to be prepared by the vaporization of graphite has been done using gentler vaporization of macroscopic samples of C_{60} and C_{70}. In light of recent developments, it seems an appropriate time for another review of molecular beam mass spectrometric investigations of the fullerenes. We will start at the very beginning with a brief review of the early history of carbon cluster research, but will rapidly move on to recent developments.

1.2   Carbon Cluster Mass Distributions and the Fullerene Hypothesis

Graphite can be laser vaporized into a carrier gas stream, the carbon allowed to cluster, and the cluster size distribution observed by mass spectrometry. When Rohlfing, Cox, and Kaldor [12] first looked at the mass spectrum of such carbon clusters, they obtained the distribution shown in Figure 1.2. It

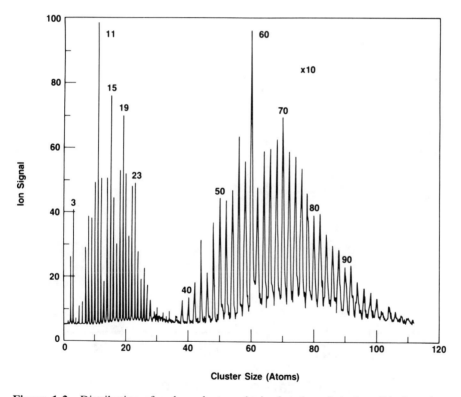

Figure 1.2   Distribution of carbon clusters obtained under relatively mild clustering conditions. (After Rohlfing, Cox, and Kaldor [12]. Reprinted by permission from *The Journal of Chemical Physics* **81**, 3322–30, 1984. Copyright American Institute of Physics.)

appears to consist of two distributions: a low-mass region in which both odd and even clusters are represented ending at about 20 atoms and a high-mass region beginning at about 40 atoms consisting of only even clusters. We now are convinced that the clusters in this second distribution are closed-cage fullerenes. Notice that C_{60} is somewhat more prominent than its neighbors in this distribution, but only somewhat. The Rohlfing, Cox, and Kaldor distribution is obtained when the clustering process is quenched rapidly.

When we began investigating carbon clusters in August 1985, at various times we hit upon clustering conditions where C_{60} was much more prominent than its neighbors [13]. A systematic study of clustering produced the mass spectra shown in Figure 1.3, where C_{60} clearly stands out. Note that C_{70} is also extra prominent. (Under some conditions C_{50} is also more prominent than its neighbors). From the chemist's point of view, there is a uniquely satisfactory structure for C_{60} that can explain this prominence, the truncated icosahedron or soccer ball structure, which in various outline forms has become very familiar.

This structure as depicted in Figure 1.4, however, contains more information

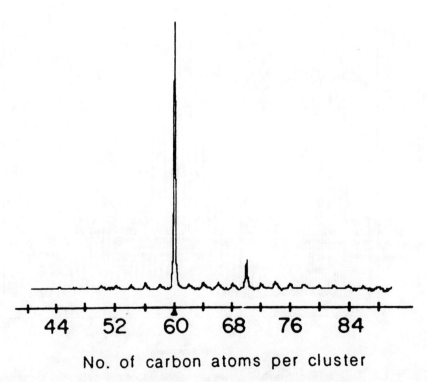

No. of carbon atoms per cluster

Figure 1.3 Distribution of carbon clusters obtained when the time allowed for the clustering process is increased by extending the flow tube. (Reprinted by permission for *Nature* **318**, pp. 162–63; Copyright (©) MacMillan Magazines Limited.)

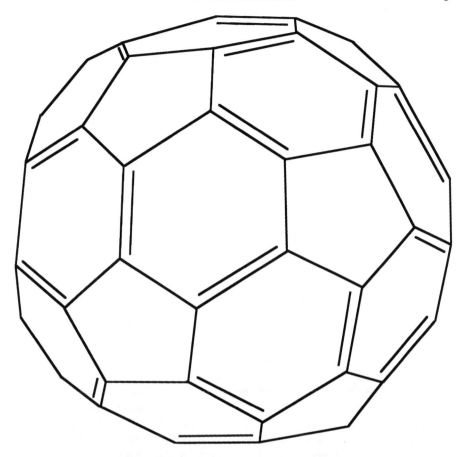

Figure 1.4 Chemical bonding of truncated icosahedron C_{60}. (Reprinted by permission for *Nature* **329**, pp. 529–31; Copyright (©) MacMillan Magazines Limited.)

than is usually displayed, because it shows the chemical bonding and demonstrates that the four bond valences of carbon are satisfied at every site, and that the molecule is a stable aromatic compound like benzene. The double bonds and single bonds can be interchanged in many ways to produce many bonding patterns, 12,500 to be precise [14]. The bonding pattern shown in the figure is the unique lowest-energy Kekulé structure and contributes somewhat more to the total electronic wave function than other Kekulé structures with double bonds in the five-membered rings. The six-membered aromatic rings would really prefer to be planar, with bonds radiating from them in the same plane, but the highly symmetrical, nearly spherical structure distributes any strain of nonplanarity uniformly. To a chemist this molecule looks very stable, and not very reactive. To our minds [13], this structure, because of its high stability and likely chemical inertness, provided a convincing explanation for the prominence of C_{60} in the carbon cluster distributions. Because the geodesic dome constructions of R. Buckminster Fuller provided the key hint that the incorporation of five-membered rings into a hexagonal sheet could lead to a closed solution for C_{60}, this truncated icosahedron isomer of C_{60} was named *buckminsterfullerene*.

The chemical inertness of the even carbon clusters of 40+ atoms, as seen by the absence of reaction when chemical reagents such as NO were injected into a extension of the molecular beam source nozzle, led to the realization that these carbon clusters could also be closed cage structures [15]. Cages consisting of twelve pentagons and $(n - 20)/2$ hexagons can be readily constructed for $n > 22$ for any even C_n species [16]. Such cages were generically called "fullerenes." Early on, an egg-shaped cage structure for C_{70} closely related to the structure of C_{60} was proposed [17] as a highly stable, chemically inert structure to explain the relative prominence of C_{70} in mass distributions such as that shown in Fig. 1.3.

1.3   Endohedral Metallofullerenes

These carbon cage molecules are hollow, and the possibility that an atom might fit into the hollow space was quickly recognized. A low-density graphite disk was soaked in $LaCl_3$, dried, and used as a target for laser vaporization. The mass distribution obtained after the molecular beam is subjected to an intense pulse of ArF excimer radiation to remove the less stable species is shown in Figure 1.5. The persistence of La-containing carbon clusters under conditions where weakly bound species are expected to photofragment strongly suggests that the metal atom is inside the cage. The absence of species such as $C_{60}La_2$ provides additional support for the incorporation of the La inside the C_{60} cage, as there is room for only one La atom, not two, inside the cage.

C_{60} has a relatively large electron affinity, 2.5 to 2.8 eV, as determined [18] from the photoelectron spectrum of C_{60}^-. Therefore, when the metal atom is placed inside the cage, electrons will be transferred from the metal to the carbon cage and the metal will be in the form of a positive ion. The van der Waals radius of a positive ion is considerably smaller than that of the corresponding neutral. There is ample room for any metal ion to fit comfortably inside the C_{60} cage. Currently, several groups are making an intense effort to produce macroscopic samples of such endohedral metallofullerenes, and considerable progress has been made [19–21].

1.4   The Diffuse Interstellar Bands and C_{60}, C_{60}^+, and C_{60}^{2+}

The nature of the absorbers giving rise to the diffuse interstellar bands has been a mystery for over 60 years. Traditionally [22], there have been 39 such fea-

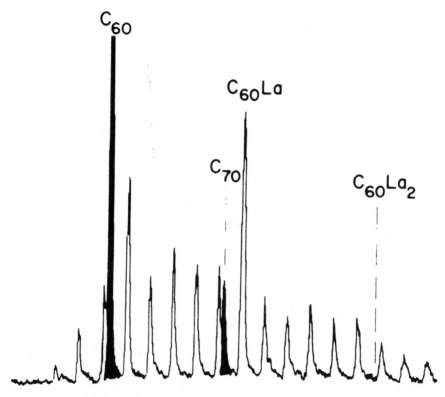

Figure 1.5 Mass distribution obtained by using a target disk of low-density graphite soaked in aqueous $LaCl_3$ and dried. The molecular beam is irradiated with $1-2$ mJ/cm^2 of ArF excimer to remove low-stability species. The shaded peaks are carbon clusters containing one La atom. (Reprinted with permission from *J. Am. Chem. Soc.* **107**, 7779–80, 1985. Copyright 1985, American Chemical Society.)

tures, but recent additions have swelled the number until now there are somewhat more than 80 features identified as diffuse interstellar bands. (The number quoted comes from a paper describing the discovery two new ir bands [23].) Many possibilities for the carriers of the diffuse interstellar bands based on either molecular species or absorbers in interstellar dust have been proposed. Here we restrict ourselves to considering molecular carriers.

All molecules have vuv absorption bands. Small molecules are unacceptable candidates for the diffuse interstellar band carriers because they all will have a photodissociative absorption in the vuv and thus will be destroyed by the vuv radiation bathing the interstellar medium at energies up to the H atom photoionization threshold at 13.6 eV. Furthermore, the rotational structure will be resolved for small molecules, giving rise to characteristic identifiable structure. Large molecules can avoid photodissociation upon vuv absorption because the molecule, if not directly photoionized, will undergo internal conversion of electronic energy to vibrational energy. Then unimolecular fragmentation competes with vibrational reradiation of the excitation in the infrared. In order for a molecule to have the requisite very high probability of surviving these processes, the rate of unimolecular decay must be very small, with excitation energies below 13.6 eV.

As molecular species candidates, C_{60} [13], C_{60}^+ [24], and C_{60}^{2+} have attractive aspects as potential carriers of the diffuse interstellar bands. The fullerenes, in particular C_{60}, might be expected to have very slow unimolecular dissociation rates under such conditions because it has a large number of vibrational modes (174) over which the energy resulting from internal conversion will be distributed and because of the strong bonding of the cage. For unimolecular decay to take place, enough energy to give fragmentation must be concentrated in rupturing a few bonds. C_{60} and its positive ions have no obvious weak point where fragmentation can preferentially occur, as all atoms are equivalent.

Of equal importance in making these species attractive carriers is that these species could be uniquely present at high concentration. Many potential molecular carriers consist of whole *classes* of molecules such as the polycyclic aromatic hydrocarbons. It seems unlikely that such a molecular class containing many members could give rise to only about 80 bands in the visible and near-uv. The situation is different for the fullerenes. C_{60} and, to a lesser extent, C_{70} are the most probable species formed by condensation of carbon. The higher fullerenes with more than 100 atoms, which in toto may be produced to a greater extent by carbon condensation than both C_{60} and C_{70}, would be expected to have such a dense spectrum that they should contribute only a weak background continuum absorption. Thus C_{60} and its positive ions and C_{70} and its positive ions seemed of high interest as possible species giving rise to the diffuse interstellar absorption bands.

1.5   Photofragmentation and "Shrink-Wrapping"

These considerations led us to undertake a photofragmentation study of the fullerenes [25] in order to obtain some estimate of the activation energy for unimolecular decay of C_{60}^+ through measurement of the fluence dependence of the photofragmentation rate. The result was that, in order for there to be a large probability for photofragmentation within 2 to 3 μsec, of C_{60}^+, at least three 6.4 eV ArF photons must be absorbed; that is, more than 19 eV of excitation energy is required for dissociation within a few microseconds. For excitation energies smaller than 13.6 eV, the rate of unimolecular decay should be quite small, confirming that C_{60}^+ might survive exposure to interstellar radiation.

The fragmentation patterns observed in this research proved to be of even greater interest. Photofragmentation of C_{60}^+ and the other fullerenes occurs by the loss of neutral fragments with an even number of carbon atoms. At low photofragmentation laser fluences, the fragment ions correspond to the loss of C_2 and C_4 neutrals. As the fluence is increased, smaller and smaller fragment ions are observed, all resulting from the loss of even carbon neutral fragments. These results are readily interpreted in terms of the loss of C_2, C_4, C_6, ... neutrals with the cage structure being reformed behind the leaving fragment. Schematic ways that this might take place are shown in Figure 1.6. Obviously these even neutral loss processes with cage shrinkage cannot continue without limit. As the cage shrinks, it becomes increasingly strained; eventually the strain will become too great, and the cage will rupture. This is precisely what is observed. As shown in Figure 1.7, the pattern of loss of an even number of carbon atoms abruptly breaks off at C_{32}^+, signaling this cage rupture.

If a metal atom is introduced inside the cage and photofragmentation experiments are carried out, the cage is now shrinking around the virtually incompressible core provided by the metal ion. Thus the cage size where rupture takes place is expected to be larger, and the rupture point is expected to increase as the metal ion size increases. This is in fact what is observed [26], as can be seen in Figure 1.8, which compares the photofragmentation patterns of $C_{60}K^+$ and $C_{60}Cs^+$. A breakoff in even carbon loss is seen for each of these species occurring at $C_{44}K^+$ when $C_{60}K^+$ is fragmented and at $C_{48}Cs^+$ when $C_{60}Cs^+$ is fragmented. These breakoff points are in excellent agreement with simple calculations in which the sizes of the cavities in the cages are compared with the van der Waals radii of the K^+ and Cs^+ ions. These experiments provided compelling evidence for the existence of endohedral metallofullerenes as well as the cage structure hypothesis.

The study of photofragmentation provides some information about activation barriers for ring rearrangement processes on the cage surface relative to the activation energy for photofragmentation. When larger species such as C_{74}^+ are chosen for photofragmentation with 2 to 3 μsec allowed for fragmentation after excitation, the C_{60}^+ fragment ion is not especially prominent, compared with C_{58}^+ and C_{62}^+ fragment ions. However, when the experiment is carried out with more time (about 100 μsec) allowed for fragmentation, the situation changes dramatically, as is shown in Figure 1.9. These results indicate that ring rearrangement processes must have a lower activation energy than the activation energy for fragmentation.

1.6 Spectroscopy of C_{60} and C_{70}

Since C_{60}^+ is highly resistant to photofragmentation, C_{60} and C_{60}^+ are even more attractive candidates as the carrier of at least some of the diffuse interstellar bands. Therefore, obtaining the electronic spectra of these species under the low-temperature, near collisionless conditions of interstellar space has been of intense interest to us. For the neutral species, C_{60}, it has proved possible to obtain electronic spectra under the desired conditions. However, the gas-phase spectrum of the ion, C_{60}^+, remains elusive.

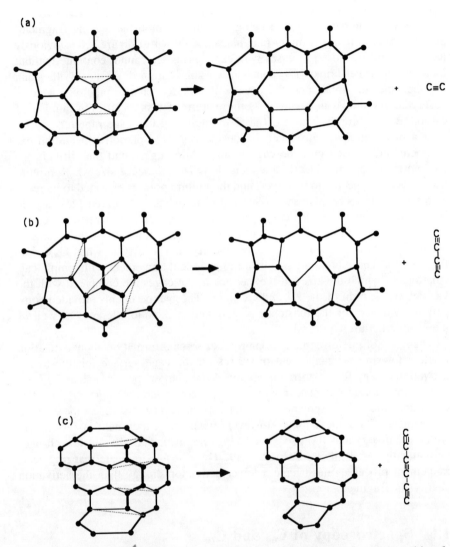

Figure 1.6 Possible mechanisms for loss of even carbon neutral fragments with reformation of the cage. Although C_{60}^+ shows C_2 loss to produce C_{58}^+, buckminsterfullerene does not have a pair of fused five-membered rings as found in the C_2 loss mechanism proposed. This seems to require ring rearrangement to one of the higher-energy C_{60} cage isomers before fragmentation. (Reprinted by permission from *The Journal of Chemical Physics* **88**, 220–30, 1988. Copyright American Institute of Physics.)

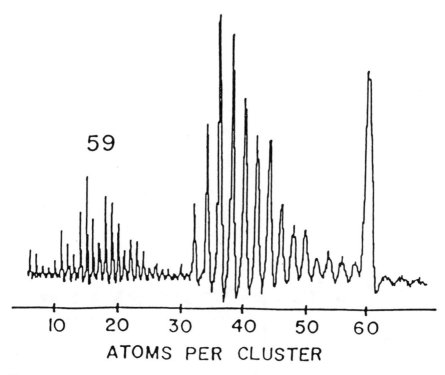

Figure 1.7 Mass distribution of carbon fragment ions upon photofragmentation of C_{60}^{+} with a time of about 2 to 3 μsec between photoexcitation and mass analysis. The C_{60}^{+} is irradiated at 193 nm with a fluence of 59 mJ/cm². Note that the even carbon loss pattern breaks off abruptly at C_{32}. (Reprinted by permission from *The Journal of Chemical Physics* **88**, 220–30, 1988. Copyright American Institute of Physics.)

Our initial scheme [27] for obtaining the spectrum of C_{60} in the cold molecular beam was based upon photofragmentation of van der Waals adducts. Normally our expansions are done with pure helium as the carrier gas, but other volatile substances can be seeded into the helium carrier. The plan chosen to obtain the desired spectrum was that a relatively innocuous adduct would be seeded into the carrier, a van der Waals adduct of C_{60} with this adduct formed in the suspersonic expansion, then the spectrum of this complex probed with a tunable pulsed dye laser. After this probe pulse, the entire beam is subjected to a pulse of 7.9 eV light from an F_2 excimer laser. As the vertical ionization potential of C_{60} is 7.65 eV [28], this results in one-photon ionization with little fragmentation of the van der Waals species. The mass spectrum of the resulting ions is then observed. If the dye laser is tuned to an absorption feature of the complex, it absorbs energy. This energy is rapidly converted to vibrational energy by intersystem crossing and internal conversion, the vibrational energy flows into the van der Waals bond, leading to fragmentation of the complex, changing the mass spectrum by depleting the adduct signal, and enhancing the

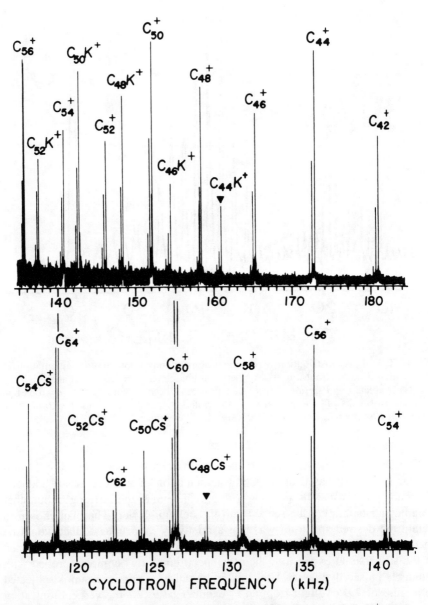

Figure 1.8 FT-ICR ArF (193 nm) photofragmentation patterns of $C_{60}K^+$ (top) and $C_{60}Cs^+$ (bottom). The bare C_n^+ clusters seen in the top panel are fragments from C_{64}^+. The bare C_n^+ seen in the bottom panel are fragments from C_{72}^+. These bare cluster signals are present because the prephotolysis mass selection is incomplete. The important point is that C_nK^+ fragmentation breaks off at $C_{44}K^+$, while the C_nCs^+ fragmentation breaks off at $C_{48}Cs^+$. (Reprinted with permission from *J. Am. Chem. Soc.* **110**, 4464–65, 1988. Copyright 1988, American Chemical Society.)

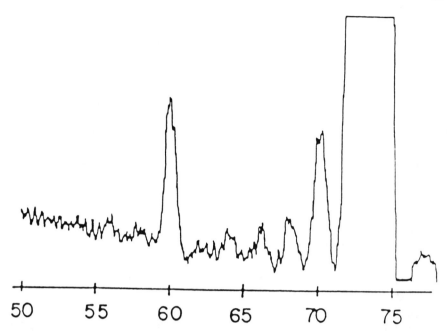

Figure 1.9 Photofragmentation pattern of C_{74}^+ when 100 μsec is allowed for fragmentation. Note that C_{60}^+ is much more prominent than its neighbors. Reprinted by permission from *The Journal of Chemical Physics* **88,** 220–30, 1988. Copyright American Institute of Physics.)

bare C_{60}^+ signal. When the dye laser is scanned, an absorption spectrum of C_{60} results, which can be detected either as a depletion of the C_{60}-adduct ion signal or an enhancement of the C_{60}^+ signal.

Of course, the absorption spectrum obtained is of the van der Waals adduct, not bare C_{60}. However, if adducts that do not themselves have absorptions in the region probed are used, the electronic transitions observed will be characteristic of C_{60} rather than the adduct. The adduct is not expected to perturb the C_{60} spectrum very seriously, and by using several adducts, the magnitude of such perturbations may be estimated. For example, if one were to use the rare gas series, He, Ne Ar, Kr, Xe, the spectrum of C_{60} might be virtually unperturbed with He, slightly perturbed by Ne, with larger changes in the spectrum as one moves to Xe.

However, such a plan is not quite feasible. It proved impossible to obtain adducts with very weak perturbers such as helium. The van der Waals binding energy of helium atoms to almost any material is quite small. While van der Waals adducts of helium with molecules have been observed [29], the expansion cooling required is extreme, as temperatures below 1 K are necessary in order to give reasonable concentrations of such adducts. In the carbon cluster expansions, even if such low temperatures are reached in the translational and rotational degrees of freedom, the vibrational modes of the carbon clusters remain relatively hot, perhaps only slightly below room temperature. Thus if a $He \cdot C_{60}$ van der Waals cluster is formed in the expansion, it will almost certainly undergo unimolecular decay in the molecular beam via conversion of C_{60} vibrational energy into the van der Waals bond. However, the van der Waals bond formed by molecules such as benzene to C_{60} are much stronger, and van der Waals adducts of C_{60} with benzene were found to survive in the molecular beam to the probing region. Probably species with several adduct molecules attached to a single C_{60} are formed in the expansion, and C_{60} is vibrationally cooled by vaporization of excess adduct molecules in the latter stages and after the expansion.

Even though this spectroscopy is limited to relatively large and thus strongly bound adducts, one can be fairly confident that the spectrum obtained closely approximates that of bare C_{60} if spectra from two very dissimilar adducts are at the same absorption wavelength. In our work, benzene and methylene chloride were chosen as adducts. After some searching, a single absorption band at 386 nm was found for both benzene and methylene chloride adducts [27].

At the time we believed that this band was the lowest allowed electronic transition of C_{60}. However, when macroscopic samples of C_{60} became available in 1990, one of the first of many physical properties investigated was the visible–uv absorption spectrum [30]. These investigations revealed a series of weak bands in the 350–400 nm region that were attributed to various vibronic bands of a single electronically forbidden, vibronically allowed electronic transition. The solution-phase spectra reported [30] did not correspond well with the 386 nm band observed by van der Waals spectroscopy. Further molecular spectroscopy investigations were needed.

Fortunately, work on macroscopic samples quickly revealed [31] that C_{60} has a relatively long-lived (about 70 µsec) first excited triplet state that is easily reached by intersystem crossing. Ground-state C_{60} is not one-photon ionized by 6.4 eV ArF photons, but the first excited triplet state is at an energy about 1.7 eV above the ground state [32, 33] and can be one-photon ionized by ArF. This permitted the determination of the collisionless lifetime of the excited triplet states of both C_{60} and C_{70}, which also has a long-lived triplet, in the molecular beam apparatus in a two-color experiment. In this experiment, C_{60} is excited to a singlet state, undergoes intersystem crossing to the triplet state in less than a few nanoseconds, and is then photoionized by ArF [33]. By varying the time between the excitation laser and the ionization laser, the triplet decay curve can be traced out.

The existence of the long-lived triplet state also permitted resonant two-photon ionization spectroscopy on C_{60} and C_{70} [34]. A narrow-band, tunable dye laser and a very weak ArF laser pulse (fluence ≈ 0.1 mJ/cm^2) are overlapped in space and time in the dc ion extraction electric field. When the dye laser is tuned to an absorption feature, the neutral molecule is excited to an electronic singlet state and rapidly undergoes intersystem crossing to the lowest triplet state, which is then one-photon photoionized by the ArF laser. Thus spectroscopic bands of C_{60} and C_{70} are detected by the appearance of C_{60}^+ and C_{70}^+ mass peaks when the dye laser is tuned over the band.

The availability of macroscopic samples of C_{60}/C_{70} mixtures is vital to the success of these experiments. Instead of vaporizing graphite to produce fullerenes, the C_{60}/C_{70} mixture is deposited on a suitable substrate (such as a roughened graphite rod or a copper rod) and used as the laser vaporization target. Far smaller vaporization laser pulse energies (≈ 0.1 mJ) are required for this than for graphite vaporization (≈ 25 mJ). Consequently, the final vibrational temperatures of C_{60} and C_{70} after the suspersonic expansion are much lower. This really makes the spectroscopy experiment possible by greatly reducing the background ionization from ArF one-photon ionization and dye laser plus ArF two-photon ionization where the dye laser excites a broad hot band background. By using a C_{60}/C_{70} mixture as the target, the spectra of C_{60} and C_{70} are obtained simultaneously by gating the signals at the C_{60}^+ and C_{70}^+ arrival times into two separate integrators.

The highest priority was given to re-examining the 375–415 nm region, where the van der Waals fragmentation band (386 nm) was observed previously [27]. After finding this band by R2PI for the free molecule at 385 nm, the rest of the region was explored. Figure 1.10 compares the molecular beam R2PI spectrum of C_{60} with the spectrum of C_{60} at 77 K in an organic matrix. The electronic transition giving rise to this series of bands is believed [30, 34] to be electric dipole forbidden. Thus the observed bands are thought to be false origins and series built from the false origins by totally symmetric and Jahn–Teller active modes. Frequencies of the normal modes of C_{60} have been predicted from theoretical calculations [35, 36], and the infrared- [37] and Raman- [38] active modes have been observed. Efforts to assign these bands using this information were somewhat ambiguous, as there appears to be more than one set of vibrational labels possible for the bands. The spectrum of C_{70} in this region is featureless.

The C_{60} matrix isolation spectrum previously reported [30] also exhibits weaker but interesting sharp features near 600 nm. Therefore, the molecular beam R2PI spectra of C_{60} and C_{70} were investigated in this region. A number of very sharp bands of both C_{60} and C_{70} (Figure 1.11 and 1.12) were observed by R2PI. The spacings between the bands observed for C_{60} are not inconsistent with the information available about normal-mode frequencies, but the number of bands observed seems to be large for this region of the spectrum to be accounted for by a single vibronically allowed electronic transition. Clearly the assignment of the spectrum of C_{70} in Figure 1.12 is at present impossible.

There is no systematic match between the observed spectrum of C_{60} and the diffuse interstellar bands. It probably can be safely concluded that C_{60} is not responsible for any of the diffuse interstellar band absorptions. However, it is interesting to note that the density of bands in the C_{60} spectrum is roughly similar to the density of interstellar absorption features in the region and that no clear pattern is discernible in either the C_{60} spectrum or the diffuse interstellar bands. To date all effort to obtain a band spectrum of C_{60}^+ by either

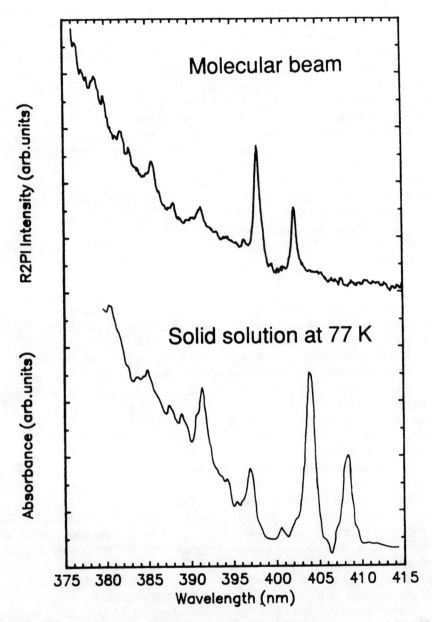

Figure 1.10 Comparison of the R2PI molecular beam electronic spectrum of C_{60} with the spectrum of C_{60} in an organic matrix (1:1 methylcyclohexane to isopentane by volume) at 77 K. There is clearly a one-to-one correspondence in the bands observed. The average matrix red shift is 351 cm^{-1}. The matrix shifts range from 333 to 362 cm^{-1}. (Reprinted by permission from *The Journal of Chemical Physics* **95**, 2197–99, 1991. Copyright American Institute of Physics.)

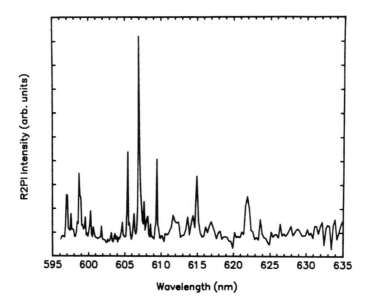

Figure 1.11 R2PI spectrum of cold C_{60} in the molecular beam in the 600 nm region. The bandwidths in this figure are limited by the number of resolution elements used in taking the spectrum. Closer examination of some of these bands shows their line-widths are limited by the bandwidth of the dye laser (≤ 0.2 cm^{-1}). (Reprinted by permission from *The Journal of Chemical Physics* **95**, 2197–99, 1991. Copyright American Institute of Physics.)

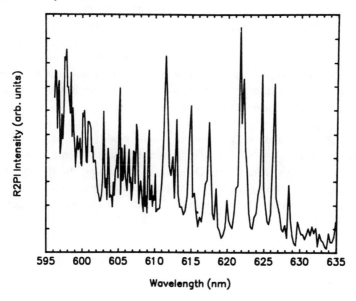

Figure 1.12 R2PI spectrum of cold C_{70} in the molecular beam in the 600 nm region. (Reprinted by permission from *The Journal of Chemical Physics* **95**, 2197–99, 1991. Copyright American Institute of Physics.)

van der Waals fragmentation or R2PI have been unsuccessful, but recently the spectrum of C_{60}^+ in argon matrices has been reported [39].

1.7 Conclusion

Mass spectrometry has been a powerful tool in the investigation of carbon clusters and the fullerenes. As we have seen, the discovery [13] that C_{60} forms spontaneously in condensing carbon vapor was made in mass spectrometry experiments. Mass spectrometry provided the initial evidence [15] that, in addition to C_{60}, many carbon cage species also form spontaneously in condensing carbon vapor. Mass spectrometry photofragmentation experiments provided [25, 26] the most compelling evidence for the fullerene hypothesis prior to the isolation of macroscopic samples. At present mass spectrometry is used as a powerful analytical tool in research aimed at producing macroscopic samples where the cage is doped with a foreign atom [40] and in research aimed at producing macroscopic samples of endohedral metallofullerenes [19].

The nature of the carriers of the diffuse interstellar bands remains a mystery. The spectroscopy experiments [34] described seem to rule out neutral C_{60} as the carrier, and the matrix isolation experiments on the spectrum of C_{60}^+, while less definitive, found an intense near-infrared band that does not seem to match any interstellar features. The major constituent of the interstellar medium is H. It is very likely that any C_{60} that finds its way into the interstellar medium is ionized and hydrogenated. At present it is unclear how far these ionization and hydrogenation processes go. It is likely that loss of an H atom on vuv photoexcitation can compete effectively with infrared reradiation and cage fragmentation processes resulting in a steady state. This simultaneously alleviates concerns about cage fragmentation removing C_{60} species from the interstellar medium and provides a means for reversing the hydrogenation process, making hydrogenated C_{60}^+ ions very attractive candidates as the carriers of the dibs. Recently Kroto and Jura [41] have discussed the possible role of fullerenes and fullerene complexes (Na, K, H, O, etc.) in explaining the diffuse bands and the ultraviolet absorption feature and calculate that the charge-transfer band of $C_{60}H^+$ should occur at a wavelength of 207 nm, in reasonable agreement with the 217 nm feature.

Acknowledgment

This work was supported by a grant from the Robert A. Welch Foundation.

References

1. T. G. Dietz, M. A. Duncan, D. E. Powers, and R. E. Smalley, *J. Chem. Phys.* **74,** 6511 (1981).

2. L. S. Zheng, P. J. Brucat, C. L. Pettiette, S. Yang, and R. E. Smalley, *J. Chem. Phys.* **83,** 4273 (1985).

3. J. M. Alford, P. E. Williams, D. E. Trevor, and R. E. Smalley, *Int. J. Mass. Spec.* **72,** 33–51 (1986).

4. J. R. Heath, Y. Liu, S. C. O'Brien, Qing-Ling Zhang, R. F. Curl, R. E. Smalley, and F. K. Tittel, *J. Chem. Phys.* **83,** 5520 (1985).

5. P. J. Brucat, L. S. Zheng, C. L. Pettiette, S. Yang, and R. E. Smalley, *J. Chem. Phys.* **84,** 3078 (1986).

6. M. D. Morse, M. E. Geusic, J. R. Heath, and R. E. Smalley, *J. Chem. Phys.* **83,** 2293 (1985).

7. J. M. Alford, F. D. Weiss, R. T. Laaksonen, and R. E. Smalley, *J. Phys. Chem.* **90,** 4480 (1986).

8. O. Cheshnovsky, P. J. Brucat, C. L. Pettiette, S. Yang, M. J. Craycraft, and R. E. Smalley, *Proceedings of International Symposium on the Physics and Chemistry of Small Clusters,* P. Jena, S. Kanna, and B. Rao, editors, NATO ASI Series, Plenum Press, New York, pp. 1–13 (1987).

9. O. Cheshnovsky, S. H. Yang, C. L. Pettiette, M. J. Craycraft, Y. Liu, and R. E. Smalley, *Chem. Phys. Letters* **138,** 119 (1987).

10. R. F. Curl and R. E. Smalley, *Science,* **242,** 1017 (1988).

11. W. Krätschmer, L. D. Lamb, K. Fostiropoulos, and D. R. Huffman, *Nature* **347,** 354–58 (1990).

12. E. A. Rohlfing, D. M. Cox, and A. Kaldor, *J. Chem. Phys.* **81,** 3322–30 (1984).

13. H. W. Kroto, J. R. Heath, S. C. O'Brien, R. F. Curl, and R. E. Smalley, *Nature* **318,** 162–63 (1985).

14. D. J. Klein, T. G. Schmalz, T. G. Hite, and W. A. Seitz, *J. Am. Chem. Soc.* **108,** 1301–02 (1986).

15. Q-L Zhang, S. C. O'Brien, J. R. Heath, Y. Liu, R. F. Curl, H. W. Kroto, and R. E. Smalley, *J. Phys. Chem.* **90,** 525–28 (1986).

16. D. E. H. Jones, *The Inventions of Daedalus,* W. H. Freeman, Oxford and San Francisco, 1982, pp. 118–19. Of course, this actually goes back to L. Euler "Elementa Doctrinae Solidorum" (1758).

17. J. R. Heath, S. C. O'Brien, Q. Zhang, Y. Liu, R. F. Curl, H. W. Kroto, F. K. Tittel, and R. E. Smalley, *J. Am. Chem. Soc.* **107,** 7779–80 (1985).

18. S. Yang, C. L. Pettiette, J. Conceicao, O. Cheshnovsky, and R. E. Smalley, *Chem. Phys. Lett.* **139,** 233–38 (1987).

19. Y. Chai, T. Guo, C. Jin, R. E. Haufler, L. P. F. Chibante, J. Fure, L. Wang, J. M. Alford, and R. E. Smalley, *J. Phys. Chem.* **95,** 7564–68 (1991).

20. M. M. Alvarez, E. G. Gillan, K. Holczer, R. B. Kaner, K. S. Min, and R. L. Whetten, *J. Phys. Chem.* **95,** 10561–63 (1991).

21. J. H. Weaver, Y. Chan, G. H. Kroll, C. Jin, T. R. Ohno, R. E. Haufler, T. Guo, J. M. Alfred, J. Conceicao, L. P. F. Chibante, A. Jain, G. Palmer, and R. E. Smalley, *Chem. Phys. Lett.* **190,** 460–64 (1991).

22. E. Herbig, *Astrophys. J.* **196,** 129–60 (1975).

23. C. Joblin, J. P. Mailard, L. d'Hendecourt, and A. Leger, *Nature* **346,** 729 (1990).

24. A. Leger, L. d'Hendecourt, L. Verstraete, and W. Schmidt, *Astr. Astrophys.* **203,** 145–48 (1988).

25. S. C. O'Brien, J. R. Heath, R. F. Curl, and R. E. Smalley, *J. Chem. Phys.* **88,** 220–30 (1988).

26. F. D. Weiss, S. C. O'Brien, J. L. Elkind, R. F. Curl, and R. E. Smalley, *J. Am. Chem. Soc.* **110,** 4464 (1988).

27. J. R. Heath, R. F. Curl, and R. E. Smalley, *J. Chem. Phys.* **87,** 4236 (1987).

28. P. Baltzer, W. J. Griffiths, A. J. Maxwell, P. A. Bruhwiler, L. Karlsson, and N. Martensson, *Phys. Rev. Lett.,* submitted (1991).

29. R. E. Smalley, D. H. Levy, and L. Wharton, *J. Chem. Phys.* **64,** 3266 (1976).

30. R. L. Whetten, M. M. Alvarez, S. J. Anz, K. E. Schriver, R. D. Beck, F. N. Diederich, Y. Rubin, R. Ettl, C. S. Foote, A. P. Darmanyan, and J. W. Arbogast, *Mat. Res. Soc. Proc.* **206,** 639–50 (1991).

31. J. W. Abogast, A. P. Darmanyan, C. S. Foote, Y. Rubin, F. N. Diederich, M. M. Alvarez, S. J. Anz, and R. L. Whetten, *J. Phys. Chem.* **95,** 11–12 (1991).

32. R. R. Hung and J. J. Grabowski, *J. Phys. Chem.* **95,** 6073–75 (1991).

33. R. E. Haufler, L.-S. Wang, L. P. F. Chibante, C. Jin, J. J. Conceicao, Y. Chai, and R. E. Smalley, *Chem. Phys. Lett.* **179,** 449–54 (1991).

34. R. E. Haufler, Y. Chai, L. P. F. Chibante, M. R. Fraelich, R. B. Weisman, R. F. Curl, and R. E. Smalley, *J. Chem. Phys.* **95,** 2197–99 (1991).

35. R. E. Stanton and M. D. Newton, *J. Phys. Chem.* **92,** 2141–45 (1988).

36. F. Negri, G. Orlandi, and G. Zerbetto, *Chem. Phys. Lett.* **144**(1), 31–7 (1988).

37. W. Kratschmer, K. Fostiropoulos, and D. R. Huffman, *Chem. Phys. Lett.* **170,** 167 (1990).

38. D. S. Bethune, G. Meijer, W. C. Tang, and H. J. Rosen, *Chem. Phys. Lett.* **174,** 219 (1990).

39. Z. Gasyna, L. Andrews, and P. N. Schatz, *J. Phys. Chem.* **96,** 1525–27 (1992).

40. T. Guo, C. Jin, and R. E. Smalley, *J. Phys. Chem.* **95,** 4948–50 (1991).

41. H. W. Kroto and M. Jura, *Astron. Astrophys.* **263,** 275–80 (1992).

CHAPTER

2

Fullerene Studies at Sussex

H. W. Kroto, K. Prassides, A. J. Stace, R. Taylor and D. R. M. Walton

In 1972–1975 at Sussex the first cyanopolyyne, HC_5N, was prepared and its microwave spectrum recorded in order to probe the dynamics of a flexing–rotating chain molecule. The project was conducted at a time when radioastronomy had revealed that the black dust clouds of the Milky Way were full of small molecules, and our results prompted a radio search for HC_5N (in collaboration with Canadian astronomers), which was successful. In addition to HC_5N, even longer chains, HC_7N and HC_9N, were discovered [1]. These observations constituted a surprising and important breakthrough in our knowledge of the carbon content of space. Further work drew attention to the fact that red giant carbon stars were pumping vast quantities of carbon molecules, as well as carbon dust, into space and that there might be some link between the carbon chains, soot formation, and these stellar processes [1, 2].

During 1980–1984 Richard Smalley and associates at Rice University, Texas, developed a powerful technique for studying, by mass spectrometry, refractory clusters generated in a plasma by focusing a pulsed laser on a solid (generally metallic) target [3]. This method not only made the study of refractory clusters feasible for the first time, but it also seemed to offer a way of simulating the chemistry in a carbon star if a graphite target were to be used. In September 1985 the experiments set up to probe this latter idea serendipitously revealed a dominant 720 amu mass peak corresponding to a C_{60} species [4]. This dominance was attributed to the exceptional stability associated with a closed truncated icosahedral cage (Fig. 2.1); the molecule was named *buckminsterfullerene* because geodesic dome concepts had played an important part in arriving

21

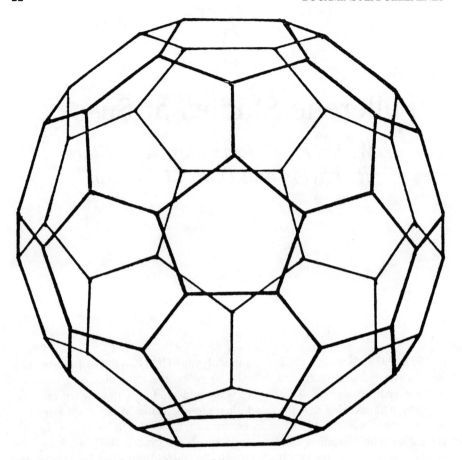

Figure 2.1 C_{60}, buckminsterfullerene [4].

at the structure. The signal for a C_{70} species, for which an oblate spheroidal structure was proposed [5], was also prominent.

Over the subsequent 5 years, a series of fundamental chemical physics studies were carried out at Sussex and Rice, both jointly and independently, which probed the behavior and properties of C_{60} and other carbon cages—the fullerenes—as they have now become known [6]. During 1985–1990 these programs and studies by other key groups such as those in Freiburg, the Naval Research Laboratory, etc. and by many theoreticians, amassed overwhelming circumstantial evidence in support of the cage proposal [5, 7–10]. The most significant observations to emerge were as follows:

1. Excellent evidence for the formation of endohedral cluster complexes, $(M)_n$, in beams was obtained [5, 8, 9, 11];
2. Pentagon isolation was a stabilizing factor [12];
3. The Pentagon Isolation Rule [13, 14] was developed and shown to be a

quantitative stability criterion in that the fullerene structural concept *necessitates* that C_{60} and C_{70} be the first and second most stable clusters, respectively;

4. Small fullerenes, such as C_{28} (Fig. 2.2 [13]), were predicted to possess extra stability;

5. C_{60} was predicted to be a possible byproduct of combustion, and it was suggested that the formation mechanism might account for the spheroidal nature of soot [15]. C_{60}^+ was subsequently found to be a dominant ion in a sooting flame [16];

6. Giant fullerenes were considered and shown to be pseudoicosahedra (Fig. 2.3 [17]), a discovery that accounted for the polyhedral shapes of certain graphite microparticles.

At Sussex, a cluster beam apparatus based on the Rice apparatus was built in order to study the gas-phase chemistry of fullerenes and to evaluate their possible implications for soot formation. Arc processing of graphite was also investigated. In late 1989 a joint Heidelberg/Tucson group—Krätschmer, Fostiropoulos, and Huffman [18]—detected weak features in the ir spectrum of arc-processed graphite that they suggested might be consistent with the presence of buckminsterfullerene. These ir observations were reproduced at Sussex. Krätschmer et al. [19] then published further striking evidence for C_{60} by monitoring vibrational isotope shifts in soot generated by processing ^{13}C-enriched graphite. The presence of C_{60} in soot generated at Sussex was confirmed mass spectrometrically (Fig. 2.4 [20]). In August 1990 the Heidelberg/Tucson team announced that they had obtained a soluble fullerene extract, the x-ray and electron-diffraction patterns of which provided unequivocal confirmation of the fullerene concept [21]. At Sussex, the solubility of C_{60} was discovered independently [20, 22]; the fullerenes were separated by column chromatography and shown to consist primarily of C_{60} and C_{70} [22]. The ^{13}C NMR spectra of these species exhibited a singlet and five lines, respectively (Fig. 2.5), conclusive proof of their fullerene structures. The C_{60} production method was improved upon by Haufler et al. [23], and the NMR line of C_{60} was also observed by Johnson et al. [24].

2.1 Fullerene Preparation

2.1.1 Arc Processing

Our first fullerene-rich soot was prepared by striking an arc between two spring-loaded carbon rods (ca. 80×5 mm) in a bell jar containing argon (ca. 100 mbar); we used helium subsequently. The current (100 A ac) was supplied from a standard welding power supply. This apparatus could only generate small amounts of soot; so a metal reactor, fitted with a screw device for keeping the carbon rods in close contact, was developed [25]. Recently, we have employed a reactor in which 305 mm rods are hand fed through water-cooled O-ring

a)

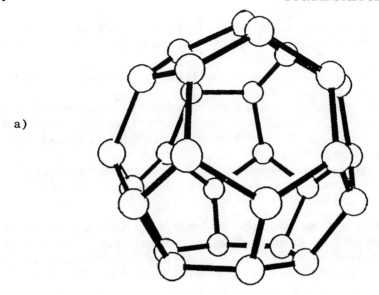

b)

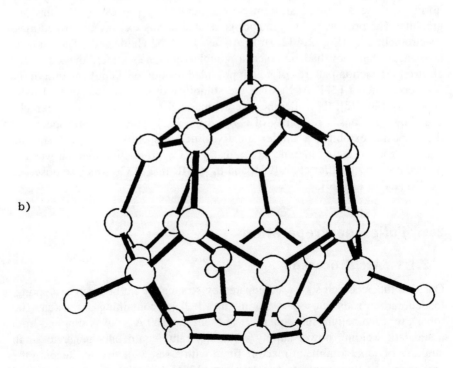

Figure 2.2 (a) C_{28}, fullerene-28 [13]. (b) $C_{28}H_4$, T_d-tetrahydrofullerene-28 [13, 20, 57].

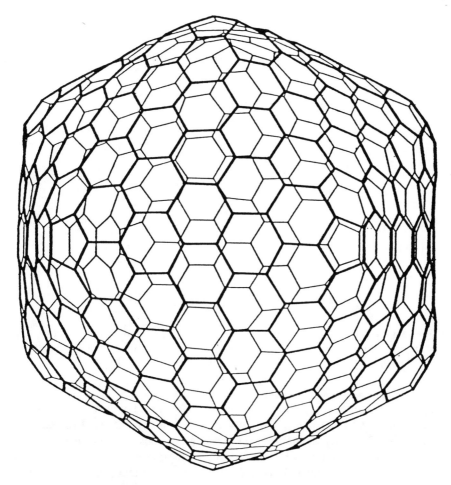

Figure 2.3 Model of the quasi-icosohedral giant fullerene C_{540} [17].

seals, a device that permits vaporization of several rods before it becomes necessary to dismantle the equipment. This design incorporates a modified end cap that minimizes the amount of unused rod.

2.1.2 Extraction and Separation

The soot was Soxhlet extracted with chloroform rather than benzene (in which the fullerenes are more soluble) for safety reasons [22]. We now use dichloromethane, which is less likely to degrade on recycling. The fullerene extract (ca. 8% of the soot) is soluble, but to widely differing extents, in most organic solvents [22]. Chromatography on neutral alumina, using hexane as elutant, yielded the first pure samples of C_{60} (magenta solution) and C_{70} (port-wine red); concentration yielded thin films that were mustard yellow and red-brown, re-

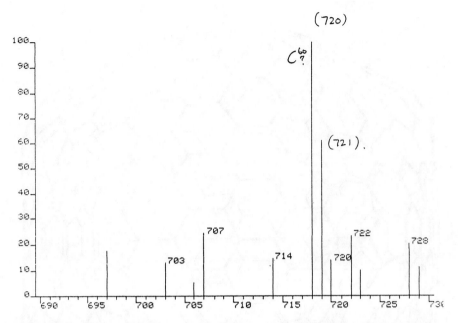

Figure 2.4 Part of the FAB mass spectrum of an arc-processed soot deposit [20, 22]. The calibration is displaced by 2 amu; however, the isotope pattern is convincing since the peaks approximate to the intensity ratio 1.0/0.66 expected for $^{12}C_{60}/^{12}C_{59}{}^{13}C$.

spectively. Fortuitously, the initial separation was carried out using an old stock of alumina that had acquired a small water content (ca. 0.5%). Both C_{60} and (especially) C_{70} partly decompose on alumina columns exposed to light [26], but the extent of decomposition is reduced in the presence of traces of water [27]. Our early work also showed C_{60} to be more soluble than C_{70} and that once both solids had been thoroughly evaporated to dryness, they could only be redissolved with extreme difficulty, especially in hexane [22].

C_{60} and C_{70} are separable by HPLC [27]. During chromatography, a greenish yellow fraction eluted after C_{70} (the amount varied from soot extract to soot extract), and preliminary mass spectra revealed the presence of C_{76} and C_{84} in this fraction. Recently, HPLC (41.4 mm i.d. preparative Rainin Dynamax Column—5 m cyano: hexane eluent; 45 bar), has yielded a mixture of C_{76} and C_{78}, which has been analyzed by ^{13}C NMR spectroscopy [28], and literature assignments corresponding to D_2-C_{76} [29] and C_{2v}-C_{78} [30] confirmed. Furthermore, a spectrum for D_3-C_{78}, different from that previously reported [30], has been obtained and a new $C_{2v'}$ isomer of C_{78} identified (Fig. 2.6 [28]). These findings have been confirmed by Kikuchi et al. [31].

2.1.3 Stability

C_{60} showed unusual properties from the outset. For example, the solid material obtained from a solution rotary evaporated to dryness can be completely and

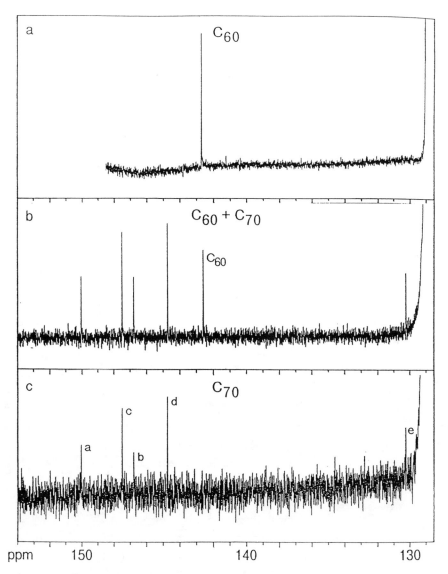

Figure 2.5 ^{13}C NMR spectra [22] obtained from chromatographically purified samples of soluble material extracted from arc-processed graphite: (a) ^{13}C NMR spectrum of a purified C_{60} sample exhibiting a single resonance; (b) spectrum of a mixed sample; and (c) spectrum of a purified sample of C_{70} from which C_{60} has been eliminated. The intense benzene solvent signal lies to the far right. This set of observations provided unequivocal evidence for the equivalence of the carbon atoms in C_{60}, in perfect agreement with the expectation for buckminsterfullerene (Fig. 2.1). The five-line C_{70} spectrum is also totally consistent with expectation for D_{5h}-fullerene-70.

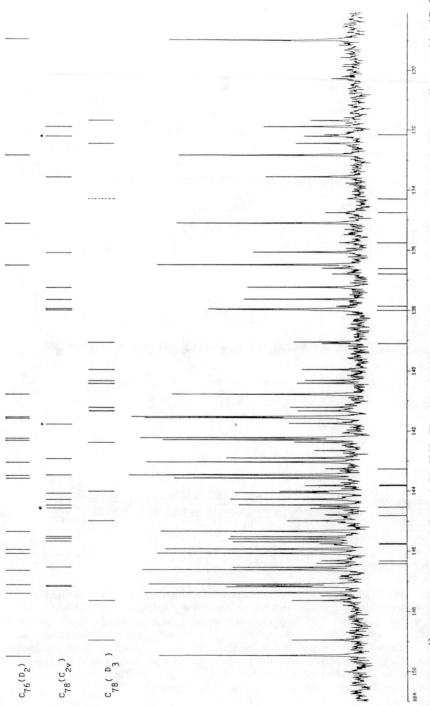

Figure 2.6 ^{13}C NMR spectrum of C_{76} and C_{78} [28]. The three half-intensity lines for C_{2v}-C_{78} are identified for D_3-C_{78}; a peak (dotted line) reported in the literature is absent. Eighteen lines, identified at the bottom of the spectrum, are ascribed to C_{2v} isomer; four other lines are not resolved from C_{76} and the background.

rapidly redissolved to give the deep magenta solution in benzene. If, however, the solid is set aside for a few days, it dissolves with difficulty, and successive benzene solutions acquire a pinkish hue, believed to be due to the presence of a beige-colored less soluble oxygen derivative of C_{60}. A recipe for purifying "old" C_{60}, utilizing the fact that this impurity is acetone soluble, has been described [32].

Solutions of C_{60} degrade slowly in light. For example, if a benzene solution is allowed to stand in Sussex sunlight (latitude 55°N), complete decomposition occurs after a few months, producing a water-soluble buff precipitate. Hexane solutions of C_{60}, irradiated with a medium-pressure uv immersion lamp, degrade completely within hours. However, this process is batch dependent, so that an artefact may either accelerate or inhibit the change; benzene solutions behave in a similar way, but at a slower rate. The possible use of photochemical irradiation for the formation of C_{60} derivatives has been emphasized [26].

2.1.4 Chemistry

C_{60} reacts as a soft electrophile, displaying behavior more characteristic of an electron-deficient polyene with relatively localized 5=5 double bonds, as is confirmed by ^{13}C NMR [33] and crystallography [33–37]. This localization arises from the avoidance of double bonds in the pentagonal rings. The presence of such double bonds would be expected to increase strain and favor antiaromaticity [38]. As a result, additions—especially nucleophilic additions—occur readily. However, these reactions tend to be reversible, and C_{60} is regenerated.

Additions broadly fall into two categories: those giving rise to eclipsing reactions of low steric requirement (e.g., when a single entity such as Ph_2C or L_2Pt spans a formerly high-order bond) and those resulting in higher eclipsing interactions. Analysis of the pattern of addition in the former has led to speculation that certain $C_{78}X_6$ compounds (i.e., involving addition across three high-order bonds) may prove to be particularly stable [39].

Fluorination.

XeF_2 rapidly fluorinates C_{60}, but the product appears to incorporate solvent fragments. With gaseous fluorine, C_{60} characteristically suddenly changes from black to dark brown, then to light brown after successive ca. 24 h intervals. Light cream material, soluble in most organic solvents, is eventually produced, and colorless plates, m.p. 287°C, exhibiting a sharp ^{19}F NMR singlet (Fig. 2.7), and ir stretching bands were isolated by solvent separation. This result was deemed to be consistent with the formation of $C_{60}F_{60}$ [40], though this interpretation is not unambiguous as other less fluorinated analogues with all-equivalent fluorines may exist. The FAB mass spectrum of less highly fluorinated material contains a range of ions in which $C_{60}F_6^+$ and $C_{60}F_{42}^+$ are dominant [41]. Microanalysis confirms the average composition: ca. $C_{60}F_{44}$. The

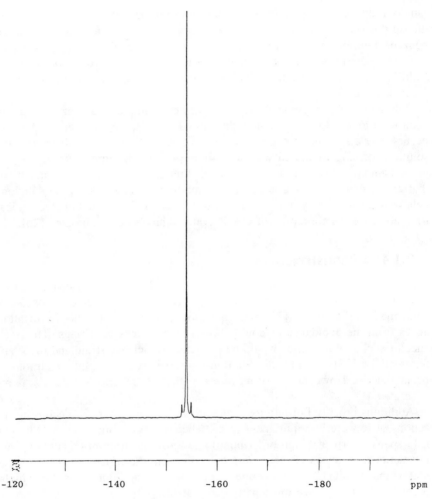

Figure 2.7 ^{19}F NMR spectrum attributed to $C_{60}F_{60}$ [40].

^{19}F NMR spectrum of fluorinated C_{60} (acetone solution) exhibits a peak at ca. 190 ppm. This fact, coupled with the observation that acetone, used as solvent for NMR spectroscopy, undergoes aldol-type condensations is attributable to the presence of HF [42]. Addition of a trace of water to these solutions leads to the immediate generation of HF [41]. Fluorinated C_{60} and water alone do not visibly react; however, addition of acetone or THF leads to a rapid evolution of heat [42]. The rate decreases as the number of fluorine atoms attached to C_{60} diminishes. Preliminary studies have been made with a wide range of nucleophiles [41, 42]. The reaction with water may well preclude the use of fluorinated C_{60} as a lubricant. Since the stability order of halogenated fullerenes appears to be: F>Cl>Br>I, it may be that if the fluorination of C_{60} can be controlled, the fluoro derivatives will turn out to be the most useful among the halegeno intermediates. The multiple nucleophilic substitution results in a continuum of FAB mass spectral lines, since the matrices are often alcohols or amines, which react with the sample [41, 42].

Bromination.

Olah et al. [43] (see Prakash et al., Chapter 12 of this volume) showed that C_{60} would brominate using neat bromine and have obtained evidence for the addition of between two and four Br atoms. The products are highly dependent on the conditions. Tebbe et al. [44] and Birkett et al. [36] have reacted C_{60} with neat bromine and obtained yellow microcrystals that analyze as ca. $C_{60}Br_{28}$. The ir spectrum (Fig. 2.8) is simple and commensurate with a highly symmetrical structure. Since it is possible to add 24 but not 28 bromines to C_{60} in a highly symmetrical fashion (Fig. 2.9), it seemed likely that two molecules of bromine were present as a solvate. Both the proposed disposition (Fig. 2.9 [36]) and the number of bromine solvate molecules have now been confirmed [45].

When a dilute solution of bromine in CS_2 is added to C_{60} in the same solvent, no reaction occurs. However, on increasing the bromine concentration, dark brown crystals of $C_{60}Br_8$ are deposited, and their structure has been determined (Fig. 2.10). This product is also obtained when C_{60} is brominated in chloroform. It turns out that repetition of the eight-fold pattern over the whole cage can lead to only one structure for $C_{60}Br_{24}$, that represented in the Schlegel diagram (Fig. 2.9). Bromination in either carbon tetrachloride or benzene yields magenta plates of $C_{60}Br_6$, whose structure has also been determined (Fig. 2.11 [36]). On warming, $C_{60}Br_6$ is converted into $C_{60}Br_8$ [36]. All three bromo compounds gradually convert into C_{60} upon heating at ca. 150°C. Like fluorinated C_{60}, the bromo compounds undergo nucleophilic substitution, but at a slower rate.

Phenylation.

When a mixture of C_{60}, bromine, ferric chloride, and benzene is heated, conventional workup followed by column chromatography yields a mixture of multiply phenylated C_{60} derivatives. The ions $C_{60}Ph_6^+$, $C_{60}Ph_8^+$, $C_{60}Ph_{12}^+$, and $C_{60}Ph_{16}^+$ predominate in the mass spectrum (Fig. 2.12). Preliminary experiments suggest that the mixture may be separable [46]. The reaction mechanism probably involves electrophilic attack on benzene by the brominated C_{60}, catalyzed by ferric chloride (Friedel–Crafts-type reaction), a view supported by the observation of the six- and eightfold substitution pattern (q.v. bromination) and multiples thereof.

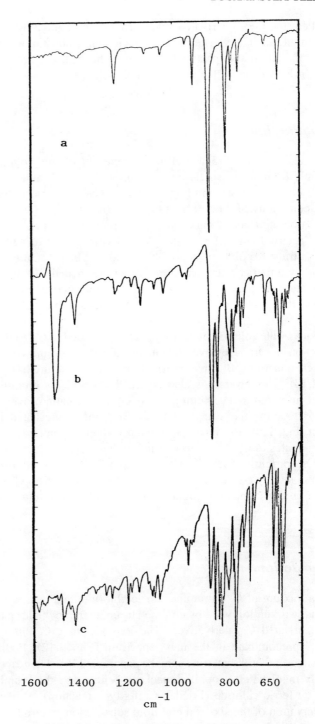

Figure 2.8 IR spectra of brominated fullerenes [36]: (a) $C_{60}Br_{24}$; (b) $C_{60}Br_8$; (c) $C_{60}Br_6$.

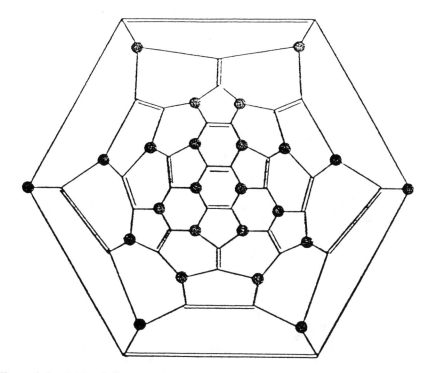

Figure 2.9 Schlegel diagram of $C_{60}Br_{24}$ [36].

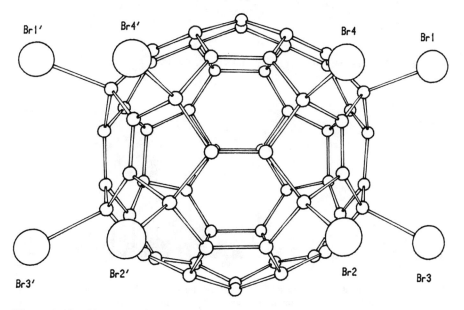

Figure 2.10 Single-crystal x-ray structure of $C_{60}Br_8$ [36].

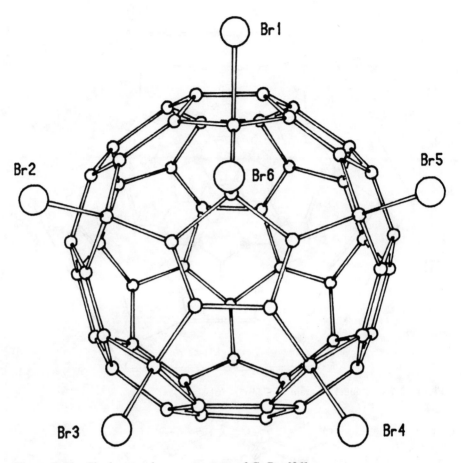

Figure 2.11   Single-crystal x-ray structure of $C_{60}Br_6$ [36].

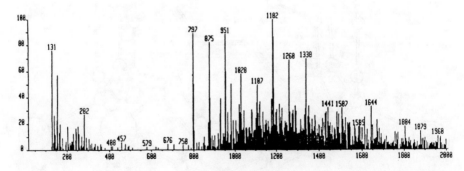

Figure 2.12   Metastable ion mass spectrum showing loss of phenyl groups from $C_{60}Ph_x$; variations from 77 amu are due to the wide slit settings used to improve sensitivity [46].

2.2 Physical Measurements

The Heidelberg/Tucson group (Krätschmer, Lamb, Fostiropoulos, and Huffman) published the ir and uv/vis spectra of the *first* fullerene extracts, which contained approximately a 5:1 C_{60}/C_{70} mixture [18, 19, 21]. At Sussex the ^{13}C NMR spectrum was recorded (Fig. 2.5 [22]), together with uv [Fig. 2.13(a),(b) [25]] and ir (Fig. 2.14 [47]) spectra of pure C_{60} and C_{70}. The MCD spectrum of matrix-isolated C_{60} [48] was obtained in collaboration with Schatz's group (Virginia), and the Raman spectra of C_{60} and C_{70} [49] were recorded with Hendra (Southampton) [Fig. 2.15(a),(b)]. The fullerene-60 NMR data, taken in conjunction with fullerene-70, proved conclusively that the C_{60} singlet was not due to a ring of identical carbon atoms or to fluxionality. The red of fullerene-70 masks the delicate magenta of fullerene-60 in the mixture. These colors are due to forbidden transitions. The numerous complex features, observed in electronic spectrum of fullerene-60, have been carefully analyzed by Leach et al. [50], who made the first detailed assignments. This study nicely complements the MCD work of Schatz et al. [48]. The strong third-harmonic ir amplification reported by Blau et al. [51] appears to have been an artefact.

TG measurements on highly purified fullerene-60 by Dworkin et al. [52, 53] revealed the presence of an order–disorder phase transition in the solid. The discovery of a second-order phase transition will be discussed here. The original powder x-ray-diffraction measurement on a mixed C_{60}/C_{70} crystal, made by Krätschmer et al. [21], revealed an hcp/fcc mixed structure. At that stage only ball-to-ball separations and dispositions were determined. As the molecules were rotating, accurate C/C bond lengths could not be determined. The results obtained by MacKay et al. on chromatographically separated C_{60} [54] are shown in Figure 2.16. Thus far, efforts to produce single crystals of pure C_{60} and C_{70} have resulted in solvated crystals [37] from solution and twinned crystals from the vapor (Liu et al. [55]). Very good structural data have been obtained from the twinned crystals [51]. An x-ray study of recently produced benzene-solvated untwinned single crystals (Meidine et al. [37]) indicate that C_{60} spheres do not appear to be spinning, allowing bond-length information to be derived.

2.2.1 Cluster Beam Studies on Smaller Fullerenes

During laser vaporization cluster beam studies at Rice [56] and Sussex [57], it was found that under certain conditions there were other magic numbers (smaller than 60) in the carbon cluster mass spectrum as well as 60 and 70. These were 28, 32, 36, and 50. A scan of part of one of these mass spectra over the C_8–C_{74} region is shown in Figure 2.17. A theoretical study based on generalized geodesic concepts and chemical stability criteria involving *singlet and multiplet* pentagon isolation concepts showed that a range of smaller fullerenes such as C_{50}, C_{36}, C_{32}, C_{28}, C_{24} should indeed exhibit extra stability over and above that of their neighbors [13]. One isomer of C_{28} (Fig. 2.2(a) [13]) in

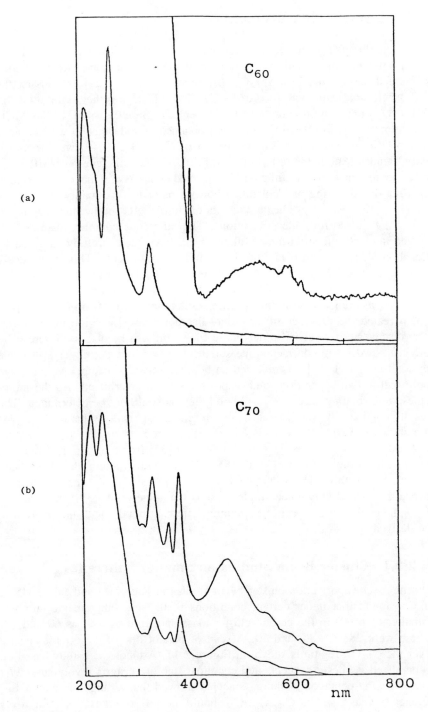

(a)

(b)

Figure 2.13 Uv-vis spectrum of (a) C_{60} and (b) C_{70} [25].

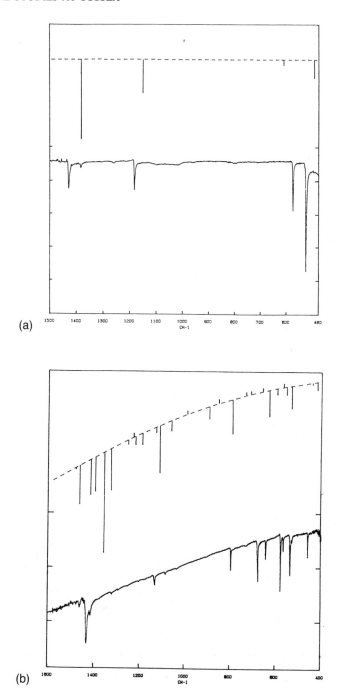

Figure 2.14 Ir spectra [47] of (a) C_{60} and (b) C_{70}; predicted band frequencies and approximate intensities (Bakowies and Thiel, *Chem. Phys.*, in press) are shown at the top.

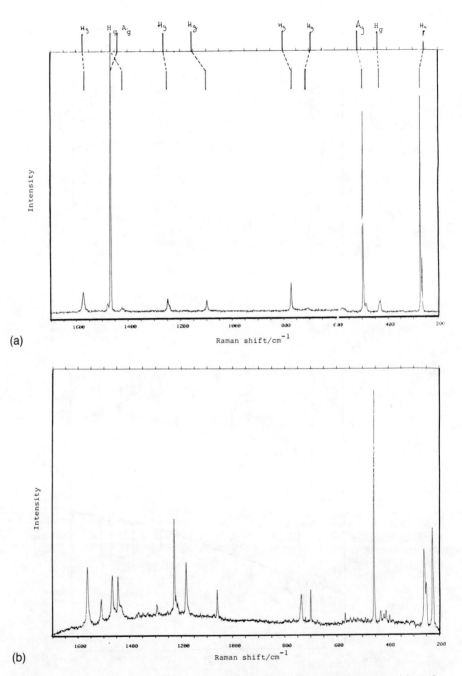

(a)

(b)

Figure 2.15 Raman spectra [49] of (a) C_{60} and (b) C_{70}; predicted band positions for C_{60} (Bakowies and Thiel, *Chem. Phys.* **151,** 309 (1991)) are shown at the top of spectrum (a).

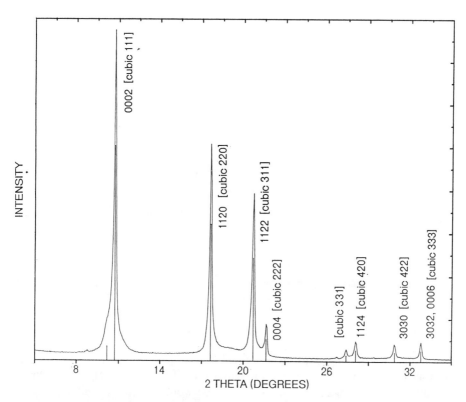

Figure 2.16 X-ray-diffraction pattern obtained by MacKay et al. [10, 54] from chromatographically purified C_{60}. The revealed structure is basically that of a strongly disordered stacking of a simple hexagonal close packing, exactly as for elemental cobalt. The hexagonal unit cell refines to $a = 10.017 \pm 0.004$ Å and $c = 16.402 \pm 0.01$ Å and contains two C_{60} spheres. The distance between centers of spheres is 10.017 Å; calculated density, 1.68 g cm^{-2}. The lines can be indexed as shown, and it is noteworthy that, because of the stacking disorder, only those reciprocal lattice rows parallel to c for which $-h + k = 3n$ are present. The c/a ratio, 1.637, is very close to the theoretical value, 1.633, and thus the pattern can also be indexed with respect to a face-centered cubic lattice ($a = 14.186$ Å) (as for copper) with stacking disorder, which removes the 200 and 400 reflections and which introduces a very weak line (the first) at a spacing of $a/(8/3)^{1/2}$ due to double diffraction arising from stacking faults. The intensity variation of the pattern as a whole corresponds to the transform of a sphere of radius 3.5 Å, giving a first minimum in the region of $2 = 25°$. The pattern is a mixture of fcc and hcp arrays, consistent with crystalline material containing solvent molecules trapped in the faults.

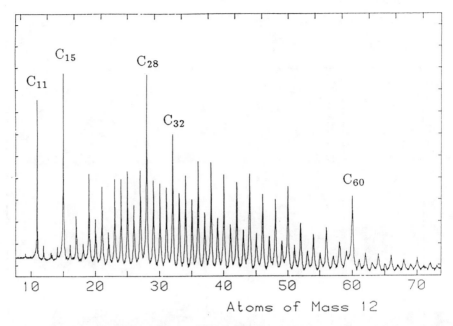

Figure 2.17 TOFMS of carbon clusters produced in a supersonic nozzle and ionized by F_2 excimer laser radiation. Under these conditions, small fullerenes—C_{28} in particular—are seen to be magic [57].

particular was predicted to have a highly symmetric tetrahedral structure. It was also predicted that such species as tetrahydrofullerene-28, $C_{28}H_4$ [Fig. 2.2(b)], should be stable, suggesting that the C_{28} cage would exhibit a shell valency of four [13, 20]. These predictions have recently gained further support from some carbon cluster beam reaction studies at Sussex [57], which indicate that C_{28} is significantly less stable towards hydrogenation than other clusters in the C_8–C_{36} range (Fig. 2.18). Furthermore, $C_{28}H_4$ also appears to be the major hydrogenated (cationic) product (Fig. 2.19). The observations also provide strong evidence for the presence of fullerene-24 and vestigial evidence for the presence of the smallest member of the family, fullerene-20 or dodecahedrene [57]. Fullerene-20, C_{20}, is of course the fully dehydrogenated analogue of dodeca-hedrane that Paquette et al. created by rational synthesis [58].

2.3 Solid-State Studies

Solid-state investigations at Sussex have concentrated on the structural and dynamic properties of C_{60} and C_{70}, and on the properties of both superconducting and insulating alkali-metal doped materials. High-resolution neutron diffraction measurements allowed us to extend earlier x-ray work [59] on the structural

H_2/He

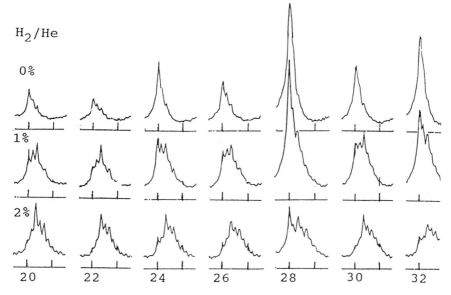

Figure 2.18 TOFMS cluster beam reaction study of *even* carbon species and hydrogenated products formed in the presence of varying concentrations of hydrogen entrained in the He pulse [57]. (Note that only the even clusters are shown so as to aid comparison.) Consecutive traces of 0, 1, and 2% H_2/He mixtures are depicted. The first marker for each cluster indicates the position of the pure carbon signal (C_n^+), [and the second marker the position of the hexahydrogenated species $(C_nH_6^+)$]. A comparison of the traces points to the enhanced stability of the C_{28} cluster. When the reaction has occurred, $C_{28}H_4$ is the dominant hydrogenated product [57].

properties of solid C_{60}, leading to a better understanding of the low-temperature crystal structure, its intermolecular bonding implications, and the structural phase transitions, accompanying orientational ordering of the C_{60} molecules. Inelastic neutron scattering measurements, on the other hand, have provided a detailed picture of the vibrational spectra of C_{60}, superconducting K_3C_{60} and insulating Rb_6C_{60}, revealing a plethora of features in addition to those observed by optical spectroscopy. Such knowledge of the experimental vibrational properties of C_{60}^{n-} ($n = 0, 3, 6$) is very important for both the theoretical description of the electronic structure of fullerenes and an understanding of the mechanism of superconductivity in the fullerides. In particular, the information provides evidence with regard to the importance of phonons in the pairing interaction.

2.3.1 Pristine Fullerenes

Powder neutron diffraction measurements have confirmed that crystalline C_{60} undergoes a first-order phase transition to a simple cubic structure ($Fm\bar{3}m \rightarrow Pa\bar{3}$) near 260 K, which is accompanied by an abrupt lattice contraction of

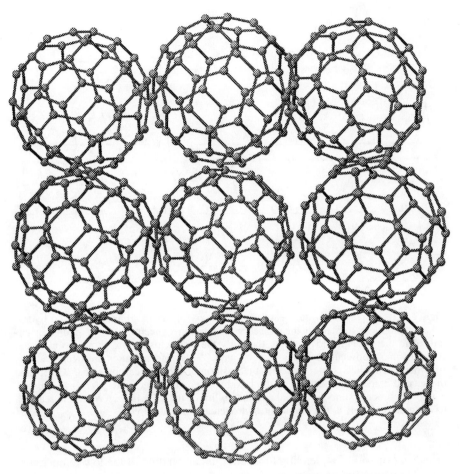

Figure 2.19 Basal-plane projection of the low-temperature structure for C_{60} [35].

0.344(8)% [60, 61]. The transition temperature, the width of the transition, and the change in lattice dimensions are sensitive functions of both the crystallinity and the purity of the solid fullerene. The simple cubic structure arises from the orientational ordering of the C_{60} molecules in the unit cell (Fig. 2.19). The solid-state packing arrangement of C_{60} molecules is subtly optimized at low temperatures [35, 61, 62], a result that cannot be explained by simple van der Waals interactions alone. Even though C_{60} is a quasispherical molecule and the intermolecular energy is dominated by the van der Waals contributions, there is a degree of bond localization that results in a significant anisotropy in the electron-density distribution. The short [0.1404(10) nm] 5=5 ring connection are associated with increased electron density. On the other hand, the long bonds [0.1448(10) nm] making up the twelve pentagons are electron deficient. The crystal structure of C_{60} at low temperatures reflects the subtle interplay between the surface electron-density domains [Fig. 2.20(a)]. Only theoretical calculations, based on our proposed model for the intermolecular interactions and explicitly taking into account the electrostatic interactions, lead to the correct global energy minimum [63, 64].

NMR studies [65, 66] of the rotational dynamics of C_{60} revealed that at the phase transition there is a dramatic change in the nature of the dynamics: The molecular motion in the low-temperature phase was described in terms of jumps between symmetry-equivalent orientations. The value of the cubic lattice constant decreases smoothly below 260 K, albeit with a distinct curvature, until at 90 K, a well-defined cusp appears (Fig. 2.21). The cusp at 90 K is the signature of a higher-order phase transition, arising from the freezing of the rotational jumping motion. The smooth contraction then continues down to 5 K. An attempt was made to refine the dynamic model proposed to explain NMR observations [65, 66] by searching for statistically preferred orientations close in energy to the global minimum configuration. For instance, a 60° rigid rotation about the [111] direction should lead to a local minimum in energy [60] since the intermolecular C_{60}–C_{60} nearest contact is now characterized by a high π-bond-order facing (ca. 159°) a hexagonal face of its neighbor [Fig. 2.20(b)]. The diffraction data in the 5–260 K temperature range were consistent with a simple cubic phase in which two coexisting orientations are present. The fraction of the energetically favorable phase increases smoothly down to 90 K and then remains constant at a value of $p = 0.835(4)$ down to 5 K. The low-temperature structure of C_{60} is best described as an orientational glass; below 90 K, a finite degree of orientational disorder is frozen in, as the thermal energies become much smaller than the rotational barriers. In the intermediate-temperature regime (90–260 K), jumps occur among inequivalent nearly degenerate positions (e.g., 60° jumps about [111] and 36° jumps about [1$\bar{1}$0]). NMR measurements, on the other hand, implied that C_{60} molecules make jumps only among symmetry-equivalent positions. An estimate of 11.4 meV for the energy difference, ΔE, between the two nearly degenerate configurations may be obtained from the temperature variation of the fraction p [61]. These conclusions have now been confirmed by sound velocity [67], dielectric relaxation [68], and thermal conductivity [69] measurements on single crystals. The latter study yielded a glass transition temperature at 85 K, an energy difference $\Delta E \approx 11.9$ meV, and the fraction of molecules at the global minimum $p \approx 0.83$. These results are in remarkable agreement with the neutron diffraction measurements. The present two-state model is, however, only an approximation to the complicated rotational dynamics of C_{60} below 260 K. The fact that the molecules perform small angular oscillations about their equilibrium orientations, with energies 20 cm^{-1} [70], should be also taken into account.

The crystal structure of C_{70} has not been studied in as much detail [71], but preliminary results reveal the existence of a fcc modification at room temperature (space group $Fm\bar{3}m$, lattice constant $a = 14.886$ Å [72]. The C_{70} molecules appear to undergo rapid reorientation [72], and the dimensions of the unit cell scale in this case with the long molecular axis. Extra peaks in neutron

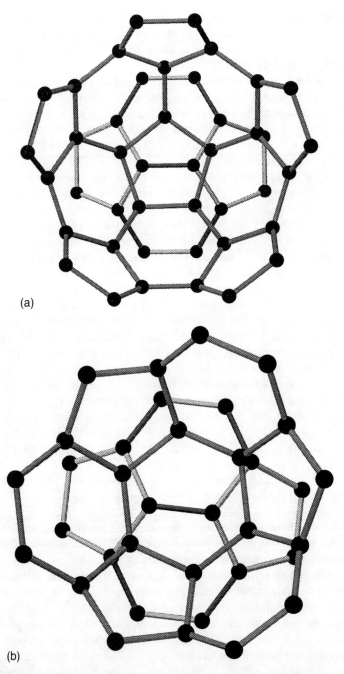

(a)

(b)

Figure 2.20 Intermolecular C_{60}–C_{60} nearest contact viewed along a center-to-center [110] direction. The front face of the far C_{60} molecule is seen through the rear face of the near C_{60} molecule [35]. The 5=5 ring connection faces the (a) pentagonal and (b) hexagonal face.

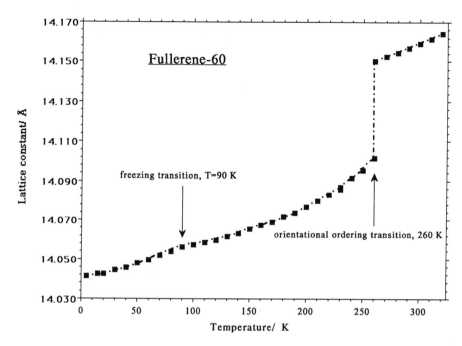

Figure 2.21   Temperature evolution of the cubic lattice constant of solid C_{60} [60].

diffraction measurements at low temperatures (not accounted for within a face-centered cell) reveal the existence of orientationally ordered phases [72].

2.3.2   Alkali-Metal-Doped Fullerides

Superconductivity in alkali-doped M_3C_{60} [73] occurs at much higher temperatures than in alkali doped graphite intercalates and has motivated efforts to understand the origin of the high T_c and the mechanism of pair formation. The fullerides are simpler materials than the high-T_c cuprates and thus are potentially more amenable to theoretical modeling. In addition to purely electronic [74] theories of the superconductivity in M_3C_{60}, phonon-induced pairing mechanisms involving low-energy (10–20 meV) M^+–C_{60}^{3-} optic [75], librational [76], and high-energy intramolecular modes [77, 78] have been considered. The electronic mechanisms for pair binding have focused on the strong intra-fullerene electron–electron repulsions. Phonon-induced mechanisms, on the other hand, ascribe the effective electron attraction to strong scattering near the Fermi surface by ion vibrations. From a BCS-type relation,

$$T_c \propto \hbar\omega_{\text{ph}} \exp[-1/VN(\varepsilon_F)] = \hbar\omega_{\text{ph}} \exp(-1/\lambda) \qquad (2.1)$$

the observed high T_cs may be understood in terms of a high average phonon frequency $\hbar\omega_{\text{ph}}$, resulting from the light carbon mass and the large force constants associated with the intramolecular modes. A high DOS at ε_F arises from the weak intermolecular interactions and strong scattering associated with intramolecular modes.

Knowledge of the vibrational density of states and the way it evolves with temperature and composition is crucial to an understanding of phonon-mediated superconductivity. As the fullerides are molecular superconducters, the electron–phonon Hamiltonian has a particularly simple form. The total electron–phonon coupling constant λ is simply expressed as a sum of partial contributions λ_n, associated with each mode n mediating the pairing interaction, $\lambda = \Sigma_n \lambda_n$. Using simple symmetry arguments [77, 78], it can be shown that, under icosahedral symmetry, only the H_g intramolecular modes can couple to the t_{1u} conduction electrons, while the totally symmetric A_g modes become active only for finite wave vectors ($Q \neq 0$). Electron–phonon coupling should produce substantial broadening and softening of the associated intramolecular H_g modes in the superconducting fullerides, as compared with pristine C_{60}. Quantitatively the effect on the phonon widths may be estimated using the expressions:

$$\lambda_n \sim (\pi/5)N(\varepsilon_F)\lambda_n\omega_n^2 \tag{2.2}$$

where λ_n is the increase in full width at half-maximum of the fivefold-degenerate nth phonon. The inelastic neutron scattering spectra of superconducting K_3C_{60} [79], insulating C_{60} [80–82], and overdoped Rb_6C_{60} [83] have all been measured at low temperatures. Phonon modes are well separated in the low-energy (30–90 meV) range, and changes in position and width can be followed with confidence as a function of the reduction level of the fullerene cage. The INS data provide evidence of strong electron–phonon coupling to the H_g modes [61]. Indeed the most remarkable feature when comparing the INS spectra of K_3C_{60} and C_{60} (Fig. 2.22) is the virtual disappearance of the $H_g(2)$ mode at 53.2 meV, due to a 4.3 meV increase in FWHM. This is consistent with strong electron–phonon coupling (Table 2.1). We find that the electron-phonon coupling constant for this mode is $\lambda(2) \sim 0.17$, ~30% of the total λ. The dominant role of this mode has also been confirmed by Raman studies on ultrathin Rb_xC_{60} ($x = 0$–3) films [84, 85]. Strong coupling is also evident for the well-separated $H_g(1)$ mode, which broadens by 1.0 meV, resulting in an estimated $\lambda(1)$ contribution to the total λ of 0.10. $H_g(4)$ is more weakly coupled, and we estimate an increase in γ of at least 3.1 meV, corresponding to a lower limit for $\lambda(4)$ of 0.04. On the other hand, the FWHM of $H_g(3)$ is virtually unchanged, even though it is a somewhat softer mode.

It is more difficult to interpret the INS spectra in the tangential-mode regime (120–200 meV). Preliminary analysis of recent high-resolution INS measurements [86] reveals that the $H_g(6)$, $H_g(7)$, and $H_g(8)$ couple strongly as the associated features that are present in C_{60} are absent in the A_3C_{60} spectrum. Our experimental results confirm uniquely the important conclusion that electron–phonon coupling strength should be distributed between buckling and tangential modes. Combining the LDA frozen-phonon λ values for the tangential modes

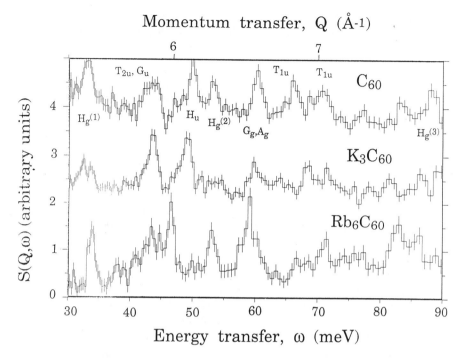

Figure 2.22 Inelastic neutron scattering vibrational spectra of the intramolecular modes of C_{60} (20 K), K_3C_{60} (5 K), and Rb_6C_{60} (20 K) in the 30–90 meV energy range [83].

calculated by Schluter et al. [78] with our estimates for the radial modes, we find $\lambda = 0.64$. With the aid of McMillan's formula, this value of λ may be combined with estimates of the density of states [$N(0) \approx 14$ states/C_{60}/eV/spin) and the Coulomb pseudopotential ($\mu^* \approx 0.15$) to give an estimate of $T_c \approx 17$ K (for $\omega_{\log} = 1072$ K), in good agreement with the experiment. A crucial point revealed by our results is the contrast with alkali-intercalated graphite, where, because of symmetry restrictions, the corresponding buckling modes do not couple to the electrons. Because of the curvature of the fullerene cage, it appears that *coupling to the radial modes accounts for roughly half of the total electron–phonon coupling strength in K_3C_{60}, which is responsible for the fact that T_c is much higher than in the lamellar graphite intercalates.* Finally, low-energy INS measurements [87] of the librational modes ($\omega = 4.1$ meV) of K_3C_{60} above and below the superconducting transition temperature reveal no anomalous behavior for these librational modes, effectively ruling out strong electron–librational coupling.

2.4 Microparticles and Nanofiber Studies

Carbon fibers have many industrial applications. Their growth mechanism and the factors that control their structure are of major strategic importance. Re-

Table 2.1 Observed positions, ω_n, of the four radial H_g modes in C_{60} and their broadenings, γ_n, on reduction to $C_{60}{}^{3-}$ together with calculated electron–phonon coupling strengths, V_n (in meV), for a density of states $N(0) = 14$ states/eV/SPIN/C_{60}

Vibrational mode	$H_g(1)$	$H_g(2)$	$H_g(3)$	$H_g(4)$
ω_n (meV)	33.1	53.2	88.1	96.2
γ_n (cm^{-1})	8	35	1	25
V_n (meV)	7.4	12.4	0.1	2.7
λ_n	0.10	0.17	0.00	0.04

cently a new type of microscopic carbon fiber has been detected that appears to consist of very small-diameter graphite tubes from 30 to 100 Å in diameter. Iijima [88] has published TEM images of such microfibers, which appear to grow on the cathode of a fullerene arc processor similar to that developed by Krätschmer et al. [21]. Similar structures have also been observed by Endo [89] in a standard carbon fiber generator. The cylinder walls may consist of only a few layers of graphite: 2–5 or maybe as many as 50 or more. Such fibers are an exciting development in micro/nano–engineering, as they may well prove to be the strongest structures so far fabricated.

A comparison between the observed TEM images of the nanofibers [88] and simulated TEM images for elongated giant fullerenes has been carried out [90]. This study is based on a detailed analysis of the structures of concentric-shell quasi-icosahedral carbon microparticles by McKay et al. (Fig. 2.22) [91]. Apart from the beautifully cylindrical structures of the nanofibers, the most striking result is the fact that the TEM images show that the ends are capped by a continuous dome of carbon [88]. The microfibers are closed by caps that are curved, polygonal, or cone shaped. The discovery that giant fullerenes have quasi-icosahedral shapes [17] readily explains these discoveries, as it shows how a closed graphite network may give rise to caps with exactly such a varying shape. Simulations of the TEM images for giant fullerenes [90] indicate how a range of shapes for such caps might occur. The pattern observed depends not only on the actual 3D shape of the fullerene but also quite critically on the orientation of the object in the electron beam (Fig. 2.22) [91]. In Figure 2.23 a computer-generated carbon network for a symmetric giant fullerene half-dome is depicted [90]. Thus it is easy to see that massive extended tubular sp^2 networks may exist that are essentially giant fullerenes—the zeppelenes (Fig. 2.24) [91].

In Figure 2.25 a TEM simulation result for a double-shell dome, similar to the structure in Figure 2.23, is shown that explains the characteristics of the TEM images observed by Iijima [88]. The new structures are overlapped in such a way that some further growth factor is required for an explanation. If a closed fullerene structure can grow by direct insertion of smaller carbon species into the graphite network, several observations on the fullerenes and the

Figure 2.23 Two TEM simulations of a concentric-shell graphite microparticle. The difference between these two images lies solely in the orientation of the particle in the electron beam [91].

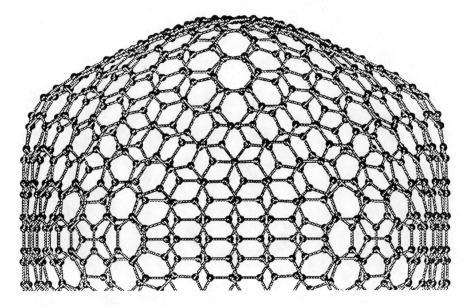

Figure 2.24   Computer simulation of a giant fullerene end cap [90]. The structure has been relaxed by molecular mechanics.

carbon microfibers are explained [90]. Iijima [88] has found that the carbon atoms in the cylindrical walls of the nanofibers are arrayed in a helical screw configuration. This can be readily understood [90] if the cap of the fullerene is nonsymmetric, which is almost certainly the general case. A simple example is shown in Figure 2.26, where the Schlegel diagram for one of the simplest (and smallest) unsymmetric fullerene caps is depicted [90]. Although this is much smaller than the caps observed so far, which have ~3 nm diameters, the result is generally valid. In Figure 2.27 one pentagon is displaced from the basic C_{5v} symmetry of a C_{60} or C_{70} half-dome. Carbon fragment insertion into the network involving screw growth of such a cap would result in the laying down of a uniform cylindrical thread of carbon atoms in a symmetrical helix [90].

2.5   Fullerene Analogues in Space?

There are several problems involving carbon where some links between C_{60}, chains, and dust are possible, and one of the most interesting relates to the carrier of the diffuse interstellar bands (DIBs). The DIBs, which are absorption bands in the spectrum of starlight due to some so-far-unidentified material in space, have been a major puzzle ever since they were identified as interstellar features in 1936. They are too broad to be atomic lines and so must be due to

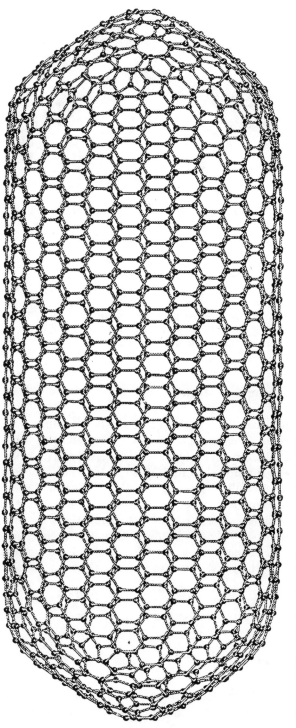

Figure 2.25 Tubular giant fullerene (zeppelene) [90].

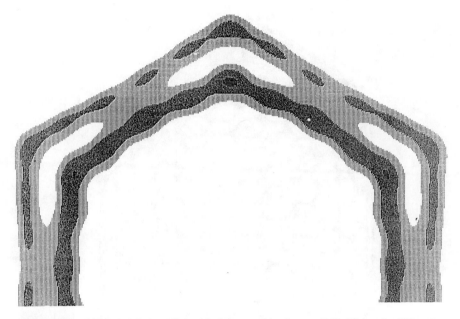

Figure 2.26   TEM simulation of double-layer grahite domes [90]. The only difference between the two layers lies in the disposition of the pentagonal cusps over the surface and the orientation of the cap in the electron beam.

some molecular aggregates. Their properties have been summarized by Herbig [92].

The problem is that this material appears to be ubiquitously abundant even in the hostile environment of space and yet, astoundingly, it has never been identified on earth. C_{60} or an analogue was suggested as a possible carrier when the molecule was first found to form spontaneously [4]. There are intriguing parallels between the fullerene-60 and the DIB carrier that have prompted the conjecture to be examined further [7]. Perhaps the most interesting analogues are *exo*hedral complexes, as it is clear that if fullerenes are formed in space then species such as $(\ ) \cdot M^+$, where M is an abundant interstellar species (e.g., Na, K, Ca, S, O, H, etc.) must form [93]. Detailed arguments that support the possibility that charge-transfer bands

$$(\ ) \cdot M^+ \overset{h\nu}{\leftrightarrow} (+) \cdot M$$

associated with such exohedral species exist have been discussed [93].

2.6   Postscript

A New Postbuckminsterfullerene World of round organic chemistry and materials science has developed overnight. Papers dealing with novel aspects of

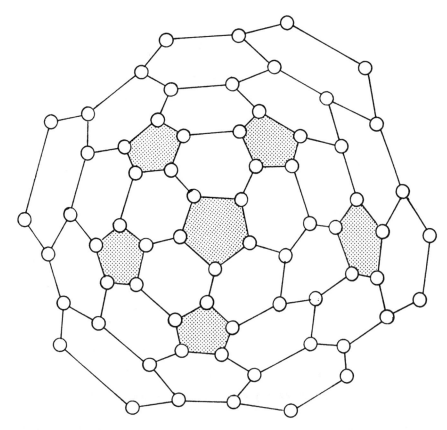

Figure 2.27 Schlegel diagram for a slightly asymmetric fullene end cap connected to a helical cylindrical array of hexagonally linked carbon atoms. The cylinder wall pattern repeats after 22 atoms have been added, that is, two revolutions of the end cap [90].

fullerenes appear almost daily. In these beleaguered times for fundamental science, particularly in the U.K., Pasteur's famous edict: "Il n'existe pas de sciences appliquees, mais seulement des applications de la science" has an ominous ring. The fullerene story is yet another shining example.

Acknowledgments

We thank our co-workers cited in the text; also Eleanor Campbell, John Copley, Bill David, Morinobu Endo, Robert Haddon, Richard Hallet, Richard Henderson, Michael Jura, D. W. Murphy, Matt Rosseinsky, Peter Sarre, Michael Schlüter, John Tom Kinson, David Wales, and Steve Wood. We also thank Alcan (U.K.), British Gas, BP, ICI, and the Royal Society for financial support.

References

1. H. W. Kroto, *Chem. Soc. Rev.* **11**, 435–91 (1982).

2. H. W. Kroto, *Internat. Rev. Chem. Phys.* **1**, 309–76 (1981).

3. T. G. Dietz, M. A. Duncan, D. E. Powers, and R. E. Smalley, *J. Chem. Phys.* **74**, 6511–12 (1981).

4. H. W. Kroto, J. R. Heath, S. C. O'Brien, R. F. Curl, and R. E. Smalley, *Nature (London)* **318**, 162–63 (1985).

5. J. R. Heath, S. C. O'Brien, Q. Zhang, Y. Liu, R. F. Curl, H. W. Kroto, and R. E. Smalley, *J. Am. Chem. Soc.* **107**, 7779–80 (1985).

6. A. Nickon and E. F. Silversmith, *Organic Chemistry—The Name Game: Modern Coined Terms and Their Origins,* Pergamon, New York, 1987.

7. H. W. Kroto, *Science* **242**, 1139–45 (1988).

8. R. F. Curl and R. E. Smalley, *Science* **242**, 1017–22 (1988).

9. R. F. Curl, *Carbon,* **30**, 1149–55.

10. H. W. Kroto, W. Allaf, and S. P. Balm, *Chem. Rev.* **91**, 1213 (1991).

11. F. D. Weiss, J. L. Elkind, S. C. O'Brien, R. F. Curl, and R. E. Smalley, *J. Am. Chem. Soc.,* **110**, 4464–65 (1988).

12. T. G. Schmalz, W. A. Seitz, D. J. Klein, and G. E. Hite, *Chem. Phys. Lett.* **130**, 203–7 (1986).

13. H. W. Kroto, *Nature (London)* **329**, 529–31 (1987).

14. T. G. Schmalz, W. A. Seitz, D. J. Klein, and G. E. Hite, *J. Am. Chem. Soc.* **110**, 1113–27 (1988).

15. Q. L. Zhang, S. C. O'Brien, J. R. Heath, Y. Liu, R. F. Curl, H. W. Kroto, and R. E. Smalley, *J. Phys. Chem.* **90**, 525–28 (1986).

16. Ph. Gerhardt, S. Loeffler, and K. Homann, *Chem. Phys. Lett.* **137**, 306–10 (1987).

17. H. W. Kroto and K. G. McKay, *Nature (London)* **331**, 328–31 (1988).

18. W. Krätschmer, K. Fostiropoulos, and D. R. Huffman, *Dusty Objects in the Universe,* E. Bussoletti and A. A. Vittone, eds., Kluwer, Dordrecht, 1990 (conference in 1989).

19. W. Krätschmer, K. Fostiropoulos, and D. R. Huffman, *Chem. Phys. Lett.* **170**, 167–70 (1990).

20. H. W. Kroto, *Angew. Chem. Internat. Edit. Engl.* **31**, 111–29 (1992).

21. W. Krätschmer, L. D. Lamb, K. Fostiropoulos, D. R. Huffman, *Nature (London)* **347**, 354–58 (1990).

22. R. Taylor, J. P. Hare, A. K. Abdul-Sada, and H. W. Kroto, *J. Chem. Soc. Chem. Commun.* 1423–25 (1990).

23. R. E. Haufler, J. Conceicao, L. P. F. Chibante, Y. Chai, N. E. Byrne, S. Flanagan, M. M. Haley, S. C. O'Brien, C. Pan, Z. Xiao, W. E. Billups, M. A. Ciufolini, R. H. Hauge, J. L. Margrave, L. J. Wilson, R. F. Curl, and R. E. Smalley, *J. Phys. Chem.* **94**, 8634–36 (1990).

24. R. D. Johnson, G. Meijer, and D. S. Bethune, *J. Am. Chem. Soc.* **112**, 8983–84 (1990).

25. J. P. Hare, H. W. Kroto, and R. Taylor, *Chem. Phys. Lett.* **177**, 394–97 (1991).

26. R. Taylor, J. P. Parsons, A. G. Avent, S. P. Rannard, T. J. Dennis, J. P. Hare, H. W. Kroto, and D. R. M. Walton, *Nature (London)* **351**, 277 (1991).

27. T. J. Dennis, J. P. Hare, H. W. Kroto, R. Taylor, D. R. M. Walton, and P. J. Hendra, *Spectrochim. Acta* **47A**, 1289–92 (1991).

28. R. Taylor, G. J. Langley, T. J. Dennis, H. W. Kroto, and D. R. M. Walton, *J. Chem. Soc., Chem. Commun.* submitted.

29. R. Ettl, I. Chao, F. Diederich, and R. L. Whetten, *Nature (London)* **353**, 149–52 (1991).

30. F. Diederich, R. L. Whetten, C. Thilgen, R. Ettl, I. Chao, M. M. Alvarez, *Science* **254**, 1768–70 (1991).

31. K. Kikuchi, T. Wakabayashi, N. Nakahara, S. Suzuki, H. Shiromaru, Y. Miyake, K. Saito, I. Ikemoto, M. Kainosho, and Y. Achiba, *Nature (London)* **357**, 142–45 (1992).

32. H. W. Kroto, K. Prassides, R. Taylor, and D. R. M. Walton, *Physica Scripta,* **T45**, 314–18 (1992).

33. C. S. Yannoni, P. P. Bernier, D. S. Bethune, G. Meijer, and J. R. Salem, *J. Am. Chem. Soc.* **113**, 3190 (1991).

34. J. M. Hawkins, A. Meyer, T. A. Lewis, S. Loren, and F. J. Hollander, *Science* **525**, 312 (1991); J. M. Hawkins S. Loren, A. Meyer, and R. Nunlist, *J. Am. Chem. Soc.* **113**, 7770–71 (1991).

35. W. I. F. David, R. M. Ibberson, K. Prassides, T. J. S. Dennis, J. P. Hare, H. W. Kroto, R. Taylor, and D. R. M. Walton, *Nature (London)* **353**, 147–49 (1991).

36. P. R. Birkett, P. B. Hitchcock, H. W. Kroto, R. Taylor, and D. R. M. Walton, *Nature (London)* **357**, 479–81 (1992).

37. M. F. Meidine, P. B. Hitchcock, H. W. Kroto, R. Taylor, and D. R. M. Walton, *J. Chem. Soc., Chem. Commun.*, 1534–37 (1992).

38. R. Taylor, *Tetrahedron Lett.* 3731–34 (1991); *J. Chem. Soc., Perkin Trans.* **2**, 3–4 (1992).

39. R. Taylor, *J. Chem. Soc., Perkin Trans.* **2**, 453–55 (1992).

40. J. H. Holloway, E. G. Hope, R. Taylor, G. J. Langley, A. G. Avent, T. J. Dennis, J. P. Hare, H. W. Kroto, and D. R. M. Walton, *J. Chem. Soc., Chem. Commun.* 966–69 (1991).

41. R. Taylor, G. J. Avent, T. J. Dennis, J. P. Hare, H. W. Kroto, D. R. M. Walton, J. H. Holloway, E. G. Hope, and G. J. Langley, *Nature (London)* **355**, 27–28 (1992).

42. R. Taylor, J. H. Holloway, E. G. Hope, A. G. Avent, G. J. Langley, J. P. Hare, T. J. Dennis, H. W. Kroto, and D. R. M. Walton, *J. Chem. Soc., Chem. Commun.* 667–68 (1992).

43. G. A. Olah, I. Bucsi, C. Lambert, R. Aniszfeld, N. J. Trivedi, D. K. Sensharma, and G. K. S. Prakash, *J. Am. Chem. Soc.* **113**, 9385 (1991).

44. F. N. Tebbe, J. Y. Becker, D. B. Chase, L. E. Firment, E. R. Holler, B. S. Malone, P. J. Krusic, E. Wasserman, *J. Am. Chem. Soc.* **113**, 9900–01 (1991).

45. F. N. Tebbe, R. L. Harlow, D. B. Chase, D. L. Thorn, G. C. Campbell, J. C. Calabrese, N. Herron, R. J. Young, Jr., and E. Wasserman, *Science* **256**, 822–25 (1992).

46. R. Taylor, G. J. Langley, M. F. Meidine, J. P. Parsons, A. K. Abdul-Sada, T. J. Dennis, J. P. Hare, H. W. Kroto, and D. R. M. Walton, *J. Chem. Soc., Chem. Commun.* 665–67 (1992).

47. J. P. Hare, T. J. Dennis, H. W. Kroto, R. Taylor, W. A. Allaf, S. P. Balm, and D. R. M. Walton, *J. Chem. Soc., Chem. Commun.* 412–13, (1991).

48. Z. Gasyna, P. N. Schatz, J. P. Hare, T. J. Dennis, H. W. Kroto, R. Taylor, and D. R. M. Walton, *Chem. Phys. Lett.* **183**, 283–91 (1991).

49. T. J. Dennis, J. P. Hare, H. W. Kroto, R. Taylor, D. R. M. Walton, and P. J. Hendra, *Spectrochimica Acta* **47A**, 1289–92 (1991).

50. S. Leach, M. Vervloet, A. Despres, E. Breheret, J. P. Hare, T. J. Dennis, H. W. Kroto, R. Taylor, and D. R. M. Walton, *Chem. Phys.* **160,** 451–66 (1991).

51. W. H. Blau, H. J. Byrne, D. J. Cardin, T. J. Dennis, J. P. Hare, H. W. Kroto, R. Taylor, and D. R. M. Walton, *Phys. Rev. Lett.* **67,** 1423–25 (1991).

52. A. Dworkin, H. Szwarc, S. Leach, J. P. Hare, T. J. Dennis, H. W. Kroto, R. Taylor, and D. R. M. Walton, *C. R. Acad. Sci. Paris,* t. 312, Ser. II, 979–82 (1991).

53. A. Dworkin, C. Fabre, D. Schutz, G. Kriza, R. Ceolin, H. Swarc, P. Bernier, D. Jerome, S. Leach, A. Rassat, J. P. Hare, T. J. Dennis, H. W. Kroto, R. Taylor, and D. R. M. Walton, *C. R. Acad. Sci. Paris,* t. 313, Ser. II, 1017–21, (1991).

54. A. MacKay, M. Vickers, J. Klinowski, J. P. Hare, T. J. Dennis, H. W. Kroto, R. Taylor, and D. R. M. Walton (unpublished results; see also Fig. 34, Ref. 9).

55. S. Liu, Y. Lu, M. M. Kappes, J. A. Ibers, *Science* **254,** 408–10 (1991).

56. J. R. Heath, S. C. O'Broen, R. F. Curl, H. W. Kroto, and R. E. Smalley, (unpublished results).

57. R. Hallet, K. G. McKay, S. P. Balm, W. A. Allaf, H. W. Kroto, and A. J. Stace, to be published.

58. L. A. Paquette, R. J. Ternansky, D. W. Balogh, and G. J. Kentgen, *J. Am. Chem. Soc.* **105,** 5446–50 (1983).

59. P. A. Heiney, J. E. Fischer, A. R. McGhie, W. J. Romanow, A. M. Denenstein, J. P. McCauley, Jr., A. P. Smith III, and D. E. Cox, *Phys. Rev. Lett.* **66,** 2911–14 (1191).

60. W. I. F. David, R. M. Ibberson, T. J. S. Dennis, J. P. Hare, and K. Prassides, *Europhys. Lett.* **18,** 219–25 (1992).

61. K. Prassides, H. W. Kroto, R. Taylor, D. R. M. Walton, W. I. F. David, J. Tomkinson, M. J. Rosseinsky, D. W. Murphy, and R. C. Haddon, *Carbon* **30,** 1277–86 (1992).

62. K. Prassides, and H. W. Kroto, *Physics World* **5,** 44–49 (1992).

63. J. P. Lu, X. P. Li, and R. M. Martin, *Phys. Rev. Lett.* **68,** 1551–54 (1992).

64. M. Sprik, A. Cheng, and M. L. Klein, *J. Phys. Chem.* **96,** 2027–29 (1992).

65. R. Tycko, G. Dabbagh, R. M. Fleming, R. C. Haddon, A. V. Makhija, and S. M. Zahurak, *Phys. Rev. Lett.* **67,** 1886–89 (1991).

66. R. D. Johnson, C. S. Yannoni, H. C. Dorn, J. R. Salem, and D. S. Bethune, *Science* **255,** 1235–38 (1992).

67. X. D. Shi, A. R. Kortan, J. M. Williams, A. M. Kini, B. M. Savall, and P. M. Chaikin, *Phys. Rev. Lett.* **68,** 827–30 (1992).

68. G. B. Alers, B. Golding, A. R. Kortan, R. C. Haddon, and F. A. Theil, Science **257,** 511–14 (1992).

69. R. C. Yu, N. Tea, M. B. Salamon, D. Lorents, and R. Malhotra, *Phys. Rev. Lett.* **68,** 2050–53 (1992).

70. J. R. D. Copley, D. A. Neumann, R. L. Cappelletti, and W. A. Kamitakahara, *J. Phys. Chem. Solids,* **53,** 1353–71 (1992).

71. G. B. M. Vaughan, P. A. Heiney, J. E. Fischer, D. E. Luzzi, D. A. Rickettsford, A. R. McGhie, Y. W. Hui, A. L. Smith, D. E. Cox, W. J. Romanow, B. H. Allen, N. Coustel, J. P. McCauley, A. B. Smith, *Science* **254**, 1350–53 (1991).

72. T. J. Dennis, C. Christides, K. Prassides, H. W. Kroto, and D. R. M. Walton, unpublished results.

73. A. F. Hebard, M. J. Rosseinsky, R. C. Haddon, D. W. Murphy, S. H. Glanim, T. T. M. Palstra, A. P. Ramirez, and A. R. Kortan, *Nature (London)* **350**, 660–62 (1991).

74. S. Chakravarty, and S. Kivelson, *Europhys. Lett.* **16**, 751–56 (1991).

75. F. C. Zhang, M. Ogata, and T. M. Rice, *Phys. Rev. Lett.* **67**, 3452–55 (1991).

76. O. V. Dolgov, and I. I. Mazin, *Solid State Commun.* **81**, 935–38 (1992).

77. C. M. Varma, J. Zaanen, and K. Raghavachari, *Science* **254**, 989–92 (1991).

78. M. Schluter, M. Lannoo, M. Needles, G. A. Baraff, and D. Tomanek, *Phys. Rev. Lett.* **68**, 526–29 (1992).

79. K. Prassides, J. Tomkinson, C. Christides, M. J. Rosseinsky, D. W. Murphy, and R. C. Haddon, *Nature* **354**, 462–63 (1991).

80. K. Prassides, T. J. S. Dennis, J. P. Hare, J. Tomkinson, H. W. Kroto, R. Taylor, and D. R. M. Walton, *Chem. Phys. Lett.* **187**, 455–58 (1991).

81. R. L. Cappelletti, J. R. D. Copley, W. A. Kamitakahara, F. Li, J. S. Lannin, and D. Ramage, *Phys. Rev. Lett.* **66**, 3261–64 (1991).

82. C. Coulombeau, H. Jobic, P. Bernier, C. Fabre, D. Schatz, and A. Rassat, *J. Phys. Chem.* **96**, 22–24 (1992).

83. K. Prassides, C. Christides, M. J. Rosseinsky, J. Tomkinson, D. W. Murphy, and R. C. Haddon, *Europhys. Lett.* **19**, 629–35 (1992).

84. M. G. Mitch, S. J. Chase, and J. S. Lannin, *Phys. Rev. Lett.* **68**, 883–86 (1992).

85. P. Zhou, K. A. Wang, A. M. Rao, P. C. Eklund, G. Dresselhaus, and M. S. Dresselhaus, *Phys. Rev. B* **45**, 10838–40 (1992).

86. C. Christides, K. Prassides, M. J. Rosseinsky, D. W. Murphy, and R. C. Haddon, unpublished results.

87. C. Christides, D. A. Neumann, K. Prassides, J. R. D. Copley, M. J. Rosseinsky, D. W. Murphy, and R. C. Haddon, *Phys. Rev. B* **46**, 12088–91 (1992).

88. S. Iijima, *Nature (London)* **354**, 56–58 (1991); see also: P. Ball, *Nature (London)* **354**, 18 (1991).

89. M. Endo and H. W. Kroto, to be published.

90. M. Endo and H. W. Kroto, *J. Phys. Chem.* **96**, 6941–44 (1992).

91. K. G. McKay, D. J. Wales, and H. W. Kroto, J. Chem. Soc., Faraday Trans. **88**, 2815–21 (1992).

92. G. H. Herbig, *Astrophys. J.* **196**, 129 (1975).

93. H. W. Kroto and M. Jura, *Astron. Astrophys.* **263**, 275–80 (1992).

CHAPTER

3

The Higher Fullerenes

Carlo Thilgen, François Diederich, and Robert L. Whetten

3.1 Higher Fullerene Isolation: From Early Enrichments to the Separation of Isomers

At the same time that macroscopic quantities of pure C_{60} and C_{70} became available through the *Krätschmer–Huffman* method, the resistive heating of graphite under inert atmosphere [1], we obtained the first indications of the formation of higher fullerenes, spherical all-carbon molecules with a number of atoms larger than 70 [2].

The evidence for the presence of higher fullerenes in the soot prepared by the method mentioned was based on mass spectrometric analysis of the crude benzene extract [3]. Besides the most prominent peaks of C_{60} ($m/z = 720$) and C_{70} ($m/z = 840$), the spectra exhibited a number of smaller peaks having a relative intensity of up to 10% and corresponding to masses up to $m/z = 1176$, which is equivalent to a molecular formula C_{98}. Thanks to a considerable improvement in the yield of extractable carbon molecules from initially 14% to 25%–30%, as well as to an optimization of the preparative separation procedure, we were able to collect enriched samples of higher fullerenes in milligram quantities [3]. Repeated gravity chromatography on neutral alumina (activity 1) using a gradient elution with toluene/hexanes afforded five fractions of slightly different colors (brown to greenish) that gave stable crystalline solids upon evaporation of the solvents. These fractions were further analyzed by high-performance liquid chromatography (HPLC) (silica gel, *n*-hexane) and by mass spectrometry.

At that time, laser-desorption time-of-flight (LD-TOF) and fast atom bombardment (FAB) mass spectrometry were the most successful detection tools in higher fullerene investigations. Electron impact (EI), desorption chemical ionization (DCI), and field desorption (FD) mass spectrometry were less useful, because with these techniques fragmentation of the higher fullerenes was very important and the peaks corresponding to C_{60} and C_{70} dominated all spectra by far.

The consecutive loss of C_2 units leading to C_{70} and ultimately to C_{60} as major fragmentation products was a general feature of all mass spectra. This observation, which had already been made with carbon ions generated by laser vaporization of graphite [4, 5], was considered to be strong evidence for the fullerene nature of the chromatographed compounds.

The first of the five fractions obtained as described (see Fig. 3.1) consisted of a highly enriched sample of C_{76} with C_{78} as a minor contaminant, and in the following one, C_{84} was the main product [3]. The third fraction was also

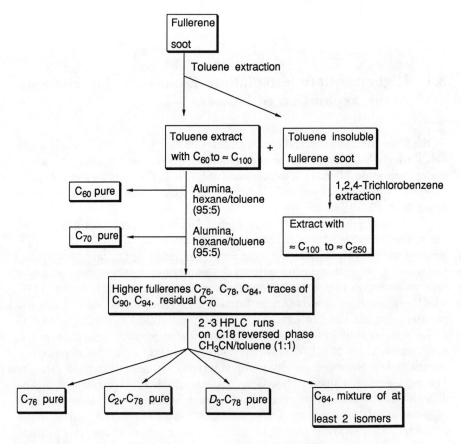

Figure 3.1   A protocol for fullerene separation and isolation.

highly enriched but did not contain a pure carbon compound. Its main constituent corresponded to the formula $C_{70}O$ and was an oxide of D_{5h}-C_{70} with a bridging oxygen atom at the outer surface of the cage. The heteroatom is easily lost in the course of the mass spectrometric experiment, especially under higher-energy conditions. The formation of $C_{70}O$ was due to trace oxygen impurities in the soot production reactor [3] and also occurred during the evaporation of carbon rods that were doped with La_2O_3 [6]. Fullerene oxides can also be synthesized starting from the carbon spheres by photooxygenation [7] or in a "wet chemistry" process by reaction with oxidizing agents, such as 2,2-dimethyl-dioxirane [8]. The last two fractions were less enriched and included C_{90}, respectively, C_{94}, as the main components.

In later procedures, [9, 10] we used gravity chromatography on alumina only for the separation of C_{60} and C_{70} in order to get an enriched higher fullerene fraction (with residual C_{70}). A major improvement of our separation techniques was possible by the use of a C_{18} reversed-phase HPLC column and a 30:70 CH_3CN/CH_2Cl_2 mixture as the eluent [11] as shown in Figure 3.2. These conditions allowed the isolation of pure C_{76}. But we were able to optimize the eluent system further and, by switching to the solvent mixture $CH_3CN/toluene$ in a 1:1 ratio, we succeeded not only in purifying C_{78}, but also in separating this fullerene into its isomers of C_{2v} and D_3 symmetry [10]. The C_{84} fraction, however, consisted of a mixture of two predominant isomers that still have to be separated [12].

3.2 Fullerene Structure Elucidation: Interplay between Theoretical Prediction and ^{13}C NMR Experiments

The number of possible isomers of carbon molecules C_n with $n > 60$ is enormous. But this number can be gradually reduced by the application of different rules that have proved to be valid for all fullerenes known so far. According to the fundamental definition, fullerenes C_{20+2m} obey Euler's rule and consist of a closed-cage-type network of C atoms having 12 pentagonal and m hexagonal faces [13, 14]. Bearing in mind this definition and applying a spiral algorithm [15], Manolopoulos and Fowler have found 21,822 possible isomers of C_{78} [16], a number that does not seem very promising in terms of structure elucidation. This number, however, can be further reduced by taking into consideration the "isolated pentagon rule" (IPR) [13, 17], which allows only fullerene structures in which all five-membered rings (5MRs) are completely separated from each other by six-membered rings (6MRs). It can be rationalized by the fact that edge-sharing pentagons—as a consequence of the enforced bond angles—induce a considerable strain in the cage-type carbon network. Furthermore, two adjacent 5MRs generate an energetically unfavorable pentalene-type 8π-electron system. Qualitative Hückel molecular-orbital (HMO) theory allows a further reduction of the number of expected isomers by favoring closed-shell or partially closed-shell (the latter have a slightly bonding LUMO)

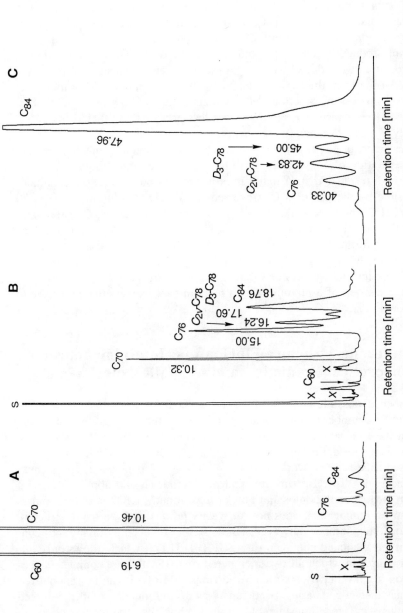

Figure 3.2 HPLC profiles (C₁₈ reversed phase, acetonitrile/toluene 1:1, 310 nm uV detection) (a) of the crude toluene extract of the fullerene soot, (b) of the higher fullerene mixture obtained after removal of all C_{60} and most of C_{70}, and (c) of a prepurified C_{84} fraction with traces of residual C_{76}, C_{2v}-C_{78}, and D_3-C_{78}. S = solvent, X = unknown impurities. (a) and (b) were obtained on a 250 × 10 mm² column (flow rate 5 ml/min, pressure 650 psi); (c) was obtained on a 250 × 25 mm² column (flow rate 6 ml/min, pressure 300 psi); retention times are given in minutes.

over open-shell structures [14, 16]. In accordance with these considerations, the only isolated isomer of C_{76} has D_2 symmetry (partially closed shell), and no T_d-C_{76} (open shell) was formed in our experiments [9]. With the observance of all the constraints discussed so far, the number of possible fullerene structures of a given C_n is reduced considerably but nevertheless increases rapidly with increasing n; Manolopoulos and Fowler found 5 for C_{78}, 24 for C_{84}, and 46 for C_{90} [16].

Predictions about relative stabilities among closed-shell isomers based on the magnitude of the HOMO–LUMO gap seem to be less reliable: We could not detected any D_{3h}-C_{78}, which is predicted to have the largest HOMO–LUMO gap among the five fullerene structures of C_{78} and should therefore be formed preferentially [14].

Theoretical considerations thus lead to a limited number of probable structures (Fig. 3.3). If they all differ in symmetry, an assignment should be easy by comparing the number of experimentally observed ^{13}C NMR lines to the number of distinct C-atom sites. In case several structures have the same symmetry, an assignment may still be possible by additionally taking into account the line intensities and relating them to the number of chemically equivalent carbon atoms.

The number of lines in different chemical shift regions provides an additional means of structure elucidation: the characteristic shift range of the fullerenes (≈ 130–150 ppm) can be subdivided into three sections typical of three different types of carbon atoms [18]. The most downfield shift region corresponds to pyracylene-type C atoms that are lying at the vertex of one five-membered ring and two 6MR, and are connected to another 5MR. The most upfield region is that of pyrene-type atoms lying at the vertex of three 6MR. Finally, the intermediate region is characteristic of corannulene-type C atoms located at the vertex of one 5MR and two 6MR, but contrary to the pyracylene sites, these are not connected to another 5MR. Even though per definition only the number of pyrene-type atoms increases with increasing molecular weight, the symmetry of a given fullerene and the position of the atoms relative to the symmetry elements has an influence on the number of chemically different sites of each type.

In an ideal case, the connectivity of all C atoms can be established by 2D NMR, as has been shown by Johnson et al. for C_{70} in a 2D inadequate ^{13}C NMR experiment [18], as shown in Figure 3.4. In the case of higher fullerenes, however, the limited amounts of pure compound available have prevented the application of this technique so far. The use of ^{13}C-enriched graphite for the production of soot is a possibility to overcome the problem.

All our ^{13}C NMR spectra have been recorded in CS_2, which, at room temperature, has proved to be a remarkable solvent for the carbon spheres, and its own resonance at 192.7 ppm is located far out of the typical fullerene shift range between 130 and 150 ppm. Due to the small quantities of high-purity material available as well as to a decrease in symmetry compared to C_{60}, the

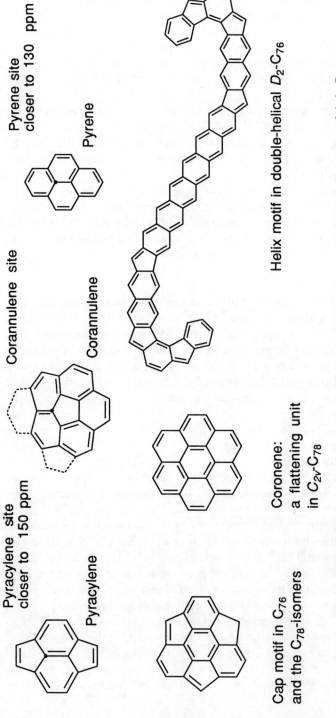

Figure 3.3   Structural motifs found in higher fullerenes and the three different kinds of atomic environments of their C atoms.

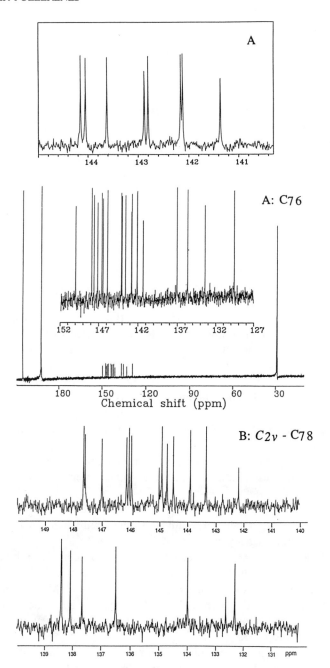

Figure 3.4 ^{13}C NMR spectra (125.6 MHz) in CS$_2$ with d_6-acetone as an internal lock. The acetone peaks are near 206 and 30 ppm; the solvent peak is at 192.7 ppm. (a) Spectrum of D_2-C$_{76}$ with an insert showing an expanded view of the fullerene region. Closely spaced lines in the region 140.3–145.0 ppm are further expanded in the box. (b) Expanded view of the fullerene region of the spectrum of C_{2v}-C$_{78}$.

accumulation of 38,000 scans for C_{76} [9] and of 71,000 scans for C_{78} [10] was necessary in order to obtain the spectra shown here.

In I_h-C_{60}, all C atoms occupy identical sites of the pyracylene type, and their resonance appears at 143.2 ppm [2, 19, 20b]. The ^{13}C NMR spectrum of D_{5h}-C_{70} shows five lines [20] that have been completely assigned by Johnson et al. [18]. The most shielded carbon resonance at 130.8 ppm corresponds to the pyrene site, whereas the three least shielded ones at 150.8, 148.3, and 147.8 ppm are assigned to the three different pyracylene-type sites. The remaining signal at 144.4 ppm is due to a corannulene-type carbon.

Comparing our spectra of C_{76} and C_{78} to the theoretically possible structures suggested by Fowler and Manolopoulos [15, 17a] we were able to identify the fullerenes as D_2-C_{76} and C_{2v}-C_{78}. Not only the overall number of NMR signals, but also their location in different shift regions (type of C site) and their relative intensities, were in accordance with this unique assignment [9, 10] (see Fig. 3.5).

Pure samples of the second, minor isomer of C_{78} have not been available in quantities large enough for NMR spectroscopy so far. However, we were able to deduce its structure from the spectrum of a mixture consisting of C_{76} and the two C_{78} isomers only. After subtracting the known resonances of D_2-C_{76} and C_{2v}-C_{78}, the isomer in question could be identified as D_3-C_{78} [10].

An interesting feature of D_2-C_{76} and D_3-C_{78} is their chirality: having been generated from achiral graphite, they were the first chiral fullerenes to be isolated, but a large number of these are conceivable, many having a similar construction scheme [21]. Their inherent chirality is not based on chiral centers but on the asymmetric, helical arrangement of their sp^2-atoms in space. They are produced as a racemic mixture, and so far no separation into enantiomers has been possible.

Besides the pyracylene, pyrene, and corannulene substructures mentioned, the higher fullerenes exhibit a number of other interesting structural motifs (Fig. 3.3) [9, 10].

A feature common to all three of the higher fullerenes discussed is the polar cap consisting of a triphenylene unit annelated with three 5MR; it gives the whole unit a strong egg-type curvature.

The whole C_{76} molecule can be considered as being made up by a spiralling double helical arrangement of two identical edge-sharing strands of annelated carbon rings. Each of these consists of a sequence of 5MR and 6MR that—when flattened out—adopts an S shape. 6MR are always annelated in a linear acene-type manner, and the curvature takes place thanks to angular annelation between 5MR and 6MR.

Whereas the corannulene and the cap motif are both bowl shaped, C_{2v}-C_{78} includes a coronene substructure that has a flattening effect. Placed on this less curved face, a model of the molecule resembles a turtle's shell.

Chiral D_3-C_{78}, on the other hand, includes a belt-shaped [9]cyclacene, a cyclic acene with nine linearly annelated benzene rings, in its equatorial region. Another belt-type substructure is encountered in C_{70} [20] and consists of a ma-

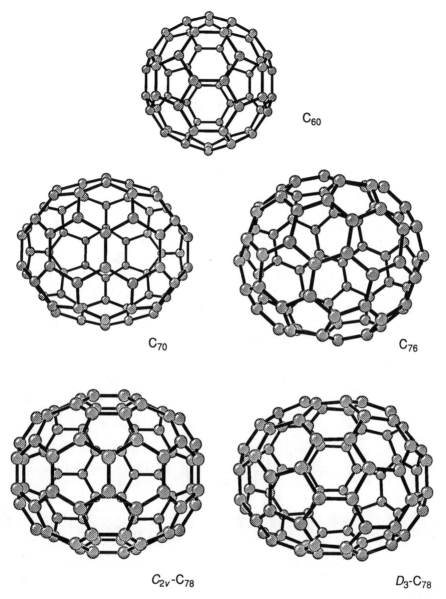

Figure 3.5 Fullerene structures as elucidated by ^{13}C NMR spectroscopy.

crocycle with 5 benzene rings connected in the para-positions called penta-*p*-phenylene.

A comparison of the NMR data reveals that in going from C_{70} to C_{84}, there is a continuous shift of the center of gravity of the ^{13}C resonances towards higher field [22]: the values observed are 145.0 ppm for C_{70}; 142.7 ppm for

C_{76}; 141.9, respectively, and 141.1 ppm for C_{2v}-C_{78} and D_3-C_{78}, and 140.3 ppm for the isomeric mixture of pure C_{84}. This can be interpreted in terms of an increase in the more benzenoid pyrene-type carbon atoms with increasing size of the carbon spheres, whereas the number of pyracylene and corannulene sites remains constant and equal to 60 in all fullerenes. C_{60}, however, with its single resonance at 143.2 ppm, does not fit into this series, and its pyracylene-type carbon atoms are more shielded than expected, a fact that might be attributed to a reduced diamagnetic ring current or some local anisotropic effects.

Even though C_{84} could not be separated into its isomers so far, we were able to record a ^{13}C NMR spectrum (Fig. 3.6) of a pure isomeric mixture of this fullerene corresponding to the large fraction in Figure 3.2(c).

The NMR spectrum of the isomeric C_{84} mixture shows 32 resonances [22] and is consistent with a spectrum recorded by Achiba et al. [23] exhibiting a much better signal-to-noise ratio. Starting from the 24 possible structures proposed by Fowler and Manolopoulos [17b] and searching for combinations of isomers leading to the observed ^{13}C NMR spectrum, Achiba et al. suggest [23] that the C_{84} fraction is a mixture of one D_{2d} isomer (11 resonances; $1 \times 4C$, $10 \times 8C$) and one D_2 isomer (21 resonances; $21 \times 4C$) in a ratio of $\approx 2:1$. Which one out of four possible D_2 isomers characterized by the same NMR pattern is present in this mixture still remains to be elucidated. Theoretical calculations seem to be largely divergent on this issue [24], but Achiba, based

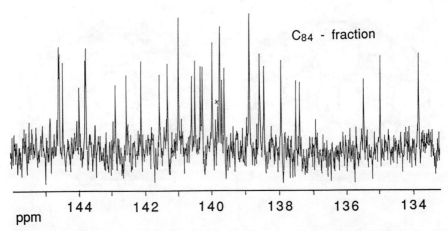

Figure 3.6 ^{13}C NMR spectrum (125.6 MHz) (32 lines) of a C_{84} fraction containing two isomers of D_{2d} (11 resonances) and D_2 symmetry (21 resonances). CS_2 was used as the solvent and d_6-acetone as an internal lock. The resonance marked by ''x'' could only be identified and assigned after receiving spectra recorded and kindly sent by Achiba et al. [23] with a much better signal-to-noise ratio. The NMR line positions are as follows: 133.84; 134.99; 135.49; 137.42; 137.52; 137.97; 138.48; 138.61; 138.90 (2); 139.65; 139.72; 139.78 (2); 139.84 (x); 140.01; 140.30; 140.36; 140.52; 140.62; 141.02; 141.35; 141.60; 142.15; 142.58; 142.91; 143.81; 143.83; 144.01; 144.50; 144.61; 144.63.

on photoelectron-spectroscopic results [25], tends to assign a structure called "round D_2-C_{84}" to it.

3.3 Electronic Absorption Spectra and Magnitude of the HOMO–LUMO Gap

All fullerenes show electronic absorption in the visible region of the spectrum [2a, 9, 10], and their solutions in apolar solvents (e.g., alkanes, aromatics, CS_2) have the following distinctive colors (Fig. 3.7): C_{60}, deep purple; C_{70}, deep orange-red; C_{76}, bright greenish-yellow; C_{2v}-C_{78}, intensely chestnut-brown; and D_3-C_{78}, golden-yellow. The isometric mixture of pure C_{84} [as shown in the large fraction of Fig. 3.2(c)] is olive green.

A trend in going from the lower- to the higher-molecular-weight fullerenes is the loss of fine structure that becomes especially apparent after C_{70}. The strongest absorptions (molar extinction coefficients ε around 10^5 l mol^{-1} cm^{-1}) of the carbon molecules are located in the ultraviolet region below 300 nm.

The absorption onset for the higher fullerenes comprises a region of weak bands (ε around 10^3), which, compared to C_{60} and C_{70} (absorption onset at 635 and 650 nm, respectively) [2a], are shifted toward longer wavelength. In the case of C_{2v}-C_{78} [10], it is located around 700 nm, but it is not sharp, and a tailing up to 900 nm is observed. Fullerene D_3-C_{78}, on the other hand, shows a set of distinct bands between 700 and 850 nm [10]. The most important shift of the longest-wavelength band into the near infrared is seen in the case of C_{76} [9]. The relatively strong absorption ($\varepsilon = 3080$ l mol^{-1} cm^{-1}) with a maximum at 715 nm has its onset around 920 nm. This corresponds to a transition energy of 1.39 eV that is much lower than the near 2.0 eV measured for C_{60} [2a]. The smaller HOMO–LUMO gap could be reproduced in a qualitative way by semiempirical molecular-orbital calculations, which predicted that C_{76} should be both a better donor and a better acceptor than C_{60} [9, 26]. These results are confirmed in principle by our cyclovoltammetric studies with the higher fullerene C_{76} [27] (Fig. 3.8).

3.4 Unexpected Electrochemical Properties of C_{76}

Up to six [20a, 28] reversible reduction (but no oxidation) steps have been found for C_{60}, which shows essentially the same redox properties as C_{70} [20a]. This exceptionally high electron affinity has been rationalized in terms of the constituent pyracylene units, which also seem to be a good model for the chemical reactivity of C_{60} [20a]. In THF, the higher fullerene C_{76}, on the other hand, not only exhibits four reversible reduction waves, but also at least one reversible oxidation step [27]. In addition, the reduction potentials are different from those of C_{60} and C_{70}. The electrochemical HOMO–LUMO gap corresponds to

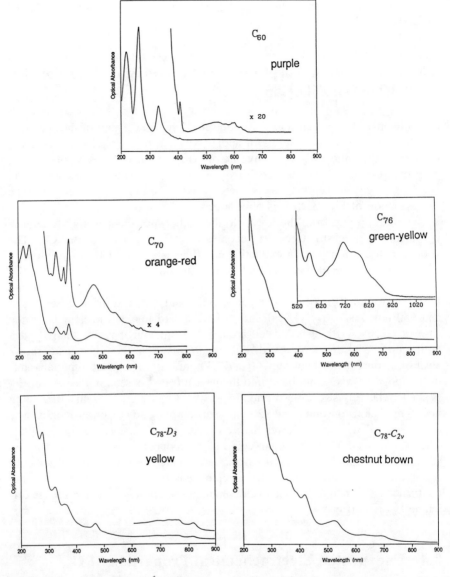

Figure 3.7 Electronic absorption spectra recorded in hexane (C_{60} and C_{70}) and in dichloromethane (C_{76} and C_{78} isomers).

≈0.4 eV; a value of 1.07 eV had been obtained by high-level density-functional theory calculations [26]. We are tempted to attribute the donor properties of chiral C_{76} to its extended linear benzo- and cyclopentadieno-annelated acene-type units, which are absent in C_{60} and in C_{70}.

A comparison between the well-known electron donor tetrathiafulvalene (TTF)

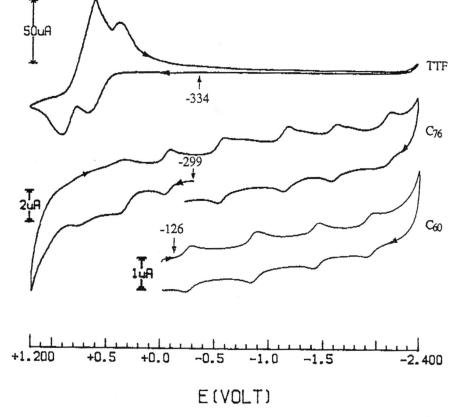

Figure 3.8 Cyclic voltammograms of TTF, C_{76}, and C_{60} in THF with 0.1 M $Bu_4N^+BF_4^-$ at ambient temperature. The arrows point to rest potentials. Working and counterelectrodes were Pt, the geometric area of the working electrode was 0.0324 cm^2, the reference electrode was Ag/AgCl, and the scan rate 1000 mV/s. The concentration of C_{76} was 0.7 mM; that of C_{60} was <0.5 mM (solubility problems prevented accurate determination of the latter).

and C_{76} reveals that, in THF, the latter is even more easily oxidized, a fact that makes it a promising potential component of acceptor-doped new materials.

3.5 Resistive Versus Arc Heating of Graphite

Even though it seems that, in the case of C_{84}, our production method yielded the same isomeric mixture as was obtained by Achiba et al. (see Section 3.2), this is not true for all C_n molecules. The higher fullerene mixtures investigated so far have been produced essentially by two methods that are related, but nevertheless nonidentical.

The first one, introduced by Krätschmer et al. in their landmark experiment [1] and used by us, [2] consists of the production of carbon soot through resistive heating of graphite electrodes having a very small contact surface (Fig. 3.9). The carbon smoke particles generated this way are deposited as a film on the walls of a glass bell jar reactor that is filled with an inert quenching gas (He) at around 150 Torr.

In the second procedure, used by Smalley et al. [5a,c] as well as by Achiba et al. [29], graphite electrodes kept at a specific distance are evaporated by the heat of an electric arc generated between their ends. This process also takes place under a similar inert gas pressure, and the soot is collected in a way comparable to that one in the resistive heating procedure (Fig. 3.10).

Whereas Achiba et al. have not yet reported the presence of C_{74} in their soot, we were able to detect traces of it through matrix LD-TOF mass spectrometry (Fig. 3.9) [30].

The most abundant species in our higher fullerene fraction was always chiral C_{76}. Furthermore, it should be mentioned that in all our experiments (run under similar conditions), the ratio of the different higher fullerenes—as detected by analytical HPLC—did not change in a noteworthy way. The same D_2-C_{76} is

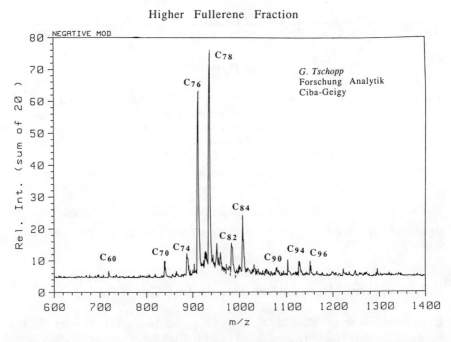

Figure 3.9 Matrix-laser-desorption time-of-flight (LD-TOF) mass spectrum of the crude toluene extract of soot obtained through resistive heating of graphite. Very mild ionization conditions were used in order to suppress fragmentation of the molecular ions as far as possible.

Higher Fullerenes: resistive (contact) heating versus arc heating

Resistive heating   Arc heating (*Achiba* et al.)

C$_{74}$	trace detected by LD-TOF	
C$_{76}$	the most abundant fraction	**C$_{76}$** equal abundance to C$_{76}$ and C$_{78}$ fractions.
C$_{78}$	second most abundant fraction containing two isomers (C$_{2v}$-C$_{78}$ and D$_3$-C$_{78}$).	**C$_{78}$** three isomers (C$_{2v}$-C$_{78}$, C$_{2v}$'-C$_{78}$, and D$_3$-C$_{78}$)
C$_{82}$	trace (LD-TOF)	**C$_{82}$** minor fraction with several isomers, a C$_2$-C$_{82}$ being the most abundant
C$_{84}$	third most abundant fraction containing two isomers, a D$_2$-C$_{84}$ (21 ^{13}C NMR resonances) and a D$_{2d}$-C$_{84}$ (11 ^{13}C NMR resonances)	**C$_{84}$** major fraction containing two isomers, a D$_2$-C$_{84}$ and a D$_{2d}$-C$_{84}$
C$_{90}$		**C$_{90}$**
C$_{94}$	minor but separable fractions	**C$_{94}$** much more abundant fractions
C$_{96}$		**C$_{96}$**

Figure 3.10 Tabulatory survey on the different product distributions as obtained by vaporization of graphite through resistive heating and arc heating, respectively.

also generated by arc heating, but in this case, together with C$_{78}$, it represents only the second most abundant fraction [29b].

C$_{78}$ obtained through resistive heating is composed of the two isomers C$_{2v}$-C$_{78}$ and D$_3$-C$_{78}$ in a ratio of ≈5:1 [10]. It is the second most abundant higher fullerene in our extracts. Achiba et al., on the other hand, analyzing a ^{13}C NMR spectrum of an isomeric C$_{78}$ mixture, can detect lines of a third isomer with C$_{2v}$' symmetry besides the two mentioned ones. The ratio C$_{2v}$'-C$_{78}$:C$_{2v}$-C$_{78}$:D$_3$-C$_{78}$ in this case is ≈5:2:2 [31a].

Whereas we can detect only traces of C$_{82}$ [30] (see matrix LD-TOF-MS, Fig. 3.9), a fullerene that plays an important role in encapsulating metal atoms [32], it is formed as a minor, but isolable, fraction during the arc-heating process [31b]. Based on a ^{13}C NMR analysis of an isomeric mixture, Achiba et al. suggest that the major component is one out of three possible C$_2$ isomers. The fraction also contains three other minor isomers that could be tentatively assigned [31b]. Very weak lines seem even to indicate the presence of other structures.

In the case of C$_{84}$, both production methods lead to the same mixture of a

D_2 and D_{2d} isomer [22, 23c]. Whereas C_{84} is only the third most abundant product of the resistive-heating process, it is the most important fraction of the fullerenes obtained through arc heating (ratio $C_{76}:C_{78}:C_{84} \approx 2:2:4-5$).

The carbon molecules C_{90}, C_{94}, and C_{96} are formed in both production processes [3, 29, 30]. Resistive heating yields only very minor, but separable, fractions. The electric arc, on the other hand, generates them in much larger quantities.

Why do the two methods described here lead to such important differences in product distribution? The obvious answer seems to be that the conditions under which the fullerenes are generated are not identical. A variety of parameters may play an important role in the fullerene formation process, such as the current density at the electrodes, their temperature, the temperature gradient between electrodes and the vapor condensation zone, the pressure, and the nature of the quenching gas, etc. These variables are not independent of each other, which makes an effective control of the reaction conditions even more difficult. In our case of the resistive-heating fullerene production process, we used an inert atmosphere of helium at a pressure of ≈ 150 Torr, a Poco graphite electrode of 1/4 and 1/8 in. diameter, respectively, and a current intensity of ≈ 40 A. These conditions generate the maximum overall fullerene yield we were able to obtain. The major difference between the resistive- and the arc-heating process seems to us the applied current intensity. Whereas we use 40 A, arc heating is performed at higher intensities, close to 80 A.

For reasons of further comparison, it would be desirable to have more detailed information about the carbon molecules, in particular fullerenes, generated by other processes like the inductive heating of graphite [33] or the burning of C-rich materials (e.g., benzene) in sooting flames [34].

Understanding the influences of different experimental parameters on the product distribution is crucial for gaining insight into the fullerene formation mechanism, which is still far from being elucidated [3, 5b, 10, 35]. An interesting contribution to our comprehension of this mechanistic problem has been made by ion-cyclotron-resonance (ICR) studies of the cyclo[n]carbons [36]. The latter are monocyclic all-carbon molecules containing n atoms that are connected in a polyyne- or a cumulene-type way [37]. In the positive ion mode ICR experiment, fascinating reactions between ions and neutral molecules of the cyclo[n]carbons lead from the all-carbon rings to fullerene ions: C_{30} forms almost exclusively C_{60}, with subsequent losses of C_2 fragments, demonstrating the fullerene nature of the product [36b]. Cyclo[24]carbon, C_{24}, undergoes di- and trimerization to C_{48} and C_{72}, respectively, and the latter, upon loss of a C_2 unit, forms almost exclusively C_{70}. No signal for C_{60}^{+} is observed [36b]. Even though caution is advised in the sense that the ion–molecule reactions just described and the fullerene formation by evaporation of graphite are two not directly related processes, the ICR studies demonstrate clearly that fullerenes can be formed in the gas phase from structurally very different precursors, probably through a myriad of rearrangements. These results might even be of practical future importance in connection with new synthetic routes to fullerenes.

3.6 Fullerene Isomerism and the Stone–Wales Pyracylene Rearrangement

The question arises why only some specific isomers of a given higher fullerene C_n (with $n \geq 78$) appear in the soot. Why is there not a single isomer—the most stable one—only, or why on the other extreme, do we not find all of them?

We tried to find an answer by having a closer look to the relatively simple case of C_{78} for which Fowler and Manolopoulos had predicted only five IPR-satisfying structures [16] (Table 3.1). As we have seen before, the fullerenes actually present in our C_{78} fraction are the C_{2v} and D_3 isomers [10] (Fig. 3.11). Qualitative HMO theory, on the other hand, taking into consideration the magnitude of the HOMO–LUMO gap, predicted the D_{3h} isomer to be the most stable one [16]. HMO theory, however, is based on electronic considerations and does not account for steric strain, another important factor interfering with molecular stability. In order to include strain in our considerations, we performed MM3 calculations [38] on the five IPR-satisfying structures, and the results [10] are shown in Table 3.1. They are in good agreement with the fact that we find the C_{2v} isomer as the main component in our C_{78} fraction. According to these calculations, our minor component, D_3-C_{78}, occupies an intermediate position between the lowest- and highest-energy structures. The isomer most favored by MM3 is the $C_{2v'}$-C_{78} found by Achiba et al. as the major isomer besides the minor D_3- and the C_{2v}-symmetric components. The arc-heating method did not yield any "high-energy" D_{3h}-C_{78} either [31a].

While generating the different structures on the computer, we realized that four of the five isomers were interconvertible by formal 90° rotation of the central bond in pyracylene units [10], an operation known as the Stone–Wales rearrangement [39] (Fig. 3.11). If considered as a pericyclic process, this concerted shift of sigma bonds leads to a four-electron transition state that is thermally forbidden by the Woodward–Hoffman rules. It was proposed as an isomerization path of C_{60} and used for the generation of nonicosahedral structures of this fullerene. Smalley et al. postulated it as the first step in the photodeg-

Table 3.1 The five ipr-satisfying isomeric structures of C_{78}, the number of independent ^{13}C nmr resonances, the gas-phase heats of formation $\Delta H°_f$ (298 K), and the differences in $\Delta H°_f$ calculated with the molecular mechanics program MM3

Structure	^{13}C NMR lines (intensity)	$\Delta H_f°$ (kcal mol^{-1})	$\Delta(\Delta H_f°)$
C_{2v}	18 (4 C) + 3 (2 C)	695.7	1.4
D_3	13 (6 C)	697.8	3.5
$C_{2v'}$	17 (4 C) + 5 (2 C)	694.3	0.0
D_{3h}	5 (12 C) + 3 (6 C)	702.3	8.0
$D_{3h'}$	5 (12 C) + 3 (6 C)	697.8	3.5

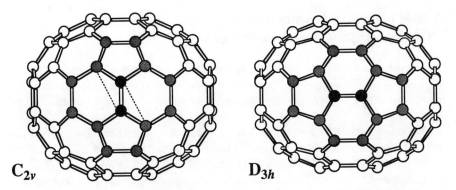

Figure 3.11 Interconversion between the C_{2v} and D_{3h} isomers of C_{78} as an example of the Stone–Wales pyracylene rearrangement.

radation of buckminsterfullerene to smaller carbon cages through consecutive C_2 losses [40].

The pyracylene rearrangement can be viewed as an interconversion mode on the C_{78} hypersurface for as many as four structures that can be transformed into each other in single steps.

$$D_{3h} \rightleftharpoons C_{2h} \rightleftharpoons C_{2v'} \rightleftharpoons D_{3h'} \tag{3.1}$$

In D_{3h}-C_{78} [10], all three central pyracylene units are oriented in such a way that they define a [9]cyclacene substructure around the equator of the carbon sphere. At the same time, three coronene motifs leading to three flattened faces separated by rather sharp edges can be distinguished on the fullerene surface. The resulting strain, in particular at the sharp edges, has to be considerable, and this is reflected by the result of the MM3 calculations. The computational results are well sustained by the fact that D_{3h}-C_{78} has not been detected in the soot of either the resistive- or arc-heating production method. Starting from the D_{3h} structure, a 90° rotation of one pyracylene unit leads to a reduction in symmetry and affords the C_{2v} structure; rotation of two such units gives the $C_{2v'}$ isomer. Rotation of all three pyracylene substructures restores the threefold symmetry in $D_{3h'}$-C_{78}.

The fifth IPR-satisfying isomer, D_3-C_{78}, however, cannot be transformed into any of the previous ones by rearrangements of the Stone–Wales type. These reflections, together with our experimental results, lead to the following hypothesis: of each family of fullerene isomers whose members are related through pyracylene rearrangements, only the most stable ones can be isolated, possibly as a result of a thermodynamic control during the cooling period of the soot production process. If the structures of a given fullerene C_n can be divided into several noninterrelated families, more isomers (the stabilomers of each group) should be isolable, because there is no energetically favorable interconversion path between these families.

The Stone–Wales analysis is well supported by the findings with the C_{78}

family. The two most stable isomers, C_{2v}- and $C_{2v'}$-C_{78} have been isolated out of a manifold of four possible structures that are interrelated by single-bond rearrangements.

In the case of fullerene C_{84} with its 24 IPR-satisfying structures [17b], Manolopoulos and Fowler have analyzed the problem of isomer interconversions computationally [41]. They took into account interconversions between IPR structures only, thus excluding high-energy intermediates with adjacent pentagons on pathways that are supposed to be slow. The calculations showed that C_{84} can be divided into two non-interrelated groups of interconvertible isomers: a small one of three members and a larger one of the remaining 21 IPR structures. A general feature of the single-step pyracylene rearrangement graphs is the fact that low-symmetry isomers are related to a larger number of other structures than high-symmetry ones.

The isomers of C_{84} actually present in the soot are the same for both production methods (resistive and arc heating of graphite): one of D_2 and one of D_{2d} symmetry [22, 31c]. Even though ^{13}C NMR spectroscopy does not allow one to decide which one of four possible IPR-satisfying structures has to be assigned to the D_2 isomer in question (all four are supposed to have 21 resonances of equal intensity), Achiba et al., on the basis of a comparison between a calculated and an experimental photoelectron spectrum, are tempted to identify it as an isomer they call "the round D_2-C_{84}" [31c]. If this assignment is correct, the two components of the C_{84} fraction belong to different isomeric families and are not interconvertible by pyracylene rearrangements [41]. This is in agreement with our hypothesis for the isomeric product distribution as stated previously. In analogy to our MM3 calculations on the structures of C_{78} [10], it will be interesting to check computationally whether the mentioned D_2 and D_{2d} isomers of C_{84} correspond to the stabilomers of each family.

3.7 Current Developments in Higher Fullerene Research

We have been showing that, despite an increasing multitude of IPR-satisfying isomers, it should be possible to determine the structure of higher fullerenes. Such elucidation is made possible by a productive interplay between theoretical calculations and predictions, on one hand, and a combination of powerful analytical tools, on the other hand. The latter involve among others an efficient HPLC-separation system, LD-TOF mass spectrometry, and very sensitive ^{13}C NMR spectroscopy. A limiting factor in the investigation of fullerenes having a size even larger than those described here are the small amounts available so far. Improvements on this problem can be obtained through modification and optimization of the production method, a goal currently under our investigation [32]. Another problem is the fact that the physical properties (e.g., solubility of fullerenes C_n with $n > 100$) seem to change so dramatically [3, 42a], that the analytical tools used so far might not be appropriate any more.

In a recent paper [42a] Lamb et al. describe the investigation of high-temperature, high-pressure toluene extracts of soot that had previously been extracted exhaustively at ambient pressure. The mass spectra thus obtained are similar to those of ambient-pressure high-boiling-point solvent extracts [3, 42b], indicating the presence of fullerenes containing from 60 to 330 atoms. In addition to reflection TOF-MS, the authors investigated the extracts by scanning tunneling microscopy (STM). The STM images show that the giant fullerenes have a roughly spherical shape and diameters ranging from 1 to 2 nm. Even though the STM data did not allow the determination of whether the giant fullerenes are hollow, the authors found no evidence of ''bucky tubes'' [21, 43] as have been observed by Ijima.

This author describes [43] the formation of carbon needles on the cathode of an arc-discharge apparatus that grow parallel to the electric field. Investigations by transmission electron microscopy and x-ray diffraction led to the conclusion that these needles, up to 1000 nm long and a few to a few tens of nanometers in diameter, consist of coaxially arranged microtubules of graphitic sheets (2–50 in number). The 6MR in these sheets are in a helical arrangement, and the tips of the needles are comparable to fullerenes cut in halves.

Finally, imagination having no limits, it is possible to think of carbon allotropes with a negative, saddle-shaped curvature instead of the positively curved fullerenes [44]. This leads to three-dimensional polymeric networks with the atoms lying on so-called triply periodic minimal surfaces. The name *schwarzites* has been suggested [44] for this type of carbon allotropes, not only because of their likely color, but also in honor of the mathematician H. A. Schwarz [45], who first explored this type of geometrical constructions that divide space into equivalent portions. Whereas the positive curvature in fullerenes is due to the incorporation of 5-membered rings into graphitic sheets, 7-, 8-membered, and larger rings are at the origin of the negative curvature of the schwarzites without introducing an excessive strain. Even though the calculated cohesive energy is smaller compared to grahite, it is significantly higher (0.2 eV per C atom) than that of C_{60}, and it might be worth looking for the described structures in the insoluble portion of the soot [44].

One of the most important problems in fullerene chemistry that needs to be solved is the elucidation of the formation mechanism. Even an explosive development of the field has not yet given a satisfying answer to this question. The studies on the higher fullerenes, with their multitude of potential structures, some of which could be detected, can provide important information, and they also should allow the prediction of new structures. Fullerene research has given insight into the structural variety of pure carbon that has been underestimated for a long time, and it has given a strong impetus to synthetic chemists and theoreticians to look for new carbon allotropes [37b].

Finally, from a more application-oriented viewpoint, the investigation of the properties of fullerenes, their derivatives, and future carbon allotropes will certainly lead to the discovery of interesting materials [46]. The superconducting alkali fullerides obtained through doping of buckminsterfullerene with alkali metals [47], the conversion of fullerenes into diamond [48], and the endohedral metal–fullerene complexes [43] with their numerous properties that remain to be studied are only a few examples of this development.

Acknowledgment

We thank the U.S. National Science Foundation, the U.S. Office of Naval Research, NATO, and the Grand-Duchy of Luxembourg for supporting this research.

References

1. W. Krätschmer, L. D. Lamb, K. Fostiropoulos, and D. R. Huffman, *Nature (London)* **347**, 354 (1990).

2. (a) H. Ajie, M. M. Alvarez, S. J. Anz, R. D. Beck, F. Diederich, K. Fostiropoulos, D. R. Huffman, W. Krätschmer, Y. Rubin, K. E. Schriver, D. Sensharma, and R. L. Whetten, *J. Phys. Chem.* **94**, 8630 (1990); (b) R. L. Whetten, M. M. Alvarez, S. J. Anz, K. E. Schriver, R. D. Beck, F. Diederich, Y. Rubin, R. Ettl, C. S. Foote, A. P. Darmanyan, and J. W. Arbogast, *Mater. Res. Soc. Symp. Proc.* **206**, 639 (1990).

3. F. Diederich, R. Ettl, Y. Rubin, R. L. Whetten, R. Beck, M. Alvarez, S. Anz, D. Sensharma, F. Wudl, K. C. Khemani, and A. Koch, *Science* **252**, 548 (1991).

4. (a) E. A. Rohlfing, D. M. Cox, and A. Kaldor, *J. Chem. Phys.* **81**, 3322 (1984); (b) H. W. Kroto, J. R. Heath, S. C. O'Brien, R. F. Curl, and R. E. Smalley, *Nature (London)* **318**, 162 (1985).

5. (a) R. F. Curl and R. E. Smalley, *Science* **242**, 1017 (1988); (b) R. F. Curl, and R. E. Smalley, *Scient. Amer.*, 54 (October, 1991).

6. R. B. Kaner, E. G. Gillan, F. Diederich, and C. Thilgen, unpublished results.

7. K. M. Creegan, J. L. Robbins, W. K. Robbins, J. M. Millar, R. D. Sherwood, P. J. Tindall, D. M. Cox, A. B. Smith, III; J. P. McCauley, Jr. D. R. Jones, and R. T. Gallagher, *J. Am. Chem. Soc.* **114**, 1103 (1992).

8. S. K. Elemes, C. Silverman, C. Sheu, M. Kao, C. S. Foote, M. M. Alvarez, and R. L. Whetten, *Angew. Chem.* **104**, 364 (1992); *Angew. Chem. Int. Ed. Engl.* **31**, 351 (1992).

9. R. Ettl, I. Chao, F. Diederich, and R. L. Whetten, *Nature (London)* **353**, 149 (1992).

10. F. Diederich, R. L. Whetten, C. Thilgen, R. Ettl, I. Chao, and M. M. Alvarez, *Science* **254**, 1768 (1991).

11. J. C. Fetzer, and E. J. Gallegos, *Polycycl. Arom. Comp.* **2**, 245 (1992).

12. F. Diederich, R. Ettl, and C. Thilgen, unpublished results.

13. R. E. Smalley, cited in: R. M. Baum, *Chem. Eng. News* **63**(51), 20 (1985).

14. (a) D. E. Manolopoulos, *J. Chem. Soc., Faraday Trans.* **87**, 2861 (1991); (b) D. E. Manolopoulos and P. W. Fowler, *Chem. Phys. Lett.* **187**, 1 (1991).

15. D. E. Manolopoulos, J. C. May, and S. E. Down, *Chem. Phys. Lett.* **181**, 105 (1991).

16. P. W. Fowler, D. E. Manolopoulos, and R. C. Batten, *J. Chem. Soc., Faraday Trans.* **87**, 3103 (1991).

17. H. W. Kroto, *Nature (London)* **329**, 529 (1987).

18. R. D. Johnson, G. Meijer, J. R. Salem, and D. S. Bethune, *J. Am. Chem. Soc.* **113**, 3619 (1991).

19. R. D. Johnson, G. Meijer, and D. S. Bethune, *J. Am. Chem. Soc.* **112**, 8983 (1990).

20. (a) P.-M. Allemand, A. Koch, F. Wudl, Y. Rubin, F. Diederich, M. M. Alvarez, S. J. Anz, and R. L. Whetten, *J. Am. Chem. Soc.* **113**, 1050 (1991); (b) R. Taylor, J. P. Hare, A. K. Abdul-Sada, and H. W. Kroto, *J. Chem. Soc., Chem. Commun.* 1423 (1990).

21. (a) P. Labastie, R. L. Whetten, H.-P. Cheng, and K. Holczer, *Chem. Phys. Lett.*, in press; (b) J. W. Mintmire, B. I. Dunlap, and T. C. White, *Phys. Rev. Lett.*, in press; (c) K. Tanaka, K. Okahara, M. Okada, and T. Yamabe, *Chem. Phys. Lett.* **191**, 46 (1992); (d) D. Robertson II, D. W. Brenner, and J. W. Mintmire, *Phys. Rev. B., Rapid Commun.*, in press.

22. F. Diederich and R. L. Whetten, *Acc. Chem. Res.* **25**, 119 (1992).

23. K. Kikuchi, T. Wakabayashi, N. Nakahara, M. Honda, S. Suzuki, H. Shiromaru, Y. Miyake, K. Saito, I. Ikemoto, M. Kainosho, and Y. Achiba, submitted for publication.

24. (a) P. W. Fowler, *J. Chem. Soc., Faraday Trans.* **87**, 1945 (1991); (b) K. Raghavachari and C. M. Rohlfing, *J. Phys. Chem.* **95**, 5768 (1991).

25. S. Hino, K. Matsumoto, S. Hasegawa, K. Kamiya, H. Inokuchi, T. Morikawa, T. Takahashi, K. Seki, K. Kikuchi, S. Suzuki, I. Ikemoto, and Y. Achiba, *Chem. Phys. Lett.* **190**, 169 (1992).

26. High-level calculations predict a small HOMO—LUMO gap of 1.07 eV for C_{76}: H. P. Cheng and R. L. Whetten, *Chem. Phys. Lett.*, in press; see also [14a].

27. Q. Li, F. Wudl, C. Thilgen, R. L. Whetten, and F. Diederich, *J. Am. Chem. Soc.* **114**, 3994 (1992).

28. (a) R. E. Haufler, J. Conceicao, L. P. F. Chibante, Y. Chai, N. E. Byrne, S. Flanagan, M. M. Haley, S. C. O'Brien, C. Pan, Z. Xiao, W. E. Billups, M. A. Ciufolini, R. H. Hauge, J. L. Margrave, R. F. Curl, and R. E. Smalley, *J. Phys. Chem.* **94**, 8634 (1990); (b) D. M. Cox, S. Bahal, M. Disko, S. M. Gorun, M. Greaney, C. S. Hsu, E. B. Kollin, J. Millar, J. Robbins, R. D. Sherwood, and P. Tindall, *J. Am. Chem. Soc.* **113**, 2940 (1991); (c) D. Dubois, K. M. Kadish, S. Flanagan, R. E. Haufler, L. P. F. Chibante, and L. J. Wilson, *J. Am. Chem. Soc.* **113**, 4364 (1991); (d) D. Dubois, K. M. Kadish, S. Flanagan, and L. J. Wilson, *J. Am. Chem. Soc.* **113**, 7773 (1991); (e) Q. Xie, E. Pérez-Cordero, and L. Echegoyen, *J. Am. Chem. Soc.* **114**, 3978 (1992).

29. (a) K. Kikuchi, N. Nakahara, M. Honda, S. Suzuki, K. Saito, H. Shiromaru, K. Yamauchi, I. Ikemoto, T. Kuramochi, S. Hino, and Y. Achiba, *Chem. Lett.* 1607 (1991); (b) K. Kikuchi, N. Nakahara, T. Wakabayashi, M. Honda, H. Matsumiya, T. Moriwaki, S. Suzuki, H. Shiromaru, K. Saito, K. Yamauchi, I. Ikemoto, and Y. Achiba, *Chem. Phys. Lett.* **188**, 177 (1992).

30. C. Thilgen, and F. Diederich, unpublished results; we thank Dr. G. Tschopp and Dr. M. Schär from the Analytical Department of Ciba-Geigy, Basel, for recording the matrix LD-TOF mass spectrum of our crude soot extract shown in Figure 3.9.

31. We thank Professor Y. Achiba for communicating some of his results to us prior to publication and for providing us with preprints of the manuscripts; (a) K. Kikuchi, T. Wakabayashi, N. Nakahara, S. Suzuki, H. Shiromaru, Y. Miyake, K. Saito, I. Ikemoto, M. Kainosho, and Y. Achiba, submitted for publicatioon; (b) Y. Achiba, K. Kikuchi, T. Wakabayashi, N. Nakahara, S. Suzuki, M. Honda, H. Shiromaru, Y. Miyake, K. Saito, I. Ikemoto, and M. Kainosho, submitted for publication; (c) K. Kikuchi, T. Wakabayashi, N. Nakahara, M. Honda, S. Suzuki, H. Shiromaru, Y. Miyake, K. Saito, I. Ikemoto, M. Kainosho, and Y. Achiba, submitted for pub-

lication. We thank Professor Achiba for communicating their ^{13}C data on C_{84} to us prior to submission.

32. Y. Chai, T. Guo, C. Jin, R. E. Haufler, L. P. F. Chibante, J. Fure, L. Wand, J. M. Alford, and R. E. Smalley, *J. Phys. Chem.* **95**, 7564 (1991).

33. G. Peters and M. Jansen, *Angew. Chem.* **104**, 240 (1992); *Angew. Chem. Int. Ed. Engl.* **31**, 223 (1992).

34. J. B. Howard, J. T. McKinnon, Y. Makarovsky, A. L. Lafleur, and M. E. Johnson, *Nature (London)* **352**, 139 (1991).

35. (a) H. W. Kroto, A. W. Allaf, and S. P. Balm, *Chem. Rev.* **91**, 1213 (1991); (b) R. E. Smalley, *Acc. Chem. Res.* **25**, 98 (1992); (c) J. M. Hawkins, A. Meyer, S. Loren, and R. Nunlist, *J. Am. Chem. Soc.* **113**, 9394 (1991); (d) D. E. Manolopoulos and P. W. Fowler, *J. Chem. Phys.*, submitted; (e) D. E. Manolopoulos and P. W. Fowler, *J. Chem. Phys.*, in press; (f) P. W. Fowler, D. E. Manolopoulos, R. P. Ryan, *J. Chem. Soc., Chem. Commun.*, in press.

36. (a) Y. Rubin, M. Kahr, C. B. Knobler, F. Diederich, and C. L. Wilkins, *J. Am. Chem. Soc.* **113**, 495 (1991); (b) M. M. Ross, S. McElvany, J. Milliken, J. Callahan, H. Nelson, A. Baronavski, L. T. Scott, N. Goroff, Y. Rubin, and F. Diederich, unpublished results.

37. (a) F. Diederich, Y. Rubin, C. B. Knobler, R. L. Whetten, K. E. Schriver, K. N. Houk, and Y. Li, *Science* **245**, 1088 (1989); (b) F. Diederich and Y. Rubin, *Angew. Chem.*, in press; *Angew. Chem. Int. Ed. Engl.*, in press.

38. (a) N. L. Allinger, Y. H. Yuh, and J.-H. Lii, *J. Am. Chem. Soc.* **111**, 8551 (1989); (b) M. Saunders, *Science* **253**, 330 (1991).

39. A. J. Stone and D. J. Wales, *Chem. Phys. Lett.* **128**, 501 (1986).

40. S. C. O'Brien, J. R. Heath, R. F. Curl, and R. E. Smalley, *J. Chem. Phys.* **88**, 220 (1988).

41. P. W. Fowler, D. E. Manolopoulos, and R. P. Ryan, *J. Chem. Soc., Chem. Commun.* 408 (1992).

42. (a) L. D. Lamb, D. R. Huffman, R. K. Workman, S. Howells, T. Chen, D. Sarid, R. F. Ziolo, *Science* **255**, 1413 (1992); (b) D. H. Parker, P. Wurz, K. Chatterjee, K. R. Lykke, J. E. Hunt, M. J. Pellin, J. C. Hemminger, D. M. Gruen, and L. M. Stock, *J. Am. Chem. Soc.* **113**, 7499 (1991).

43. S.Ijima, *Nature (London)* **354**, 56 (1991).

44. T. Lenosky, X. Gonze, M. Teter, and V. Elser, *Nature (London)* **355**, 333 (1992).

45. H. A. Schwarz, *Gesammelte Mathematische Abhandlungen*, Springer, Berlin, 1890.

46. J. E. Fischer, P. A. Heiney, A. B. Smith, III, *Acc. Chem. Res.* **25**, 112 (1992).

47. R. C. Haddon, *Acc. Chem. Res.* **25**, 127 (1992).

48. (a) M. N. Regueiro, P. Monceau, and J.-L. Hodeau, *Nature (London)* **355**, 237 (1992); (b) see also: R. Meilunas, R. P. H. Chang, S. Liu, and M. M. Kappes, *Nature (London)* **354**, 271 (1991).

CHAPTER

4

Fullerene Structures

T. G. Schmalz and D. J. Klein

The proposal in 1985 by Kroto et al. [1] that a prominent C_{60} peak in the mass spectrum of laser-vaporized graphite was due to a hollow, closed-cage carbon molecule with the shape of a truncated icosahedron attracted the immediate attention of theorists. Although it was recognized immediately that the other even-numbered clusters in the mass spectrum might also be closed-cage molecules, and a structure was soon suggested for C_{70} [2], most early theoretical work was focused on the highly symmetrical C_{60} structure, which was named *buckminsterfullerene* in recognition of R. Buckminster Fuller's geodesic domes. Naturally enough, the early work concentrated on establishing the reasonableness of the hollow three-dimensional form, and on calculating physical and spectroscopic properties of buckminsterfullerene that might be useful for experimental verification of the structure. An alternative form for C_{60}, called *graphitene*, was suggested early on by Haymet [3], and in 1986 we examined the five most symmetrical closed-cage C_{60} isomers with five- and six-membered rings only, including both buckminsterfullerene and graphitene, primarily with a view to establishing the special stability of buckminsterfullerene [4, 5].

As the hollow-cage form of buckminsterfullerene began to become generally accepted on the basis of indirect evidence, despite its resistance to isolation in macroscopic quantities, we became interested in a more systematic study of fullerene structures, both for C_{60} and for carbon clusters with other numbers of atoms. We define a "fullerene" as a three-dimensional network of sp^2 hybridized carbon atoms in which each atom is connected by a bond to exactly three neighbors. The tetravalency of carbon is then satisfied by the formation of a π bond between each atom and one of its neighbors. The possibility that these π bonds might be delocalized, and so lead to an aromaticlike resonance stabilization of the molecule was, of course, one of the initial reasons for the suggestion that the buckminsterfullerene structure might be stable.

If the carbon atoms of a cluster are considered to be the vertices of a closed trivalent net, with the σ bonds forming the edges, then it is a simple consequence of Euler's theorem that the face counts of the net must satisfy the following equation

$$3f_3 + 2f_4 + f_5 = 12(1 - \gamma_s) + \sum_{n \geq 7} (n - 6)f_n$$

where f_n is the number of n-sided faces and γ_s is the genus of the closed surface S defined by the net. For clusters that can be deformed into a sphere (i.e., those forming surfaces with no "holes"), γ_s is equal to zero, and one concludes that faces of five or fewer sides are required to provide the curvature necessary to close the fullerene in three-dimensional space. We argued that since four- and three-membered rings of sp^2 carbon atoms are more highly strained, as well as chemically less favorable in aromatic systems, the most stable structures would likely result when $f_3 = f_4 = 0$, and $f_n = 0$ for $n > 6$ to minimize the number of five-membered rings [6]. This leads immediately to the result that all fullerenes composed of only five- and six-membered rings and homeomorphic to a sphere must have exactly 12 pentagonal faces. It is known constructively that polyhedra satisfying these conditions can be produced for any number of six-sided faces except one [7], leading to the result that fullerenes are possible for any even number of atoms greater than or equal to 20 except 22. The restriction to five- and six-membered faces seems now to be generally accepted as part of the definition of a fullerene.

We originally believed that stable fullerenes were likely to be of relatively high symmetry, and so we began our studies by adapting a transfer-matrix methodology, which had originally been developed for studying polymers [8]. In this method, the fullerene was divided into (rotational) unit cells by slicing the molecule into wedges around a rotation axis. The transfer-matrix method permits the calculation of graphically related properties of the molecule such as the Hückel energy-level spectrum or its set of Kekulé structures. By identifying all possible rotational unit cells, we were able to generate and treat all 33 fullerenes with up to 84 atoms having a five- or sixfold rotation axis or T_d or T_h symmetry [6], as well as all icosahedral symmetry cages with up to 240 atoms [9]. While this produced attractive structures for several prominent mass spectral peaks, it produced no structures at all for some mass numbers and only relatively unattractive structures for others. It became apparent that any really thorough systematic study of fullerene structures would have to deal with all isomers, even those without symmetry.

With the announcement by Krätschmer et al. [10] in 1990 of a simple way to produce bulk quantities of fullerenes, the likelihood that a variety of larger fullerenes could be subjected to direct experimental study increased significantly. While the fullerene-rich soot produced in the electric arc generation process is dominantly C_{60} and C_{70}, observable quantities of many other fullerenes are produced as well. A number of these have been separated chromatographically [11], and several have now been characterized [12, 13]. We therefore set out to develop a method for the systematic generation of all possible isomeric fullerene structures, and to develop a suite of theoretical tools to screen the structures so generated. We have focused primarily on theoretical methods that are simple (and fast) enough to be applied routinely to thousands of possible structures. Our goal is both to identify likely candidates for experimental isolation and to try to identify structural trends associated with stability. Ultimately one might hope that an understanding of the structural features leading to stability could throw some light on the mechanism by which the fullerenes form, grow, and react.

In the remainder of this chapter we describe the method by which we generate fullerene structures, first in graphical form, and then as a set of three-dimensional coordinates. We will also describe some of the techniques that have proved useful in analyzing these structures. After that we will discuss some of the structural trends that have emerged from these studies. Next we extend these ideas to some more exotic fullerenes or fullerenelike materials, the buckytubes and buckytori. Finally we close with some brief speculation concerning formation and growth mechanisms.

4.1 Construction and Analysis of Fullerenes

4.1.1 Graphical Isomer Generation

The basic scheme [14] we have developed to generate a planar graph for each fullerene isomer is essentially a brute force approach, in which we begin with a ''seed'' graph and systematically add a pentagon or hexagon at any one of the possible locations around the periphery of the ''growing'' cage and then iterate the process. Of course, this is a rapidly branching tree process since there are usually several locations where a ring may be added and either a five- or six-membered ring may be added at each point. The number of growing cage structures can be kept manageable by exploiting constraint relations as soon as possible to eliminate structures that cannot close into appropriate cages. A simple example of a constraint is that the number of pentagons cannot exceed 12. If the number of vertices v and the number p of edges shared between two pentagons along with the number q of vertices shared between a triple of pentagons are specified, a series of additional contraints can be found [14].

The advantage of this approach to isomer generation is that if one starts with a seed graph guaranteed to be present in all cages, one must generate every possible fullerene isomer. The disadvantage, other than time and storage demands on the computational resource, is that even with maximal exploitation of constraints there is no guarantee that the same cage may not be generated more than once. It is therefore necessary to apply some sort of graph iso-

morphism check to identify unique structures. The method we have developed
[15] to compare two graphs is guaranteed to give one of three results: (1) the
two graphs are isomorphic; (2) the two are nonisomorphic; or (3) the algorithm
is indecisive. So far, we have found no such indecision for fullerene cages.
Indeed we have found no fullerene cages that are isospectral so that so far
simply diagonalizing the adjacency matrix and comparing eigenvalues has been
sufficient to recognize multiple copies of any isomer. To give some feeling for
the redundancy in our generation method, a total of 130,139 completed graphs
were generated to produce the 1,812 isomers of C_{60}. During the generation
process 82,205,454 partial graphs were produced at least temporarily before
being eliminated.

   The output of the generation program is the adjacency matrix for each fullerene isomer, in other words, just a list of which vertices are bonded to each
given vertex by sigma bonds. Diagonalization of the adjacency matrix yields
directly the Hückel energy-level spectrum of the molecule, and other molecular
characteristics such as the equivalence classes of atoms in the NMR spectrum
can also be extracted from it. However, to characterize a molecule fully, it is
necessary to convert its graphical structure into actual physical coordinates for
the atoms. Ultimately, this requires the application of some sort of quantum-
chemical or molecular-mechanics approach to produce an optimized structure,
but at least an approximate picture of the molecule can be obtained directly
from its graph. We describe that technique briefly in the next section before
returning to the analysis of cage structures.

4.1.2   Topological Coordinates

The method we have recently begun to employ to generate approximate atomic
positions in fullerenes was proposed by Manolopoulos and Fowler [16], based
upon the approximate symmetry of certain of the π electron energy levels available from a Hückel calculation. In particular, because of the topologically
spherical shape of fullerenes, the lowest π orbitals resemble S, P, D, . . .
surface harmonics [17] in terms of their nodal structure, and the P-type molecular orbitals can be used to produce x, y, z Cartesian coordinates. This method
of generating coordinates takes no account of strain, so it does not properly
predict deviations from standard bond lengths or bond angles, but it does generate atomic positions that respect the overall point group symmetry of the
molecule, while preserving the face structure of the cage. These coordinates
are termed *topological coordinates* to indicate that they contain mathematical,
but not chemical, structural information.

   Nevertheless, we find that they provide a reasonable picture of molecular
shape. The associated topological moments of inertia can provide an immediate
indication of asphericity, which we shall argue is likely to be associated with
high strain. We are currently investigating whether other meaningful strain
measures can be extracted directly from the topological coordinates. In any
event, we find that these topological coordinates provide reasonable starting

locations for subsequent geometry optimization using semiempirical quantum chemistry or empirical interatomic potentials. The topological coordinates also provide direct information as to the ir/Raman activity of the vibrational modes of the molecule [16].

4.1.3 Resonance Stabilization

The relative stability of a fullerene isomer is in general determined by a balance between two competing factors. Once dangling bonds have been eliminated by the formation of a fullerene, the factor that tends to stabilize one structure relative to another is π-electron resonance—in other words, delocalization of the π-like electrons on the surface of the molecule. The factor that tends to destabilize a structure is strain in the σ-bond framework. A certain amount of strain is the necessary price paid for forming a closed three-dimensional cage, but additional strain may result from particular arrangements of the atoms. In our early work [6] we argued that for molecules whose shape does not deviate much from a sphere, the strain energy should be nearly independent of the arrangement of the five- and six-membered rings, reflecting simply the 4π Gaussian curvature of the closed cage and the angle strain of the 12 pentagons. (To produce an essentially spherical shape, the pentagons must be well separated. If they touch, there are obviously additional local strain contributions.) In this case, relative stability should be dominated by π resonance energy, which depends strongly on the arrangement of the five- and six-membered rings.

This leaves open the question of just how much deviation from spherical shape constitutes a significant deviation. It is now apparent that relatively small distortions from sphericity can introduce strain energy that can be as important as resonance in determining the most stable structures. We will return to the question of strain in the next section, but in this section we will consider methods of estimating π resonance stabilization. For fullerenes that are sufficiently spherical, we still believe that a large resonance stabilization will be one hallmark of unusual stability.

As mentioned earlier, our isomer generation program gives as output the adjacency matrix for each fullerene that when diagonalized gives the Hückel energy-level spectrum. One possibility for assessing resonance stabilization is to use the Hückel π-electron energy found by assigning electrons to the lowest-energy Hückel orbitals. Unfortunately, we have not found this very useful. It turns out that almost all fullerene isomers with a given number of carbon atoms have nearly the same Hückel total energy, even though the orbital energy-level pattern may be quite different. This is most likely related to the well-known failure of Hückel theory to recognize the instability of fused five-membered rings.

On the other hand, the Hückel energy-level structure around the Fermi level may be of some utility in assessing kinetic stability. A zero HOMO–LUMO gap is likely to be an indicator of high reactivity and/or a tendency to distort to a lower symmetry. We expect that stable fullerenes will have a significant

HOMO–LUMO gap, though the converse is not true: A large gap does not necessarily imply stability. It is interesting to note that buckminsterfullerene itself has one of the largest gaps of any fullerene, indicating that it may be exceptionally resistant to photo- or thermochemical activation.

The method we prefer to use to assess π resonance stabilization, which has its roots in valence bond theory, was developed originally for planar aromatic hydrocarbons. It was introduced originally by Herndon [18] under the name of *quantitative resonance theory,* and has been widely and successfully applied [19] to a variety of aromatic systems. It was later formulated by Randić [20] in the mathematical form that we use under the name of *conjugated circuit theory.* The method is based on an analysis of the Kekulé structures of the molecule. A conjugated circuit is a cycle of alternating single and double bonds extending completely around the periphery of a ring of atoms of a particular size. Figure 4.1(a) shows the three Kekulé structures of naphthalene, with dark lines representing double bonds and light lines representing single bonds. Structure I has two conjugated 6-circuits, one around each of the two hexagons. Structures II and III have only one 6-circuit, around the left and right hexagons, respectively, but they each have a conjugated 10-circuit around the perimeter of the molecule. Conjugated 6-circuits are taken to be strongly stabilizing when they occur in a Kekulé structure. 10-circuits are also considered to be stabilizing, but significantly less so than 6-circuits.

By contrast, in either of the two Kekulé structures of pentalene shown in Figure 1(b), the only conjugated circuits involve eight bonds around the perimeter of the molecule. According to Hückel $4n/(4n + 2)$ ideas, conjugated circuits of 4, 8, 12, . . . bonds are taken to be destabilizing. Conjugated circuit theory assigns a destabilizing influence to such cycles, thus accounting directly for the instability of fused 5-membered rings. The final formula for the resonance stabilization energy is then

$$E = \frac{1}{K} \sum_{n \geq 1} (R_n \#^{(4n+2)} + Q_n \#^{(4n)})$$

where $\#^{(m)}$ is the total number of conjugated m-cycles in all the Kekulé structures of the molecule, K is the number of Kekulé structures, and the Q_n and R_n are oppositely signed parameters that decrease roughly geometrically with the size of the circuit, and whose values are available from studies of planar aromatic systems [18, 19].

Conjugated circuit theory shares with Hückel theory the feature that it can be completely evaluated from the graph of the molecule. Starting from Kasteleyn's [21] result that the determinant of an appropriately signed adjacency matrix gives the square of the number of Kekulé structures, it is possible to show [22] that counting conjugated circuits just involves evaluating appropriate small minors of the inverse of the adjacency matrix. Conjugated circuit computations are thus not significantly more time consuming than ordinary Hückel calculations. We have found that they are much more sensitive than Hückel

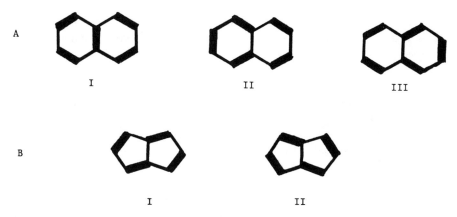

Figure 4.1 The Kekulé structures of (a) naphthalene and (b) pentalene, illustrating the types of conjugated circuits that may occur in fullerenes.

calculations for picking out favorable fullerene structures as long as strain is not an important factor. They clearly show the unusual stability of C_{60} and C_{70}. Such calculations are always possible [23], since all fullerenes (regardless of sizes of rings) have Kekulé structures.

4.1.4 Strain

Strain is intimately related to the distortions required to form a nominally planar benzenoid sheet into a closed surface S. The strain is naturally anticipated to correspond to the sum of the squares of the principal curvatures, K_1 and K_2, for all points in the surface. That is,

$$\varepsilon(\text{strain}) \sim \sum_{\text{vertices}} (K_1^2 + K_2^2)$$

$$= 2 \sum |K_1 K_2| + \sum (|K_1| - |K_2|)^2$$

and we call these two resultant terms the total absolute isotropic (or Gaussian) curvature and the total absolute anisotropic curvature. For a closed surfaces S each is bounded in terms of the genus γ_s,

$$\sum |K_1 K_2| \geq 4\pi(\gamma_s + 1)$$

$$\sum (|K_1| - |K_2|)^2 \geq A\gamma_s$$

where the value of A ($\approx 8\pi^2 - 16\pi$) is a surmise of what might be called the "Willmore" conjecture [24]. The first of these relations becomes an equality for "nonlumpy" surfaces, such as convex surfaces (or polyhedra). Indeed both terms disfavor lumpy surfaces, and the second term favors a uniformity of absolute curvature (i.e., it favors $|K_1| \approx |K_2|$). Thence polyhedra that are more nearly spherical (rather than prolate or oblate) are favored. The proportionality in the strain equation may be related [6] to molecular force-field results.

The continuum-limit differential–geometric view of the preceding paragraph interrelates to the combinatorial–graphical structure of the embedded network of bonds. This network implicates a planar-faced closed surface S, for which Descartes' relation holds

$$\sum_{\text{vertices}} \delta = 4\pi(1 - \gamma_s)$$

where the δ are each vertex's angle defect—the amount by which the sum of the bond angles at a vertex is less than the planar value of 2π. In fact, this is a limiting case of the celebrated Gauss–Bonnet theorem, since δ also is the total Gaussian curvature at a vertex if the sharp (polyhedral) corners are rounded off smoothly. Now for an "ideal" vertex with n_1-, n_2-, and n_3-sided regular polygons meeting there, the associated angle defect is

$$\delta_0 = 2\pi - \sum_i \left(\pi - \frac{2\pi}{n_i} \right) = 2\pi \left(\frac{1}{n_1} + \frac{1}{n_2} + \frac{1}{n_3} \right) - \pi$$

If all vertices have such angle defects, Descartes' relation becomes

$$\sum_{\text{vertices}} 2\pi \left(\frac{1}{n_1} + \frac{1}{n_2} + \frac{1}{n_3} \right) - \pi = 4\pi(1 - \gamma_s)$$

or upon conversion to a face sum and multiplication through by $3/\pi$, one obtains

$$\sum_{\text{faces}} (6 - n) = 12(1 - \gamma_s)$$

which duplicates the Euler relation of our introduction. But from this exercise one sees that $\pi(6 - n)/3$ is a "combinatorial" Gaussian curvature to be identified to each face. The formula given for δ_0 is idealized to accommodate the least possible angle strain in each polygon neglecting their interconnections with others. Thence the actual geometric values will deviate from these combinatorial values. Indeed the mismatch of these two types of curvature strain may be viewed to give an internal strain

$$\varepsilon(\text{mismatch}) \sim \sum_{\text{vertices}} |\delta - \delta_0|$$

Notably the truncated icosahedral structure of buckminsterfullerene is again singled out as exceptional. The absolute anisotropic and mismatch strains are 0, while the absolute isotropic strain reaches its minimum value. Among all fullerene polyhedra the only other that may be expected to be so singular as regards strain is the dodecahedron, though here the curvature per atom is excessive (and the species is open shell).

An alternative approach to the internal strain is to view it as resulting from deviations of bond lengths and/or bond angles from their idealized values. One could then attempt to represent it as

$$\varepsilon(\text{mismatch}) = \varepsilon(\text{pent} + \varepsilon(\text{dist})$$

where $\varepsilon(\text{pent})$ is the bond angle distortion introduced by the pentagons (which should asymptotically approach a constant for well-separated pentagons) and $\varepsilon(\text{dist})$ represents the contribution from bond-length compression or extension. Presumably $\varepsilon(\text{dist})$ could be expressed in terms of conventional force constants. How best to extract estimates of these various types of strain from the graphical or topological representations of the molecule remains a topic for further study.

4.2   Structural Trends in Stability

4.2.1   Isolated Pentagon Rule

Probably the most firmly established structural feature associated with stability of fullerenes is what has come to be known as the *isolated pentagon rule*. This rule states that fullerenes will be most stable if no two pentagonal faces share an edge, that is, if each pentagon is completely surrounded by hexagons. We originally suggested this rule [5] partly on the basis of the conjugated circuit ideas discussed in Section 4.1.3. If no two pentagons share an edge, only conjugated 6- and 10-circuits can occur in a fullerene, all of which lead to resonance stabilization. If, however, two or more pentagons abut, conjugated 8-circuits like those in pentalene may occur, and this leads to resonance destabilization. Placing the pentagons close to each other is also likely to increase the anisotropy of the molecule, leading to an increase in strain energy. Kroto [25] has further justified the rule on the basis of the increased local strain that results when a carbon atom is forced to be part of two or more pentagons. Evidently here we have a case where resonance energy and strain work in parallel to produce the same result.

The isolated pentagon rule now seems firmly established both experimentally and theoretically. All of the fullerenes that have been isolated in macroscopic quantities—C_{60}, C_{70}, C_{76}, and two isomers of C_{78}—are isolated pentagon structures. The preponderance of C_{60} in both laser and electric arc production of fullerenes is almost certainly related to the fact that buckminsterfullerene is the smallest fullerene that can satisfy the rule. The next most prominent product, C_{70}, is also the next smallest isolated pentagon structure [14]. In Figure 4.2 we show the π resonance energy per site versus the number of edges shared between pentagons for a variety of fullerenes. The declining trend is obvious. (The strange double line appearance results because we have included the larg-

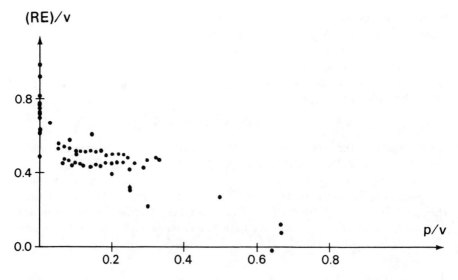

Figure 4.2 The resonance energy per site, computed by conjugated circuit theory, as a function of the ratio of edges shared between pentagons to number of sites, for a variety of fullerenes.

est and smallest value for each p-class of C_{60}'s while omitting points in between.) Raghavachari and Rohlfing [26] have recently found similar trends in total energy, using both Hartree–Fock and MNDO methods, for isomers of C_{60} containing up to four pairs of abutting pentagons.

Kroto [25] has provided a very interesting extension to the isolated pentagon rule. He has argued that strain should be least for an isolated pentagon, greater for two pentagons sharing a single edge, greater still for three pentagons meeting at a vertex, and still greater for a contiguous group of four pentagons. In laser vaporization studies C_{50} is often a prominent peak, which is rationalized by the observation that C_{50} is the smallest fullerene that can contain only pairs of abutting pentagons [6]. All of the larger fullerenes photofragment by loss of C_2 to form the next smaller fullerene [27] down to C_{32}, which fragments into a variety of small components. This may well be related to the fact that all C_{30} isomers contain groups of more than three fused pentagons. Finally Kroto notes that $(T_d)C_{28}$, which is a prominent metastable peak under certain clustering conditions, does contain only (four) groups of three fused pentagons. The four carbon atoms at the apices of the tetrahedron are in an environment quite similar to the central carbon in the stable Gomberg triphenylmethyl radical, and the ground state of C_{28} is almost certainly not closed shell.

This last observation takes on additional significance in light of recent experimental results [28], showing that while empty fullerenes photofragment only down to C_{32} before "exploding," the endohedral fullerenes $U@C_n$ retain a fullerene structure down to C_{28}. Superficially, one would expect an endohedral fullerene to require a larger cage than an empty one. However, this apparent paradox is neatly resolved in terms of formation of U^{4+} and a stable $C_{28}{}^{4-}$.

In summary, the isolated pentagon rule seems to be a reliable guide to stable fullerene structures. It and its extensions have proved useful in rationalizing a great deal of the observed fullerene behavior.

4.2.2 Symmetry and Sphericity

In some sense buckminsterfullerene is the most symmetrical of all molecules. It has the largest possible point group (I_h), and all of its atoms are equivalent (though the bonds are not). The striking symmetry of buckminsterfullerene certainly was one of the factors that first attracted attention to the fullerene hypothesis, and it was natural to assume that other stable fullerenes might also be highly symmetric. However, it now appears that the high symmetry of buckminsterfullerene may also be viewed to be a consequence of other structural characteristics. For larger fullerenes these other structural features can often be achieved without high symmetry.

One of these structural requirements is the isolated pentagon rule discussed in the previous section, and for C_{60} this requirement alone forces the I_h symmetry. Likewise for C_{70}, there is only a single isolated pentagon structure that has relatively high symmetry. But for fullerenes of 76 or more atoms, multiple isolated pentagon structures are possible [29]. At this point it appears that sphericity, rather than symmetry, becomes the most important structural requirement. As we argued previously, a spherical shape distributes the strain as evenly as possible, minimizing the anisotropic contribution to strain energy. Because bond lengths in the molecule can be different, the most spherical shape need not result in high symmetry. This is evidently the reason why the observed isomers of C_{78} are of C_{2v} and D_3 symmetry [13], rather than the higher symmetry D_{3h} predicted from analysis of electronic factors only [29, 30]. It thus appears that for larger fullerenes a reasonably accurate determination of molecular shape will be important in identifying favorable structures.

4.2.3 Clar Sextet Isomers

One other structural feature that may be associated with resonance stability can be seen from Figure 4.3. This figure shows the π resonance stabilization energy, computed via conjugated circuit theory, for the most stabilized isolated pentagon cage of each size, plotted versus the number of atoms in the cluster. As can be seen clearly, unusually large stabilization energies occur for at least one isomer of C_{60}, C_{72}, C_{78}, C_{84}, C_{90}, and C_{96}, and in fact for any number of atoms divisible by six from there on. Except for the missing point at C_{66} where there are no isolated pentagon isomers, this suggests some sort of structural regularity. Investigation shows that what these fullerenes have in common is a special sort of Kekulé structure possessed by no other fullerenes. The importance of this sort of Kekulé structure in leading to resonance stabilization

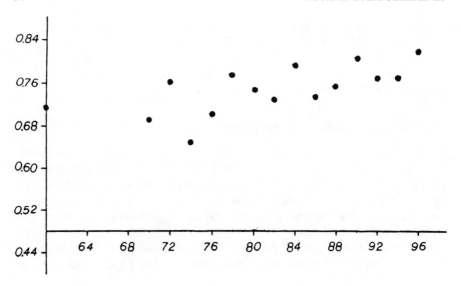

Figure 4.3 The resonance energy per site, computed by conjugated circuit theory, of the most stabilized isolated pentagon isomer for fullerenes of 60 through 96 atoms.

is closely related to Clar's ideas [31] concerning aromaticity in planar aromatic hydrocarbons, and so we have called these *Clar sextet fullerenes* [32].

The special Kekulé structure is shown in Figure 4.4 for C_{78}, where bold lines represent double bonds and light lines represent single bonds. In this Kekulé structure every double bond is seen to take part in two conjugated 6-circuits. In conjugated circuit language, this represents the maximum possible stabilizing contribution from a Kekulé structure. Buckminsterfullerene itself possesses a Kekulé structure with such a maximum of "Clar sextets," and it is possible to construct such isomers for any fullerene having $60 + 6n$ atoms, for $n > 1$ [32]. When, as for 84 carbon atoms, more than one such isomer is possible, all are found to have resonance energies larger than any non-Clar sextet isomer.

The experimental importance of the Clar sextet criterion at this point remains unclear. Buckminsterfullerene is, of course, highly stable. But C_{78} has been isolated as a mixture of two isomers, neither of which is the Clar sextet form [13]. As Diederich et al. have shown [13], this is due to the fact that the Clar sextet isomer is quite flat and thus highly strained. Here the strain energy simply swamps out the resonance stabilization. The same thing seems to happen at C_{84}, where again the Clar sextet isomers are highly aspherical. We are currently investigating larger Clar sextet fullerenes to see if we can identify isomers in which resonance stabilization will not be associated with high strain.

The case of C_{72} is more subtle. Although this structure is also slightly flattened, its anisotropic strain does not appear to be inordinate. However, C_{72} is formed in very low abundance in both laser vaporization and carbon arc ex-

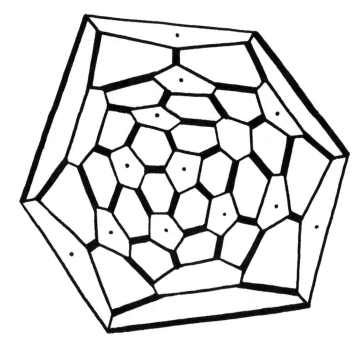

Figure 4.4   The special Kekulé structure of the Clar sextet isomer of C_{78}, in which each double bond participates in two conjugated 6-circuits.

periments, and has not been isolated. It is possible that the low abundance reflects some sort of kinetic barrier to formation, but Raghavachari [33] has explained the instability of the molecule as being due to lengthening of some bonds around the perimeter of the "polar" hexagons, presumably to relieve strain, to the point that they become essentially single. This may act to lock the molecule into a single Kekulé structure, preventing resonance stabilization. This should serve as a salutory warning that structural trends, though very valuable as a means of identifying potentially stable isomers, are just guides. Interesting structures ultimately need to be subjected to thorough analysis.

4.3   Buckytori, Buckytubes, Etc.

Beyond the polyhedral fullerene structures there are a number of other possibilities. The next higher genus ($\gamma_s = 1$) gives tori. From the Euler-related equation of the introduction, one sees that tori need not have any rings of size other than hexagons, since the net amounts of positive and negative Gaussian curvatures exactly cancel. But of course, the absolute isotropic curvature is at least 8π (as indicated in Section 4.1.4), and additionally the absolute anisotropic curvature contributes at least $A \approx 8\pi^2 - 16\pi$. Further, with only hexagons, the energy of mismatch of (geometric and combinatoric) curvatures must be notable. Of course such mismatch might be attempted to be minimized with (negative combinatoric curvature) heptagons around the inside of a torus' annulus, along with (counterbalancing positive combinatoric curvature) pentagons around the torus' outer periphery. However, such pentagons and heptagons should tend to diminish the otherwise often large resonance energies. In Refs. [6] and [34], plus a great deal of unpublished work on tori composed solely of hexagonal rings, there have been found many cases of exceptionally high Kekulé structure counts and associated resonance energies, though of course in the large-torus limit the Kekulé structure count per site approaches that for graphite.

In a suitable limit, buckytori become buckytubes. Indeed buckytubes with the standard theoretical construct of cyclic boundary conditions are actually tori. The structures of buckytubes composed solely of hexagons may be completely characterized [35]. All hexagons are equivalent in any such tube so that any hexagon may be chosen as a "seed" from which one may imagine three trigonally oriented acenic strips propagating away from the seed, as in Figure 4.5. One of two things happens as one follows the paths of these three acenic strips until an intersection first occurs: Either one propagating strip hits the seed, after say t_+ steps, or else none of the propagating strips hits the seed but rather two strips intersect one another, after say t_+ and $t_- \leq t_+$ steps. These

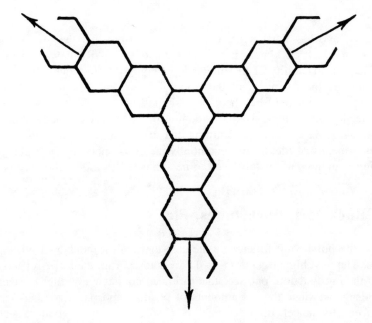

Figure 4.5   Three acenic strips that may be used to provide a complete characterization of any buckytube.

two circumstances are indicated in parts (a) and (b) of Figure 4.6. In the first case one takes $t_- = 0$, and it emerges that the pair of integers t_+ and t_- (with $t_+ > 0$ and $0 \leq t_- \leq t_+$) provide a complete characterization of buckytube (graphical) structures. If $t_-(t_+ - t_-) \neq 0$, then the tubes have a helical structure, and two possible mirror image embeddings in space are possible. The $t_+ = t_-$ case has been studied [36] in the local density-functional approximation, and Hückel MO-theoretic results are available [35] for all buckytubes. One finds [35] that one out of three of the structures have a zero band gap, whose occurrence is related to a similar circumstance in earlier studied [37, 38] graphitic strip polymers.

The buckytubes found experimentally [39, 40] have rather large diameters, are helical, and are of finite length with caps. The caps of course must involve some rings smaller than hexagons, since there is needed a net amount of 2π Gaussian curvature to close off each end of a tube. That is, if no three- or four-membered rings are used, one needs exactly 6 five-membered rings for each cap. Such bicapped tubes with a longitudinal five- or sixfold axis follow from the construction in Appendix F of Ref. [6] and are considered in detail in Ref. [41]. These correspond to the nonhelical buckytubes with $t_-(t_+ - t_-) = 0$, but with the (so-called) buckytube period $t_+ + t_-$ a multiple of 5 or 6.

Finally structures of genus higher than 2 are possible. Perhaps those of greatest interest are those where the mismatch and absolute anisotropic curvatures of Section 4.1.4 are 0. With these higher genuses negative curvature is required so that to avoid absolute anisotropic curvature, one must avoid positive curvatures. Indeed, surfaces with negative Gaussian curvature and zero anistropic curvature are termed [42] "minimal." They are of infinite extent with probably the most interesting being periodic. Indeed some initial work on such bucky species have been reported [43, 44], but we go into this no further now.

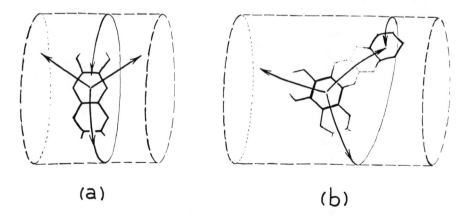

(a)   (b)

Figure 4.6 The two ways in which the strips of Figure 4.5 may intersect: (a) a $t_- = 0$ tube; (b) a $t_- \neq 0$ tube.

4.4   Formation Mechanism

It should be stated at the outset that almost everything in this section falls into the category of (we hope not wild) speculation. Next to nothing is known about how fullerenes actually grow and form. Nevertheless, a reasonably attractive picture can be put together, which rationalizes a lot of what is known about the occurrence of fullerenes. This picture has been presented in brief outline by us in an earlier work [45], and much more fully by Smalley in an excellent review [46]. It assumes that fullerenes grow by accretion of single atoms or other small fragments. For small clusters, chains or rings are the most stable conformation, but at some point, perhaps around 20 atoms, two- and three-dimensional structures become more stable. Growing clusters try to minimize dangling bonds by forming five-membered rings when possible in what would otherwise be a graphitic net, thus causing the fragment to curl. However, the isolated pentagon rule is respected during the growth process. The clusters are assumed to have enough excess energy to rearrange into a stable conformation at each growth step before adding another ring.

In this view, buckminsterfullerene results from a perfect growth history. If five-membered rings are introduced at every opportunity during the growth process that does not violate the isolated pentagon restriction, they will be separated by precisely one hexagon, and buckminsterfullerene will be the inevitable result. Of course, during the actual growth process accreting atoms are unlikely to arrive in the perfect location, so it must be possible for fragments to rearrange to get back on the perfect path. Smaller fullerenes result from fragments that do not succeed in escaping from locally formed adjacent pentagons to a lower-energy isolated pentagon structure, while larger fullerenes form when a fragment passes up one or more opportunities to close dangling bonds with pentagons. For C_{60} to be the dominant product, it is not necessary for it to be more stable than graphite or other larger fullerenes, which it is not. It is only necessary that it be the most stable arrangement of 60 carbon atoms, which it almost certainly is, and for the partially open isolated pentagon structures to be at least locally stable arrangements for smaller fragments from which they can grow.

How likely is this scenario? It is at least not completely unreasonable. It has recently been shown that corannulene is highly flexible even at room temperature [47], so at the elevated temperatures involved in fullerene growth it is not inconceivable that growing fragments can undergo very efficient annealing. Since only a few of the many possible isomers of the fullerenes are actually formed, annealing is apparently efficient even after the fullerene has closed. In addition, the fact that fullerenes can retain their closed form while losing C_2 during high-energy photofragmentation [27] indicates that they must be capable of accommodating enormous amounts of internal energy. Whether the isolated pentagon rule can play such an important role in hot open fragments is something for which we do not yet have any direct evidence.

One of the most interesting open questions concerns the rearrangement pathways in fragments, closed fullerenes, and during photofragmentation. Stone and Wales [48] introduced a rearrangement mechanism for closed fullerenes, known as the *pyracylene rearrangement,* which rotates a bond linking two pentagons by 90°, changing the two pentagons into hexagons and the originally adjacent hexagons into pentagons. Curl and O'Brien [27] proposed a related mechanism for C_2 photoejection. Diederich and Whetten [13] have rationalized the occurrence of two C_{78} isomers on the basis of the Stone–Wales mechanism by noting that four of the five isolated pentagon isomers can interconvert in this fashion, while the fifth cannot. However, these mechanisms have not been studied in any detail. The barriers to rearrangement and the minimum energy paths are not known, nor is it known if there may be other competitive, or even superior, mechanisms available. Understanding the mechanisms of fullerene growth, interconversion, and ultimately chemical reativity promises to provide a fertile field of study for years to come.

References

1. H. W. Kroto, J. R. Heath, S. C. O'Brien, R. F. Curl, and R. E. Smalley, *Nature* **318,** 162–63 (1985).

2. J. R. Heath, S. C. O'Brien, Q. Zhang, Y. Liu, R. F. Curl, H. W. Kroto, and R. E. Smalley, *J. Am. Chem. Soc.* **107,** 7779–80, (1985).

3. A. D. J. Haymet, *J. Am. Chem. Soc.* **108,** 319–21 (1986).

4. D. J. Klein, T. G. Schmalz, W. A. Seitz, and G. E. Hite, *J. Am. Chem. Soc.* **108,** 1301 (1986).

5. T. G. Schmalz, W. A. Seitz, D. J. Klein, and G. E. Hite, *Chem. Phys. Lett.* **130,** 203–7 (1986).

6. T. G. Schmalz, W. A. Seitz, D. J. Klein, and G. E. Hite, *J. Am. Chem. Soc.* **110,** 1113–27 (1988).

7. B. Grunbaum and T. S. Motzkin, *Can. J. Math.* **15,** 744–51 (1963).

8. D. J. Klein, G. E. Hite, and T. G. Schmalz, *J. Comput. Chem.* **7,** 443–56 (1986).

9. D. J. Klein, W. A. Seitz, and T. G. Schmalz, *Nature* **323,** 703–6 (1986).

10. L. D. Krätschmer, K. Lamb, K. Fostiropoulos, and D. R. Huffman, *Nature* **347,** 354–58 (1990).

11. F. Diederich, R. Ettl, Y. Rubin, R. L. Whetten, R. Beck, M. Alvarez, S. Anz, D. Sensharma, F. Wudl, K. C. Khemani, and A. Koch, *Science* **252,** 548–51; (1991). D. J. Parker, P. Wurz, K. Chaterjee, K. R. Lykke, J. E. Hunt, M. J. Pellin, J. C. Hemminger, D. M. Gruen, and L. M. Stock, *J. Am. Chem. Soc.* **113,** 7499–503 (1991); H. Shinoshara, H. Sato, Y. Saito, M. Takayama, A. Izuoka, and T. Sugawara, *J. Phys. Chem.* **95,** 8449–51 (1991).

12. R. Ettl, I. Chao, F. Diederich, and R. L. Whetten, *Nature* **353,** 149–53 (1991).

13. F. Diederich, R. L. Whetten, C. Thilgen, R. Ettl, I. Chao, and M. M. Alvarez, *Science* **254,** 1768–70 (1991).

14. X. Liu, D. J. Klein, T. G. Schmalz, and W. A. Seitz, *J. Comput. Chem.* **12**, 1252–59 (1991).

15. X. Liu and D. J. Klein, *J. Comput. Chem.* **12**, 1243–51 (1991).

16. D. E. Manolopoulos and P. W. Fowler, *J. Chem. Phys.* **96**, 7603–14 (1992).

17. A. J. Stone, *Inorg. Chem.* **20**, 563–71 (1981).

18. W. C. Herndon, *J. Am. Chem. Soc.* **95**, 2404–6 (1973).

19. W. C. Herndon and M. L. Ellzey, Jr., *J. Am. Chem. Soc.* **96**, 6631–42 (1974).

20. M. Randić, *Chem. Phys. Lett.* **38**, 68–70 (1976); M. Randić, *J. Am. Chem. Soc.* **99**, 444–50 (1977).

21. P. W. Kasteleyn, *J. Math. Phys.* **4**, 287–93 (1963).

22. D. J. Klein and X. Liu, *J. Comput. Chem.* **12**, 1260–64 (1991).

23. D. J. Klein and X. Liu, *J. Math. Chem.*, to appear.

24. J. Langer and D. Singer, *Bull. London Math. Soc.* **16**, 531–34 (1984).

25. H. W. Kroto, *Nature* **329**, 529–31 (1987).

26. K. Raghavachari and C. M. Rohlfing, *J. Phys. Chem.* **96**, 2463–66 (1992).

27. S. C. O'Brien, J. R. Heath, R. F. Curl, and R. E. Smalley, *J. Chem. Phys.* **88**, 220–30 (1988).

28. R. E. Smalley, *Materials Research Society Symposium on Novel Forms of Carbon*, San Francisco, 1992.

29. X. Liu, T. G. Schmalz, and D. J. Klein, *Chem. Phys. Lett.* **188**, 550–54 (1992).

30. P. W. Fowler, D. E. Manolopoulos, and R. C. Batten, *J. Chem. Soc. Faraday Trans.* **87**, 3103–4 (1991).

31. E. Clar, *The Aromatic Sextet*, John Wiley, London, 1972.

32. X. Liu, D. J. Klein, T. G. Schmalz, C. A. Folden, *J. Amer. Chem. Soc.*, to be submitted.

33. K. Raghavachari, *Materials Research Society Symposium on Novel Forms of Carbon*, San Francisco, 1992.

34. R. Tosi and S. J. Cyvin, *J. Math. Chem.*, submitted for publication.

35. D. J. Klein, W. A. Seitz, and T. G. Schmalz, *J. Phys. Chem.*, in publication.

36. J. W. Mintmire, B. I. Dunlap, and C. T. White, *Phys. Rev. Lett.* **68**, 631–34 (1992).

37. M. Bradburn, C. A. Coulson, and G. S. Rushbrooke, *Proc. Roy. Soc. Edinburgh* **62**, 336–49 (1948).

38. D. J. Klein, T. G. Schmalz, W. A. Seitz, and G. E. Hite, *Intl. J. Quantum Chem.* **S19**, 707–18 (1986); D. J. Klein, *Rep. Molec. Theory* **1**, 91–94 (1990).

39. S. Iijima, *Nature* **354**, 56–58 (1991).

40. S. Iijima, T. Ichihashi, and Y. Ando, *Nature* **356**, 776–78 (1992).

41. P. W. Fowler, *Chem. Soc. Faraday Trans.* **86**, 2073–77, (1990).

42. R. Osserman, *A Survey of Minimal Surfaces*, Dover, New York, 1986.

43. D. Vanderbilt and J. Tersoff, *Phys. Rev. Lett.* **68**, 511–13 (1992).

44. T. Lenosky, X. Gonge, M. Teter, and V. Elser, *Nature* **355**, 333–35 (1992).

45. D. J. Klein and T. G. Schmalz, Buckminsterfullerene, Part A: Introduction. In *Quasicrystals, Networks and Molecules of Five-fold Symmetry*, I. Hargittai, Ed., VCH Publishers, New York, 1990.

46. R. E. Smalley, *Acc. Chem Res.* **25,** 98–105 (1992).

47. L. T. Scott, M. M. Hashemi, M. S. Bratcher, *J. Am. Chem. Soc.* **114,** 1920–21 (1992); A. Borchardt, A. Fuchicello, K. V. Kilway, K. K. Baldridge, and J. S. Siegel, *J. Am. Chem. Soc.* **114,** 1921–23 (1992).

48. A. J. Stone and D. J. Wales, *Chem. Phys. Lett.* **128,** 501–3 (1986).

Ab Initio Theoretical Predictions of Fullerenes

Gustavo E. Scuseria

Theoretical calculations have played a significant role in the recent development of the fullerene field. The proposal of an icosahedral soccer ball structure of 60 carbon atoms [1] (buckminsterfullerene) to interpret the remarkably stable peak appearing in the mass spectra of carbon clusters produced by laser vaporization of graphite was quickly followed by a number of calculations [2–7] that later proved crucial in the identification of C_{60} in the soot produced by resistive heating of graphite [8–10]. The theoretical determination of the number of active infrared bands and their positions, as well as the optical spectrum of C_{60}, proved critical in the early developments of the "second chapter" (mass production) of the fullerene story.

Since the fall of 1990, when fullerenes became widely available in macroscopic quantities, the number of papers dealing with theoretical aspects of these fascinating new molecules has been growing steadily. Kroto et al. [11] have recently reviewed the fullerene bibliography including both theoretical and experimental aspects.

Theoretical contributions in the fullerene field have been hampered by the large molecular size of the systems of interest, making conventional ab initio quantum-chemistry techniques very expensive to apply. However, the recent advances in computer speed combined with methodological improvements have dramatically extended the range of applicability of ab initio quantum-chemistry techniques. In particular, the direct self-consistent field (SCF) Hartree-Fock method originally proposed by Almlöf et al. [12] has been central to this development. This method is based on the simple but powerful idea that beyond a certain molecular size, it is no longer practical to store the electron-electron repulsion integrals required by the SCF procedure. Therefore, the only possible alternative is to simply recalculate them "on the fly," as required by each iteration. In contrast, the conventional SCF method stores on disk all the integrals and retrieves them when necessary. In very large memory machines it would be possible to hold in core all matrices and integrals, and in high-symmetry cases this is the approach employed for C_{60} by Pitzer and co-workers [13].

Several efficient computer implementations of the direct SCF ideas have been achieved in the past few years. In particular, the TURBOMOLE suite of programs [14], has proved very efficient for fullerene research. For example, we have carried out calculations on C_{240} employing a double-zeta basis set. This calculation, which includes 2,400 basis functions, is unthinkable from a conventional SCF perspective. By taking advantage of "locality" in a rigorous way, the numerical values of the two-electron integrals can be estimated using mathematical bounds, and only those that are larger than a given threshold are actually calculated. This prescreening of integrals greatly reduces the computational effort in a direct SCF calculation. Employing TURBOMOLE in large fullerene systems, we have observed that the CPU time required for a given calculation approximately increases as the square of the number of basis functions (n^2), rather than the theoretical factor of n^4.

In this paper, I will review the theoretical calculations carried out in our research group in the past few years. Although some of this work has already been published, it will be presented here in light of the most recent experimental results. In other cases, the material discussed herein has not been published yet. Most of the calculations discussed in this work have been carried out at the SCF level of theory employing TURBOMOLE. In a few cases, correlation effects were included employing second-order Moller-Plesset perturbation theory (MP2) [15].

5.1 The Equilibrium Structure of C_{60}

Before fullerenes were available in macroscopic quantities [8], semiempirical [16, 17] and ab initio SCF [18–20] predictions of the equilibrium structure of C_{60} had been performed. These calculations demonstrated that the two symmetry-distinct carbon-carbon bond lengths in the icosahedral C_{60} geometry were different. The predicted bond distances were consistent with an alternating single- and double-bond description of an aromatic structure for the C_{60} shell. The ab initio predictions were limited to the SCF method employed in conjunction with minimum and double-zeta quality basis sets [18–20].

In 1990, with faster computers and more efficient programs, it was possible to extend the range of these ab initio calculations to include both polarization functions and electron correlation effects [21, 22]. The theoretical predictions are presented in Table 5.1.

With the availability of C_{60} in macroscopic quantities, several groups un-

Table 5.1 The equilibrium structure of C_{60} (Å)

Basis	Method	C-C	C=C
$4s2p^a$	SCF	1.451	1.368
$4s2p^b$	MP2	1.470	1.407
$4s2p1d^a$	SCF	1.453	1.372
$5s3p1d^a$	SCF	1.448	1.370
$5s3p1d^b$	MP2	1.446	1.406
Exp^c	ED	1.458(6)	1.401(10)
Exp^d	NMR	1.45	1.40
Exp^e	X ray	1.432(9)	1.388(5)
Exp^f	X ray	1.445(30)	1.388(30)
Exp^g	X ray	1.467(21)	1.355(9)

[a]Scuseria, Ref. 21.
[b]Häser, Almlöf, and Scuseria, Ref. 22.
[c]Gas-phase electron diffraction, Hedberg et al., Ref. 23.
[d]NMR, Yannoni et al., Ref. 24.
[e]Osmium derivative, Hawkins et al., Ref. 25.
[f]Platinum derivative, Fagan et al., Ref. 26.
[g]C_{60} crystal, Liu et al., Ref. 27.

dertook measuring the C_{60} bond lengths under different molecular environments and using different techniques [23–27]. The experimental results are compared with the theoretical predictions in Table 5.1. It should be pointed out that large-basis-set SCF predictions [21] preceded the experimental measurements and, in some cases, were helpful in guiding them. Included in Table 5.1 are gas-phase electron diffraction results [23], an NMR determination of the bond lengths [24], and X-ray structures obtained from an osmium derivative [25], a platinum derivative [26], and a C_{60} crystal [27]. The most meaningful comparison of the ab initio theoretical predictions is with the experimental results obtained in the gas phase by Hedberg et al. [23].

In general, the quality of the SCF predictions is fairly good. Even the minimum basis STO-3G results are of acceptable quality. However, a quick glance at Table 5.1 demonstrates that "to get the right answer for the right reason," *both* large basis sets and correlation effects are needed for C_{60}. Our best theoretical predictions for the C_{60} bond lengths, 1.446 and 1.406 Å, obtained with a triple-zeta plus polarization basis set ($5s3p1d$) at the MP2 level of theory, are in excellent agreement with the gas-phase experimental values of 1.458(6) and 1.401(10) Å. It should be pointed out that for C_{60} with a $5s3p1d$ basis set (1140 basis functions), MP2 was the only affordable and practical correlated technique we could use. Nevertheless, this calculation is one of the largest MP2 calculations reported to date. With a more modest $4s2p$ basis set, and employing the SCF method, the agreement between theory and experiment is still qualitatively good.

In synthesis, results presented in Table 5.1 demonstrate that small-basis-set

SCF calculations give qualitative correct results, but large basis sets and correlation effects are necessary to fine tune the bond-length predictions.

5.2   The Equilibrium Structure of C_{70}

A natural continuation of the C_{60} work discussed previously was to calculate the equilibrium structure of C_{70}. For this molecule, however, the computational effort is much greater, since the number of floating operations in an SCF calculation scales as the inverse of the order of the point symmetry group. In essence, a D_{5h} (20 symmetry elements) calculation of C_{70} is at least six times more expensive than an I_h (120 symmetry elements) calculation of C_{60}, just because of symmetry.

Although C_{70} is the second most important peak in the mass spectrum of fullerenes, theoretical predictions for this molecule are much more scarce than for C_{60}. Actually, the lack of *any* calculations on C_{70} prompted us to carry out an SCF study of its equilibrium geometry [28], which followed soon after the first macroscopic preparation of C_{70} was reported [8]. Employing basis sets of double-zeta plus polarization quality, in conjunction with analytic energy gradient techniques, the geometry of C_{70} was optimized within the D_{5h} point symmetry group. Several other studies soon followed [29–34], and in all cases their results have been in excellent agreement with the ab initio bond-length predictions [28].

Recently, the equilibrium structure of C_{70} was measured using thin-film electron diffraction in conjunction with a simulated annealing method [35]. The theoretical predictions and the experimental values for the 8 symmetry-distinct bond lengths of C_{70} are compared in Table 5.2. The labeling of the bond lengths employed in Table 5.2 is reported in Figure 5.1.

Table 5.2   The equilibrium structure of C_{70} (Å)

Bond type	Experiment[a] thin-film ED	Theory[b] dzp/SCF
2	1.39 ± 0.01	1.407
4	1.46 ± 0.01	1.457
5	1.37 ± 0.01	1.361
6	1.47 ± 0.01	1.446
7	1.37 ± 0.01	1.375
8	1.464 ± 0.009	1.451
1	$1.41^{+0.03}_{-0.01}$	1.475
3	$1.47^{+0.01}_{-0.03}$	1.415

[a]Thin-film electron diffraction, McKenzie et al., Ref. 35.
[b]Scuseria, Ref. 28.

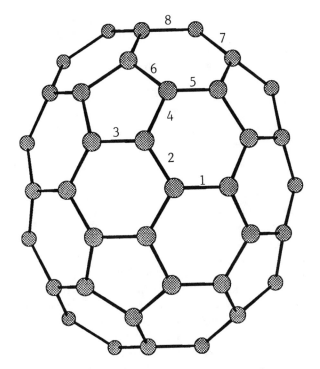

Figure 5.1   Bond type definitions employed in Table 5.2 for the C_{70} molecule.

Table 5.2 is organized in a way that highlights the discrepancies between theory and experiment. Although most of the theoretical predictions fall within the error bars of the experimental results, our calculated values for bond lengths type 1 and 3 are clearly in contradiction with experiment. It is unlikely that our $4s2p1d$/SCF predictions could be in error by as much as 0.06 Å. Moreover, it is even more difficult to imagine that two bond lengths may be in error by that much *and* in opposite directions (see Table 5.2). At the $4s2p1d$/SCF level of theory, such large errors would be unusual, as demonstrated by the theoretical predictions for C_{60} in Table 5.1.

Evidently, better calculations and other experimental measurements would be extremely helpful in settling this controversy. Unfortunately, recent gas-phase electron diffraction studies carried out by Hedberg and collaborators [36] have not been conclusive and have been plagued by difficulties in deconvoluting the experimental data for a complex molecule like C_{70}.

5.3   Hydrogenated C_{60}

One of the promising avenues in C_{60} research is the synthesis of new compounds originating from the addition of atoms or radicals to the interior or the exterior of C_{60}. In particular, $C_{60}H_{36}$ has been reported as a product in the hydrogenation of C_{60} employing the Birch reduction mechanism [37]. Hydrogenated C_{60} could have potential applications as a hydrogen storage device, and has also been suggested as one of the infrared emission sources in interstellar space.

Only a few theoretical studies have appeared in the literature on hydrogenated C_{60}. Using large basis sets including polarization functions, we reported the first SCF calculation of $C_{60}H_{60}$ some time ago [21]. In that study, an icosahedral I_h symmetry structure for $C_{60}H_{60}$ was proposed, but no partially hydrogenated cases were analyzed.

Employing the local-density-functional (LDF) method, Dunlap et al. [38] have calculated the electronic and equilibrium structure of $C_{60}H_{36}$ and $C_{60}H_{60}$. For $C_{60}H_{60}$, these authors found that introducing one hydrogen atom inside the C_{60} cage gives a more stable configuration than the one obtained from fully exo-hydrogenated C_{60}.

Saunders [39] has also studied several isomers of $C_{60}H_{60}$ with hydrogen atoms inside the C_{60} cage using molecular mechanics [40]. The lowest-energy isomer was predicted to have ten hydrogen atoms inside the cage, a result that may be rationalized in terms of energetically favorable chair conformations for the H–C–C–H fragments at the 10 distinct C-C edges shared between hexagons [39]. In each of the 10 possible H-C-C-H fragments, one hydrogen atom is inside the cage while the other remains outside.

Recently, semiempirical MNDO, AM1, and PM3 calculations were also carried out by Bakowies and Thiel [41] on $C_{60}H_{36}$ and $C_{60}H_{60}$. Their predictions for the equilibrium bond lengths are consistent with the ab initio [21, 42] and LDF results [38].

We have recently carried out a series of ab initio theoretical calculations to study both exo- and endo-hydrogenation of C_{60} [42]. Employing basis sets of double-zeta quality at the SCF level of theory, a series of $C_{60}H_{12n}$ ($n = 1$–5) molecules in tetrahedral (T_h) symmetry were analyzed and their equilibrium geometries and relative stabilities predicted.

In all exo- and endo-hydrogenated C_{60} molecules studied, the geometrical structures were optimized employing analytic energy gradient techniques. The predicted equilibrium structures of the $C_{60}H_{12n}$ ($n = 1$–5) molecules are presented in Figure 5.2. The equilibrium interatomic distances were in all cases consistent with normal C–C and H–H bond lengths [42].

Based on the relative energies of these molecules, the main conclusion of this study was that for the exo-hydrogenated case, $C_{60}H_{36}$ and $C_{60}H_{48}$ are the most stable molecules in the $C_{60}H_{12n}$ series. If hydrogen atoms are allowed inside C_{60}, it was also found that only for $C_{60}H_{60}$ does a lower-energy isomer exist that contains 12 hydrogen atoms inside the cage. Endo-hydrogenation of the other $C_{60}H_{12n}$ species did not produce isomers lower in energy than the exohydrogenated ones. The total energy of the 12-endo-hydrogenated form of $C_{60}H_{60}$ ($H_{12}@C_{60}H_{48}$) is much lower than that of the previously predicted I_h structure [21]. However, the barrier for hydrogen penetration of the C_{60} cage through a

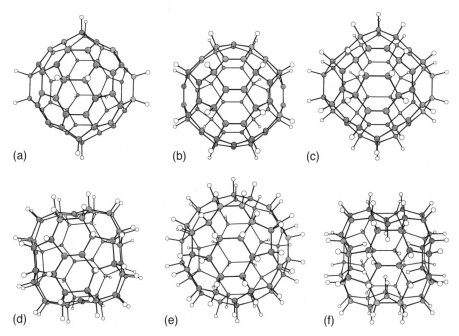

Figure 5.2 Equilibrium geometries of hydrogenated C_{60} molecules as predicted by the dz/SCF level of theory: (a) $C_{60}H_{12}$, (b) $C_{60}H_{24}$, (c) $C_{60}H_{36}$, (d) $C_{60}H_{48}$, (e) $C_{60}H_{60}$, and (f) $H_{12}@C_{60}H_{48}$.

six-membered ring is very large, approximately 2.7 eV/atom. Thus, hydrogen atoms will not attach to the inside of the C_{60} cage unless trapped when the cage closed, or unless special experimental conditions are set up (i.e., enough energy in a collision process is given to the hydrogen atoms).

5.4   Fluorinated C_{60}

The initial report of the C_{60} molecule [1] contained a reference to $C_{60}F_{60}$, a compound presumed to have interesting superlubricant properties. Fluorinated derivatives of C_{60} have now been studied both theoretically [21, 41, 43–45] and experimentally [46, 47].

Ab initio calculations at the SCF level of theory employing basis sets of double-zeta plus polarization quality (dzP) indicated the existence of a stable, I_h symmetry $C_{60}F_{60}$ molecule [21]. Experimentally [46, 47], a number of fluorinated C_{60} intermediates have been identified. In particular, $C_{60}F_{36}$ and $C_{60}F_{38}$ are found to be the most prominent peaks in the mass spectrum of fluorinated fullerenes [47]. One experimental group has associated a single line observed in the ^{19}F NMR spectrum with an icosahedral $C_{60}F_{60}$ species [46]. However,

$C_{60}F_{60}$ has not been observed in the mass spectrum of fluorinated fullerenes by other authors [47].

From a theoretical viewpoint, Fowler, Kroto, Taylor, and Walton [44] (FKTW) have pointed out that the experimental evidence of a single NMR line is compatible with both I_h and I symmetry $C_{60}F_{60}$ isomers. These authors further suggested that a twisted $C_{60}F_{60}$ I isomer, while preserving the 60-fold equivalence of the F and C nuclei, would allow for a relief of steric strain from the "pincushion" I_h isomer [see Fig. 5.2(e) for an analogous $C_{60}H_{60}$ isomer]. Recent semiempirical MNDO calculations by Bakowies and Thiel [41], as well as our ab initio SCF results [45], do not support this interpretation. The theoretical evidence clearly indicates that the I_h $C_{60}F_{60}$ isomer is preferred over the I species. Calculations carried out employing basis sets of double-zeta quality at the SCF level of theory showed that the energy is a steep uphill function of the FCCF torsional angle with a minimum at the I_h conformation [45]. Full geometry optimization starting from different I conformations lead us back to the I_h isomer. In all cases, no indication of a stationary point for an I equilibrium structure was found. It was concluded that externally fluorinated $C_{60}F_{60}$, if observed experimentally, must have I_h symmetry, as originally suggested [21].

Two competing effects are at work in $C_{60}F_{60}$; one is the F-F steric repulsion, and the other is the stiffness of the C_{60} cage. As FKTW suggested [44], fluorine steric strain can be reduced by moving the F atoms out from their eclipsed positions in the I_h isomer to gauche conformations in the I symmetry species. However, carbon hybridization is affected in the process, and the stiffness of the C_{60} cage should be taken into consideration. The theoretical evidence strongly suggests that the stiffness of the bucky cage is much more important than the steric repulsion of the fluorine atoms. Hence a $C_{60}F_{60}$ equilibrium structure of I_h symmetry is the most likely candidate.

The question of the lowest-energy structure of $C_{60}F_{60}$ when endo- and exo-fluorination is permitted is also interesting. Following the study on hydrogenated C_{60} [42], we analyzed an analogous structure to that of the lowest-energy isomer of $C_{60}H_{60}$ containing 12 fluorine atoms inside the C_{60} cage ($F_{12}@C_{60}F_{48}$). The fluorine atoms were attached to carbon atoms that form double bonds in C_{60}, and the geometry was optimized employing a double-zeta basis at the SCF level of theory. The equilibrium structures of the I_h $C_{60}F_{60}$ and T_h $F_{12}@C_{60}F_{48}$ isomers are similar to those of the hydrogenated species presented in Figures 5.2(e) and 2(f). The 12 endo-fluorinated T_h isomer of $C_{60}F_{60}$ is predicted to be 372 kcal/mol lower in energy than the exo-hydrogenated I_h structure. As with $C_{60}H_{60}$, the barrier for fluorine penetration of the cage is expected to be very large. We therefore do not expect endo-fluorinated compounds to appear under normal fluorination conditions.

5.5   The Giant Fullerenes: Spheres or Tubes?

The recent experimental discovery [48] that a new type of structure consisting of needlelike microtubules of graphitic carbon can be grown using an apparatus similar to that used to mass produce fullerenes challenges the conventional view that big fullerenes such as C_{120}, C_{180}, and C_{240} have spheroidal shapes [49]. The carbon needles detected by Iijima using high-resolution electron transmission microscopy [48] were described as being up to 1 μm in length and ranging from 20 to 300 Å in diameter. These needles are usually found in concentric cylindrical layers separated by ~3.4 Å, the distance between graphitic sheets in bulk graphite. These so-called "buckytubes," if grown to macroscopic length, might constitute vastly superior carbon fibers. Iijima has also detected the presence of seven-membered rings in the fullerene soot and has assigned them an important role in the growth of the buckytubes [50].

Experimentally, the current techniques for mass spectroscopy permit a resolution up to ~C_{300}, so very little is known about the mass spectrum distribution of the largest fullerenes. Recently, Lamb et al. [51] characterized the extract of soot produced by the Krätschmer-Huffman technique by mass spectroscopy and imaging by scanning tunneling microscopy (STM). The STM images showed that the giant fullerenes in these samples were roughly spherical in shape, corresponding to fullerenes containing up to 330 carbon atoms. No evidence of buckytubes was found.

On the other hand, employing high-resolution electron transmission microscopy, Ugarte [52] has recently reported that the soot extracted from the surface of the electrodes in the arc generation machine contains a large number of buckytubes and "buckyonions." The tubes are similar to those reported by Iijima [48, 50]. Ugarte's buckyonions consist of concentric spherical fullerenes, and although similar to those photographed by Iijima in 1980 [53, 54], they are much sharper, rounder, and bigger [52]. Ugarte has observed giant buckyonions up to 500 Å in diameter [52].

We have recently carried out a series of SCF calculations to test the energetics of monolayer capped buckytubes versus the large spheroidal fullerenes. In particular, we have focused on C_{80}, C_{120}, C_{180}, and C_{240}. The isomers studied, as well as their point-symmetry groups and relative energies, are presented in Figures 5.3–5.6 and Table 5.3. The equilibrium geometry of all isomers was obtained employing analytic energy gradient methods in minimal-basis SCF calculations. The optimized geometries are precisely those presented in Figures 5.3–5.6. In some cases, single-point SCF energy calculations employing a double-zeta (dz) basis at the STO-3G/SCF optimized geometries were carried out, but no significant changes in the energy differences between isomers were observed. The dz basis employed for carbon is identical to that reported in Refs. [21] and [28]. Given that the low-symmetry SCF calculations of the large fullerenes are computationally demanding, the geometry optimizations were stopped prior to full convergence in a few cases, once it became clear that a particular isomer was not the most stable one.

The first fullerene for which we can make a distinction between spheres and tubes is C_{80}. The I_h C_{80} isomer has an open-shell electronic configuration (h_u^4), and it is subject to Jahn–Teller distortion. Lowering the symmetry to T_h, the h_u orbital splits into $e_u + t_u$, and it is then possible to close the e_u orbital

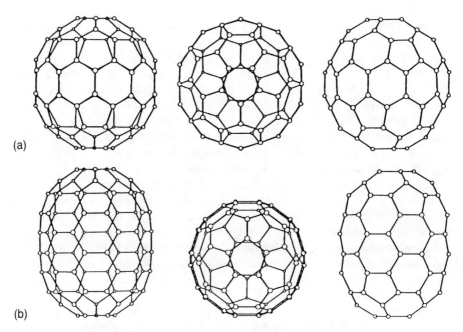

(a)

(b)

Figure 5.3 Equilibrium geometries of two isomers of C_{80} as predicted by the STO-3G/SCF level of theory: (a) D_{5d} (SF), (b) D_{5d} (VF).

to obtain a ^1A_g electronic state. Relaxing the symmetry to D_{5d}, the h_u orbital transforms into $a_{1u} + e_{1u} + e_{2u}$, so at least two Jahn–Teller stable states may be formed by closing e_{1u} or e_{2u}. Ultimately, we found that for this C_{80} topology, the lowest-energy closed-shell electronic state ($17a_{1g}^2\ 7a_{2g}^2\ 24e_{1g}^4\ 24e_{2g}^4\ 6a_{1u}^2$ $18a_{2u}^2\ 24e_{1u}^4\ 24e_{2u}^4$) is not related to the (h_u^4) I_h electronic state through Jahn–Teller distortion. This D_{5d} isomer [Fig. 5.3(a)] achieves a very small deformation from the icosahedral shape, with diameters of 7.97 and 8.03 Å.

A second D_{5d} isomer is also possible for C_{80} [Fig. 5.3(b)]. In this cage the pentagons are "vertex-facing" (VF) each other, unlike the other $I_h \rightarrow D_{5d}$ isomer where the pentagons are "side-facing" (SF) one another. C_{80} has a total of 7 isomers within the isolated pentagon rule (IPR) [55]. (C_{60} and C_{70} are the first two fullerenes to satisfy IPR [56].) Only two out of the seven possible IPR isomers for C_{80} were considered in this study. The VF D_{5d} C_{80} isomer has a distinct cylindrical shape with calculated diameters of 7.1 and 9.0 Å at the STO-3G/SCF level of theory. Its energy is lower than that of the SF isomer by 132 kcal/mol.

For C_{120}, the most rounded-shape fullerene or buckysphere is the pear-shaped object of T_d symmetry in Figure 5.4a. Unfortunately, C_{120} has several hundred isomers within the IPR. In our work, only three different classes of tubes were analyzed for C_{120}. They correspond to three closed cylinders (or capsules) capped by half of C_{72}, C_{80}, and C_{60}. Their symmetries are D_{6d}, D_{5d}(SF), and D_{5d}(VF),

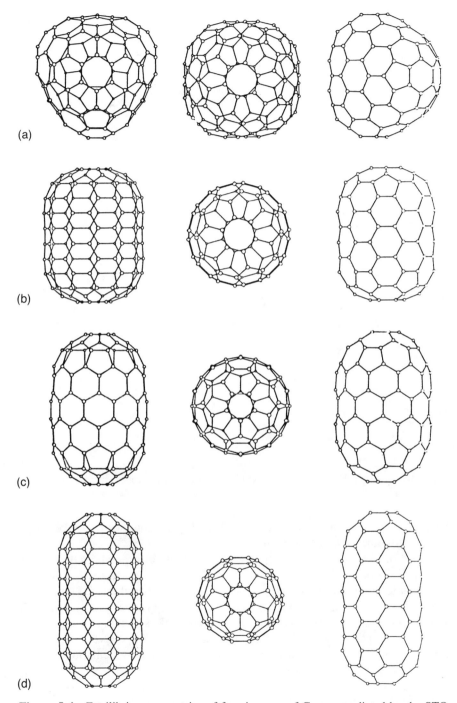

Figure 5.4 Equilibrium geometries of four isomers of C_{120} as predicted by the STO-3G/SCF level of theory: (a) T_d, (b) D_{6d}, (c) D_{5d} SF, (d) D_{5d} VF.

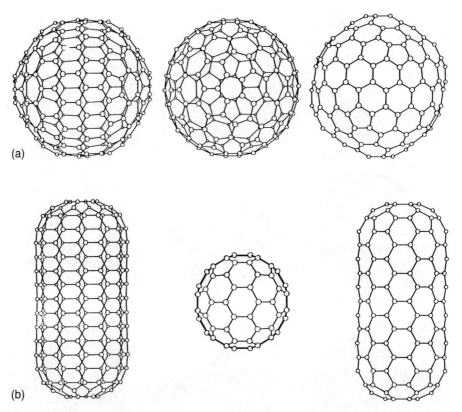

Figure 5.5 Equilibrium geometries of two isomers of C_{180} as predicted by the STO-3G/SCF level of theory: (a) I_h, (b) D_{6h}.

respectively. In all cases, the buckytubes were found to be higher in energy than the pear-shaped T_d structure (see Fig. 5.4). The most stable buckytube (D_{6d}) was found to be 32 kcal/mol higher in energy than the T_d isomer at the STO-3G/SCF level of theory. As with C_{80}, VF pentagons seem to give lower-energy structures than SF pentagons.

Based on the C_{120} results, only the D_{6h} tube that corresponds to the lowest-energy D_{6d} tube in C_{120} was studied for C_{180}. At the STO-3G/SCF level, this tube was found to lie 208 kcal/mol higher in energy than the I_h C_{180} fullerene. The C_{180} fullerenes are shown in Figure 5.5. In I_h symmetry, C_{180} has three symmetry-distinct classes of atoms. They contain 60 symmetry-equivalent atoms each in spheres of radii 6.29, 6.09, and 6.03 Å, respectively. These figures are in good agreement with a local-density-functional calculation employing effective core potentials [57].

Finally, for C_{240} the I_h structure was tested against the D_{5h} tube made of C_{60} caps and shown in Figure 5.6. Again, the icosahedral fullerene was much more stable than the cylindrical buckytube. In I_h symmetry, C_{240} has three symmetry-

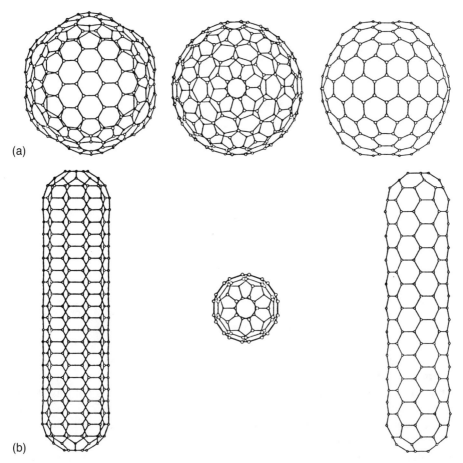

Figure 5.6 Equilibrium geometries of two isomers of C_{240} as predicted by the STO-3G/SCF level of theory: (a) I_h, (b) D_{5d} VF.

distinct classes of atoms, containing 120, 60, and 60 symmetry-equivalent carbon atoms. The radii of the spheres for these classes were 7.33, 7.08, and 6.95 Å, respectively, in good agreement with the LDF results [57].

The main conclusion of this study is that for sizes up to 240 carbon atoms, there is no indication of tubes being energetically more favorable than spheroidal fullerenes. This conclusion is consistent with a recent study by Adams et al. [58] employing a different theoretical approach and with the semiempirical MNDO predictions of Bakowies and Thiel [59]. However, local-density-functional calculations suggest a possible thermodynamic preference for tubular structures rather than cage structures [60]. Other buckytube studies have focused on different aspects of these fascinating objects [61, 62].

It should be pointed out that an important factor in the mechanism of stabilization of both buckytubes and buckyonions is the dispersion or van der

Table 5.3 Total and relative energies for large fullerene isomers (spheres versus tubes) calculated at the SCF level of theory (all closed-shell electronic states)

Fullerene	$E(\text{h})$	ΔE (kcal/mol)
C_{80} ($I_h \rightarrow T_h$)	$-2992.010\ 65$	286
C_{80} (D_{5d} SF)	$-2992.256\ 87$	132
C_{80} (D_{5d} VF)	$-2992.467\ 75$	0
C_{120} (T_d)	$-4489.207\ 53$	0
C_{120} (D_{6d})	$-4489.157\ 06$	32
C_{120} (D_{5d} VF)	$-4489.036\ 03$	108
C_{120} (D_{5d} SF)	$-4488.773\ 26$	272
C_{180} (I_h)	$-6734.466\ 61$	0
C_{180} (D_{6h})	$-6734.134\ 87$	208
C_{240} (I_h)	$-8979.723\ 32$	0
C_{240} (D_{5d} VF)	$-8978.527\ 50$	750

Waals forces between different graphitic layers. In our monolayer fullerenes this effect is not present. A multilayer SCF calculation would not be meaningful since van der Waals forces are not accounted for at this level of theory. Correlated methods are necessary to take dispersion forces into account. In particular, bulk graphite is ~1.5–2.0 kcal/mol per carbon atom more stable than infinite two-dimensional graphitic sheets, precisely because of this effect. It should also be pointed out that the observed buckytubes [48, 50] are always helical, while the capsules considered in our study are not. Helical tubes can fine tune their diameters, bringing the interlayer distance conveniently close to the bulk graphite value.

In conclusion, although our theoretical results are consistent with the existence of buckytubes under special kinetic conditions, they clearly indicate that they are higher in energy than the icosahedral carbon structures. An important consequence of this prediction is that under high-energy experimental conditions, amorphous carbon should anneal to spheroidal rather than cylindrical shapes, a conclusion consistent with Ugarte's observations [52].

5.6 C_{28}, An Inside/Outside Tetravalent Fullerene

One of the interesting aspects of the original C_{60} story is that under the experimental conditions that make C_{60} and C_{70} *not* special, C_{28} is a very prominent peak [63]. Given that under special circumstances C_{60} can be made *very* special, the question of whether the relatively large C_{28} peak corresponded to a fullerene remained unanswered until very recently.

Topologically the smallest possible fullerene is C_{20}, formed by 12 pentagons. In fact, despite some theoretical evidence that C_{20} may be a fullerene [64], no indication has ever been found in carbon cluster beams for an abundant bare C_{20} cluster. Nevertheless, the fully hydrogenated $C_{20}H_{20}$ molecule, dodecahedrane [65], does turn out to be a highly stable hydrocarbon.

A possible closed-cage T_d structure for C_{28} that contains four six-membered rings has been mentioned by Kroto as being particularly stable [66]. Photofragmentation of C_{60} gave, however, products that abruptly terminated at C_{32}, prompting the interpretation that C_{32} could be the smallest stable fullerene [67]. Moreover, mass selection of C_{32} followed by photofragmentation resulted in loss of C_3 and other carbon chains [67], a clear departure from the C_2 loss mechanism central to the fullerene hypothesis [68].

A recent series of experiments and calculations carried out at Rice [69] has dramatically changed our understanding of C_{28}. When using a pellet made of a mixture of UO_2 and graphite in the laser vaporization machine, a very large peak in the mass spectrum was observed corresponding to U@C_{28}. Other U@C_n fullerenes are also formed and when mass selected for a shrinkwrapping laser experiment [69], the endpoint is consistently found to be U@C_{28}. A series of unsuccessful attempts to break this product apart gave firm support to the interpretation that very strong bonds between uranium and carbon are being formed from the inside of C_{28}.

A series of SCF calculations done in our group gave further support to the hypothesis that uranium is forming four covalent bonds with the C_{28} cage. As a fullerene, C_{28} has only two possible isomers of which the open-shell $^5A_2$ T_d structure is the ground state. The T_d structure is the most symmetrical possible structure for 28 atoms: a triplet of pentagons arranged at each vertex of a tetrahedron. This T_d structure has been mentioned in the literature before [66, 70].

Direct SCF and analytic energy gradient methods were applied to optimize the geometry of C_{28} shown in Figure 5.7. The $^5A_2$ ground state has valence electronic occupations $8a_1 14t_2^3$, corresponding to a dangling bond in each of the four tetrahedral vertices where the triplet of pentagons join. Evidently, this open-shell system is highly reactive, and is not experimentally observed unless special experimental conditions are set up [69]. If the clusters are "chemically cooked" long enough, the C_{60} becomes the prominent peak.

The other possible isomer of C_{28} is a closed-shell D_2 fullerene that has an elongated shape. Our calculations indicate that it lies much higher in energy than the T_d isomer [71]. The $^5A_2$ electronic state found in our work to be the ground state of the T_d isomer had not been considered before by other groups [70]. All the theoretical evidence indicates that uranium (in a positive oxidation state) is forming four covalent bonds to the four six-membered rings in tetrahedral positions of T_d C_{28}. This uranium-carbon bonding is similar to that of uranocene [72, 73].

The open-shell electronic structure of C_{28} suggests that external addition of four hydrogen atoms at the dangling bonds may stabilize this species. The methane analogue $C_{28}H_4$ was mentioned by Kroto as a potentially stable species [66].

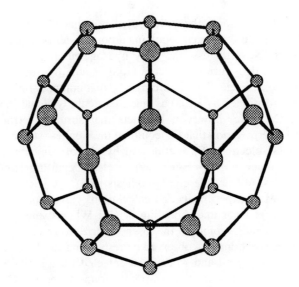

Figure 5.7 Equilibrium geometry of the ground state $^5A_2$ T_d isomer of C_{28} calculated at the dz/SCF level of theory.

Indeed, according to our SCF calculations, the molecule $C_{28}H_4$ does turn out to be a remarkably stable hydrocarbon with a very large HOMO-LUMO gap. This particular situation makes C_{28} the first molecule ever that has the capability of covalent bonding both from the inside and outside; that is, it has ''inner'' and ''outer'' valences of 4.

The question of whether C_{28} is a fullerene should be reformulated as: What can we put inside or attach outside C_{28} to stabilize it? Although it may be complicated to set up a particular experiment for every element of the periodic table, it is very simple to test different hypotheses with our theoretical methods. Our calculations quickly indicated that $X@C_{28}$ for X = Ti, Zr, and Hf were all ''happy'' molecules. Later on, the experiments showed a large peak in the mass spectrum for $Zr@C_{28}$. Both $Ti@C_{28}$ and $Hf@C_{28}$ have also been detected, but their signal was small and not special [69].

Considered as a tetrahedral unit, one can imagine C_{28} connected in a diamond lattice to form yet a new form of pure carbon. Eventually, some of the links may be replaced by metal atoms. C_{28} has also shown the capability of bonding both to the outside and inside simultaneously [74], giving rise to innumerable possibilities for stabilizing this remarkable fullerene.

5.7 Negative Curvature and Fullerenes

Seven-membered carbon rings have received much attention recently [75–77]. Carbon networks composed of six- and seven-membered rings have been proposed and their heats of formation per carbon atom estimated to be much lower than that of C_{60} [76]. Iijima [50] has observed seven-membered rings in his electron transmission micrographs and assigned them a central role in the growth pattern of buckytubes. Ugarte [52] has also observed seven-membered rings in the soot extracted from the electrodes of the arc-generation machine.

Normal fullerenes are made of 12 five-membered and an arbitrary number of six-membered carbon rings that form a closed, spheroidal cage. Seven-membered rings introduce negative curvature, and greatly increase the number and type of objects that may be built out of carbon networks. The negative curvature analogue of C_{60} is an object formed by 12 heptagons, each of them surrounded by 7 hexagons. The resulting structure has tetrahedral symmetry and does not close on itself (Figure 5.8). These negative-curvature units may be joined in a diamond lattice that divides space into disjoint regions. A few different forms of these infinite lattices have recently been described [75–77].

From a theoretical perspective it is an interesting question as to whether we can build closed carbon cages out of negative curvature units. The answer to this question is affirmative [78] but requires the use of a different negative-

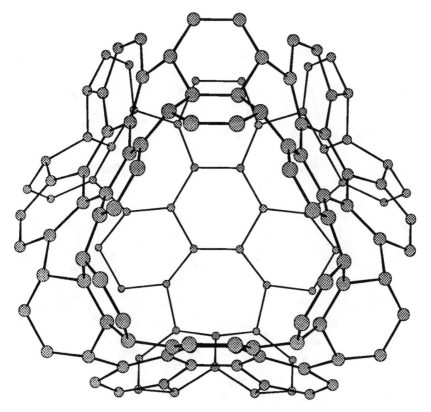

Figure 5.8 An sp^3 schwartzon of T_d symmetry containing 12 seven-membered rings.

curvature unit than those previously discussed [75–77]. The sp^2 "schwartzon" of D_{3h} symmetry shown in Figure 5.9 is the junction of 3 tubes at 120°, and contains only six seven-membered rings. A given number of them may be joined in a fullerene network that will only contain six- and seven-membered rings. From a topological perspective, this "hyper-fullerene" will have a large number of handles. Euler's theorem,

$$n_5 - n_7 = 6\chi = 6(2 - 2g)$$

where χ is the Euler characteristic, g is the genus (number of handles), n_5 the number of pentagons, and n_7 the number of heptagons, gives $n_5 = 12$ for a normal fullerene. A fullerene is defined as an spheroidal carbon network ($g = 0$) containing no seven-membered rings ($n_7 = 0$) and an arbitrary number of hexagons (except $n_6 = 1$). A torus ($g = 1$) may be formed with $n_5 = n_7$ polygons. The sp^3 schwartzon (Fig. 5.8) has an Euler characteristic of $\chi = -2$, while the sp^2 schwartzon (Fig. 5.9) has $\chi = -1$. We note that hyper-C_{60} has $\chi = -60$ (i.e., a genus of 31), which is different from the number of hyper-σ bonds (90). Mathematically, when a handle is cut, the object should not split

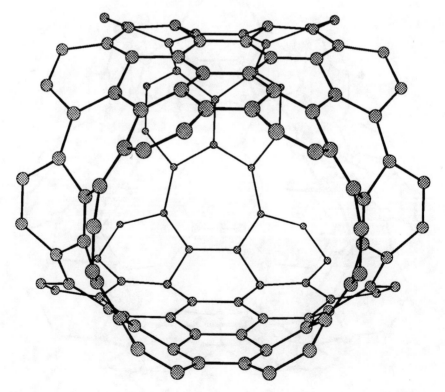

Figure 5.9 An sp^2 schwartzon of D_{3h} symmetry containing 6 seven-membered rings.

into pieces. Thus, not all possible cuts of a hyper-fullerene σ bond should be counted as handles.

Employing the sp^2 schwartzon, one may actually build a fractal of fullerenes, as previously discussed [78]. A given number of sp^2 schwartzons are connected into a network that resembles the sp^2 schwartzon itself (self-similarity). This process may be repeated indefinitely. At any step in the construction, the network can be closed employing a fullerene topology. The name *hyper-fullerenes* was used in connection with these topological constructions.

Although hyper-fullerenes are hypothetical structures, a detailed understanding of the electronic structures, equilibrium geometries, and energetics of negative-curvature fullerenes will help elucidate many of the interesting aspects of soot formation in the arc-generation experiments. If feasible, the synthesis of negative-curvature carbon networks will greatly enhance the possibility of building a realistic carbon-based nanotechnology [79].

Acknowledgments

Support by the National Science Foundation (Grant No. CHE-9017706), The Welch Foundation, and the Petroleum Research Fund of the American Chemical Society is gratefully acknowledged.

Note Added in Proof

After this manuscript was submitted for publication, two experimental reports [80, 81] have confirmed the theoretical predictions [28] for bond types 1 and 3 of C_{70} (see Table 5.2 and Fig. 5.1) discussed in this work.

References

1. H. W. Kroto, J. R. Heath, S. C. O'Brien, R. F. Curl, and R. E. Smalley, *Nature* **318,** 162 (1985).

2. S. Larsson, A. Volosov, and A. Rosen, *Chem. Phys. Lett.* **137,** 501 (1987).

3. Z. C. Wu, D. A. Jelski, and T. F. George, *Chem. Phys. Lett.* **137,** 291 (1987).

4. R. E. Stanton and M. D. Newton, *J. Am. Chem. Soc.* **92,** 2141 (1988).

5. S. J. Cyvin, E. Brendsal, B. N. Cyvin, and J. Brunvoll, *Chem. Phys. Lett.* **143,** 377 (1988).

6. D. E. Weeks and W. G. Harter, *Chem. Phys. Lett.* **144,** 366 (1988).

7. Z. Slanina, J. M. Rudzinski, M. Togasi, and E. Osawa, *J. Mol. Struct. (Theochem)* **202,** 169 (1989).

8. W. Krätschmer, K. Fostiropoulos, and D. R. Huffman, *Chem. Phys. Lett.* **170,** 167 (1990).

9. W. Krätschmer, L. D. Lamb, K. Fostiropoulos, and D. R. Huffman, *Nature* **347**, 354 (1990).

10. D. R. Huffman, *Physics Today*, pp. 22–29, November 1991.

11. H. W. Kroto, A. W. Allaf, and S. P. Balm, *Chem. Rev.* **91**, 1213 (1991).

12. J. Almlöf, K. Faegri, Jr., and K. Korsell, *J. Comp. Chem.* **3**, 2469 (1986).

13. A. H. H. Chang, W. C. Ermler, and R. M. Pitzer, *J. Chem. Phys.* **94**, 5004 (1991).

14. R. Ahlrichs, M. Bär, M. Häser, H. Horn, and C. Kölmel, *Chem. Phys. Lett.* **162**, 165 (1989).

15. C. Moller and M. S. Plesset, *Phys. Rev.* **46**, 618 (1934).

16. D. S. Marynick and S. Estreicher, *Chem. Phys. Lett.* **125**, 383 (1986).

17. J. M. Rudzinski, Z. Slanina, M. Togasi, and E. Osawa, *Thermochim. Acta* **125**, 155 (1988).

18. R. L. Disch and J. M. Schulman, *Chem. Phys. Lett.* **125**, 465 (1986).

19. H. P. Lüthi and J. Almlöf, *Chem. Phys. Lett.* **135**, 357 (1987).

20. J. Almlöf and H. P. Lüthi, University of Minnesota, Supercomputer Institute Report 87/20, March 1987.

21. G. E. Scuseria, *Chem. Phys. Lett.* **176**, 423 (1991).

22. M. Häser, J. Almlöf, and G. E. Scuseria, *Chem. Phys. Lett.* **181**, 497 (1991).

23. K. Hedberg, L. Hedberg, D. S. Bethune, C. A. Brown, H. C. Dorn, R. D. Johnson, and M. de Vries, *Science* **254**, 410 (1991).

24. C. S. Yannoni, P. P. Bernier, D. S. Bethune, G. Meijer, and J. R. Salem, *J. Am. Chem. Soc.* **113**, 3190 (1991).

25. J. M. Hawkins, A. Meyer, T. A. Lewis, S. D. Loren, and F. J. Hollander, *Science* **252**, 312 (1991).

26. P. J. Fagan, J. C. Calabrese, and B. Malone, *Science* **252**, 1160 (1991).

27. S. Liu, Y-J Lu, M. M. Kappes, and J. A. Ibers, *Science* **254**, 408 (1991).

28. G. E. Scuseria, *Chem. Phys. Lett.* **180**, 451 (1991).

29. K. Raghavachari and C. M. Rohlfing, *J. Phys. Chem.* **95**, 5768 (1991).

30. J. Baker, P. W. Fowler, P. Lazzeretti, M. Malagoli, and R. Zanasi, *Chem. Phys. Lett.* **184**, 182 (1991).

31. J. J. P. Stewart and M. B. Coolidge, *J. Comp. Chem.* **12**, 1129 (1991).

32. M. Froimowitz, *J. Comp. Chem.* **12**, 1157 (1991).

33. F. Negri, G. Orlandi, and F. Zerbetto, *J. Am. Chem. Soc.* **113**, 6037 (1991).

34. W. Andreoni, F. Gygi, and M. Parrinello, *Chem. Phys. Lett.* **189**, 241 (1992).

35. D. R. McKenzie, C. A. Davis, D. J. H. Cockayne, D. A. Muller and A. M. Vassallo, *Nature* **355**, 622 (1992).

36. K. Hedberg, results presented at the 14th Austin Symposium on Molecular Structure, Austin, Texas, March 2–4, 1992.

37. R. E. Haufler, J. Conceicao, L. P. F. Chibante, Y. Chai, N. E. Byrne, S. Flanagan, M.

M. Haley, S. C. O'Brien, C. Pan, Z. Xiao, W. E. Billups, M. A. Ciufolini, R. H. Hauge, J. L. Margrave, L. J. Wilson, R. F. Curl, and R. E. Smalley, *J. Phys. Chem.* **94,** 167 (1990).

38. B. I. Dunlap, D. W. Brenner, J. W. Mintmire, R. C. Mowrey, and C. T. White, *J. Phys. Chem.* **95,** 5763 (1991).

39. M. Saunders, *Science* **253,** 330 (1991).

40. N. L. Allinger, Y. H. Yuh, and J.-H. Lii, *J. Am. Chem. Soc.* **111,** 8551 (1989).

41. D. Bakowies and W. Thiel, *Chem. Phys. Lett.* **193,** 236 (1992).

42. T. Guo and G. E. Scuseria, *Chem. Phys. Lett.* **191,** 527 (1992).

43. J. Cioslowski, *Chem. Phys. Lett.* **181,** 68 (1991).

44. P. W. Fowler, H. W. Kroto, R. Taylor, and D. R. M. Walton, *J. Chem. Soc. Faraday Trans.* **87,** 2685 (1991).

45. G. E. Scuseria and G. K. Odom, *Chem. Phys. Lett.* **195,** 531 (1992).

46. J. H. Holloway, E. G. Hope, R. Taylor, G. J. Langley, A. G. Avent, T. J. Dennis, J. P. Hare, H. W. Kroto, and D. R. Walton, *Chem. Comm.* 966 (1991).

47. H. Selig, C. Lifshitz, T. Peres, J. E. Fischer, A. R. McGhie, W. J. Romanow, J. P. McCauley, and A. B. Smith *J. Am. Chem. Soc.* **113,** 5475 (1991).

48. S. Iijima, *Nature* **354,** 56 (1991).

49. H. W. Kroto, *Nature* **91,** 1213 (1991).

50. S. Ijima, T. Ichihashi, and Y. Ando, *Nature* **356,** 776 (1992).

51. L. D. Lamb, D. R. Huffman, R. K. Workman, S. Howells, T. Chen, D. Sarid, and R. F. Ziolo, *Science* **255,** 1413 (1992).

52. D. Ugarte, *Nature* **359,** 707 (1992).

53. S. Iijima, *J. Cryst. Growth* **50,** 675 (1980).

54. S. Iijima, *J. Phys. Chem.* **91,** 3446 (1987).

55. T. G. Schmalz, W. A. Seitz, D. J. Klein, and G. E. Hite, *Chem. Phys. Lett.* **130,** 203 (1986).

56. T. G. Schmalz, W. A. Seitz, D. J. Klein, and G. E. Hite, *J. Am. Chem. Soc.* **110,** 1113 (1988).

57. B. I. Dunlap, D. W. Brenner, J. W. Mintmire, R. C. Mowrey, and C. T. White, *J. Phys. Chem.* **95,** 8737 (1991).

58. G. B. Adams, O. F. Sankey, J. B. Page, M. O'Keeffe, and D. A. Drabold, *Science* **256,** 1792 (1992).

59. D. Bakowies and W. Thiel, *J. Am. Chem. Soc.* **113,** 3704 (1991).

60. D. H. Robertson, D. W. Brenner, and J. W. Mintmire, *Phys. Rev. B* (in press).

61. J. W. Mintmire, B. I. Dunlap, and C. T. White, *Phys. Rev. Lett.* **68,** 631 (1992).

62. J. W. Mintmire, D. H. Robertson, B. I. Dunlap, R. C. Mowrey, D. W. Brenner, and C. T. White, MRS Symposia Proceedings No. 247, Materials Research Society, Pittsburgh PA, 1992.

63. R. E. Smalley, personnal communication.

64. V. Parasuk and J. Almlöf, *Chem. Phys. Lett.* **184,** 187 (1991).

65. L. A. Paquette, R. J. Ternansky, D. W. Balogh, and G. J. Kentgen, *J. Am. Chem. Soc.* **105,** 5446 (1983).

66. H. W. Kroto, *Nature* **329,** 529 (1987).

67. S. C. O'Brien, J. R. Heath, R. F. Curl, and R. E. Smalley, *J. Chem. Phys.* **88,** 220 (1988).

68. Q. L. Zhang, S. C. O'Brien, J. R. Heath, Y. Liu, R. F. Curl, W. H. Kroto, and R. E. Smalley, *J. Phys. Chem.* **90,** 525 (1986).

69. T. Guo, M. D. Diener, Y. Chai, M. J. Alford, R. E. Haufler, S. M. McClure, T. Ohno, J. H. Weaver, G. E. Scuseria, and R. E. Smalley, *Science* **257,** 1661 (1992).

70. M. Feyereisen, M. Guotowski, J. Simons, and J. Almlöf, *J. Chem. Phys.* **96,** 2926 (1992).

71. L. P. Delabroy and G. E. Scuseria, to be published.

72. A. H. H. Chang and R. M. Pitzer, *J. Am. Chem. Soc.* **111,** 2500 (1989).

73. R. M. Pitzer, personal communication.

74. T. Guo, G. E. Scuseria, and R. E. Smalley, to be published.

75. A. L. Mackay and H. Terrones, *Nature* **352,** 762 (1991).

76. T. Lenosky, X. Gonze, M. Teter, and V. Elser, *Nature* **355,** 333 (1992).

77. D. Vanderbilt and J. Tersoff, *Phys. Rev. Lett.* **68,** 511 (1992).

78. G. E. Scuseria, *Chem. Phys. Lett.* **195,** 534 (1992).

79. P. Ball and L. Garwin, *Nature* **355,** 761 (1992).

80. A. L. Balch, J. W. Lee, and M. M. Olmstead, (preprint).

81. G. Roth and P. Adelmann, *J. Phys. I France* **2,** 1541 (1992).

6

Predicting Properties of Fullerenes and their Derivatives

C. T. White,* J. W. Mintmire, R. C. Mowrey,
D. W. Brenner, D. H. Robertson, J. A. Harrison,
and B. I. Dunlap

Kroto, Curl, Smalley, and co-workers' observation of uniquely abundant, preeminently stable C_{60} clusters in supersonic carbon cluster beams produced by the laser vaporization of graphite led to their suggestion of a beautiful, truncated icosahedral structure for this special cluster [1]. This chemically sound cage of spherically arranged carbon atoms, named *buckminsterfullerene* in honor of Buckminster Fuller and his geodesic domes [1], has no interior carbon atoms and no dangling bonds, as depicted at the top of Figure 6.1.

Buckminsterfullerene (Bf) is the most prominent member of a family of structurally related, hollow, carbon cages composed solely of five- and six-membered rings formed from threefold-coordinated carbon atoms. This family, known as the *fullerenes*, was originally invoked by Heath et al. [2] to explain the predominance of high-mass (C_N; $N \gtrsim 30$), even-membered carbon clusters first observed in cluster distributions produced by the laser vaporization of graphite [3]. Although there is no obvious upper limit on size, for complete closure each fullerene must contain $2(10 + M)$ carbon atoms corresponding to exactly 12 five-membered and M six-membered rings [4]. Hence the smallest fullerene is C_{20} and all fullerenes must contain an even number of carbon atoms [5]. Also, starting at C_{20} any even-membered carbon cluster, except C_{22}, can be recast as a somewhat aromatic fullerene [6–11], consistent with the observed greater abundance of the even-membered clusters [5].

Bf is unique among the fullerenes because it is the smallest that can avoid the local strains [8–11] and undesirable antiaromatic effects [8] associated with abutting pentagons [8–11]. Additionally, all carbons are equivalent in this spe-

* To whom all correspondence should be addressed.

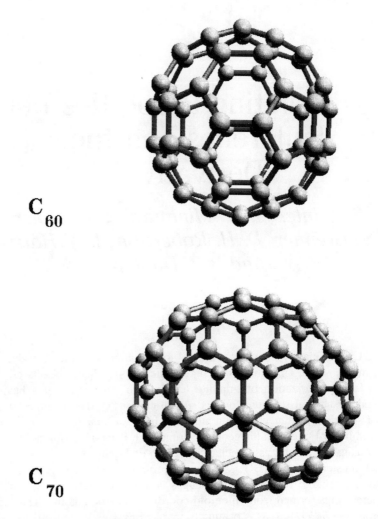

C_{60}

C_{70}

Figure 6.1 The special fullerenes Bf and (D_{5h}) C_{70}.

cial fullerene so that the strain of closure is equally shared with no self-evident reactive sites [1, 8–11]. The next larger fullerene with all pentagons isolated is (D_{5h}) C_{70} [8, 9, 11], depicted at the bottom of Figure 6.1. C_{70} was also seen in unusual, but less pronounced, relative abundance in the early cluster-beam experiments [1]. Although chemically and aesthetically appealing and indirectly supported by a convincing body of experimental [2, 5, 10, 12–15] and theoretical [16–36] evidence, including a striking series of "shrink-wrapping" studies [15], the proposed structures for C_{60} [1] and for C_{70} [8, 9, 20] had to wait several years to be verified directly by experiments made possible by

Krätschmer, Huffman, and co-workers' stunning breakthrough yielding macroscopic amounts of these special fullerenes [37, 38]—the fullerites.

Begun in cluster science experiments designed to study the nature of interstellar dust [1], fullerene chemistry has grown at a breathtaking pace with the production and purification of gram quantities of C_{60}. A battery of spectroscopic techniques have definitively probed the character of both C_{60} [38–51] and C_{70} [39, 41, 42, 52, 53], making detailed comparisons between theory and experiment possible for these to two special fullerenes. Higher fullerenes such as C_{76} and C_{78} have also been isolated in lesser amounts and their properties characterized [54–56], yielding convincing evidence for the first fullerene isomers [55]. Spectacular STM images not only of C_{60} [43, 57] but also of giant spheroidal fullerenes up to two nanometers in diameter have been obtained [58]. Condensed-phase samples of C_{60} have been crushed to diamond [59], shocked to graphite [60], and burned in air [61], but C_{60} itself often rebounds when collided with steel, graphite, and semiconductor surfaces [62–64] at hypervelocities (>2 km/sec). Fullerene projectiles have been found to entrap individual target atoms in high-energy collisions with small target gases [65–69], while C_{82} fullerenes with a La atom inside, denoted by La@C_{82}, have been synthesized in milligram quantities [70, 71], yielding the first of an anticipated large set of endohedral precursors for chemistry in the "near round." Rather than being chemically inert, C_{60} has proved reactive [48, 72–74], leading to a host of new chemicals that are only now beginning to be characterized. The discovery of superconductivity at 18 K in potassium-doped C_{60} [75] followed by the observation of superconductivity in Rb_xC_{60} [76, 77], Cs_xC_{60} [78], and $Cs_xRb_yC_{60}$ [79] with superconducting transition temperatures near 30 K establishes the alkali fullerides as by far the best organic-based superconductors yet synthesized. Graphitic tubules less than 3 nm in diameter but up to 1 μm in length have been observed [80], and a family of such fullerene tubules shown to be excellent synthetic targets for novel lightweight metals [81–84]. Clearly, a revolution has begun in chemistry and materials science based on one of the most familiar and important of all elements, carbon.

In this chapter, we will review some of the results obtained by the Theoretical Chemistry Section at the Naval Research Laboratory from theoretical studies of the fullerenes and related structures [35, 81, 82, 85–105]. These results include: predictions for the geometries, electronic structures, and photoelectron line shapes of several pure fullerenes and their derivatives [35, 85–91]; molecular-dynamics (MD) simulations of Bf during hypervelocity impacts with surfaces [92, 93], in high-energy collisions with small rare-gas target atoms [94, 95], and under compression [96]; studies of the dynamic curling and closure of graphitic ribbons as an early step to fullerene formation [97]; and results for the structural, electronic, and elastic properties of graphitic tubules [81, 82, 98, 99]. A discussion will be provided at the beginning of each section that introduces and motivates the topic, while connecting it to other experimental and theoretical studies.

In Section 6.1, we will give a brief sketch of an array of computational methods developed by members of our group over more than a decade that were well suited to the study of fullerene structures and dynamics [106–113]. We will begin with an outline of our first-principles, linear-combination-of-atomic-orbitals (LCAO), local-density-functional (LDF) method [106–109] that is ideal for calculating the properties of large symmetric molecules such as the fullerenes and some of their derivatives. Next, we will discuss our related LCAO-LDF band-structure method originally developed to study quasi-one-dimensional polymers [110] with translational symmetry but recently extended to treat chain polymers with helical symmetry [111] such as the fullerene tubules. We will also briefly discuss our all-valence, tight-binding Hamiltonian [112] that is useful in studying a range of larger fullerene-derived structures [82, 98, 100]. Finally, we will conclude this section with an outline of our many-body empirical hydrocarbon potential [113] that allows for the dynamic making and breaking of bonds and the associated hybridization changes that can occur, for example, during fullerene collisions with surfaces.

In Section 6.2, we will review our results for the pure fullerenes and some of their simpler hydrogenated and fluorinated derivatives [35, 85–91]. We will begin by showing that our calculated results for the geometry [35, 85, 86], ionization potentials [86, 87], electron affinity [86, 87], and photoelectron line shapes [88] for Bf are all in good agreement with the corresponding experimental results [44–51, 63]. Next, we will discuss our results for some of the larger fullerenes [86–89], which include not only (D_{5h}) C_{70} and some higher fullerenes [87–89], but also first-principles studies of the giant icosahedral fullerenes C_{180} and C_{240} [86]. These studies reveal that carbon is less tightly bound in C_{60} than in many larger fullerenes, suggesting the importance of kinetics in understanding the remarkably high yield of this new allotrope of carbon. We will then discuss the recently synthesized partially hydrogenated fullerene derivative $C_{60}H_{36}$ [72]. We find that a suggested structure for this molecule [72] yields a closed-shell species with a large electric-dipole-forbidden HOMO-LUMO (highest occupied molecular orbital-lowest unoccupied molecular orbital) gap [85, 90]. Finally, we will conclude this section with a brief discussion of the so far hypothetical, fully saturated species $C_{60}H_{60}$ and $C_{60}F_{60}$ [85, 91]. Our results for $C_{60}H_{60}$ suggest that if this fully hydrogenated fullerene derivative is prepared, then isomers with some of the hydrogens bonded to the interior cage wall are likely to be the lower in energy [85].

In Section 6.3, we will discuss some of our MD simulations involving Bf [92–96]. We will begin with a discussion of simulations of hypervelocity impacts of Bf with hydrogen-terminated (111) diamond surfaces [92, 93]. Consistent with experiments [62–64], our results suggest that Bf is highly resilient in energetic surface impacts. Next, we will discuss simulations of the trapping of He inside the Bf cage during high-energy gas-phase collisions [94, 95]. The computed cross sections for He entrapment during scattering with Bf follow the same trends as seen experimentally [95]. Finally, we will conclude this section with a discussion of MD simulations of Bf under anisotropic compression [96]. The flattening of the cage we observe during these simulations could relate to the recent shock compression of C_{60} fullerites to graphite [60].

In Section 6.4, we will discuss our recent studies involving early steps that might be on the way to fullerene formation [97]. We will begin by pointing out that size is one of the key issues involved. Next, we will discuss our MD simulations of the behavior of graphitic fragments that could condense in the cooling carbon vapor prior to fullerene formation. We find these fragments, modeled as ribbons, exhibit large instantaneous deviations from planarity that often result in the formation of open-ended hollow carbon structures representing good fullerene precursors. Finally, we will conclude this section by discussing the possible consequences of these simulations for fullerene formation.

In Section 6.5, we will review some of our results for the fullerene tubules [81, 82, 98, 99]. We will begin by showing how all graphitic tubules defined by rolling up a single graphite sheet can also be defined in terms of their helical and rotational symmetry [98]. This result, together with their translational symmetry, allows these tubules to be classified in a fashion familiar in the description of chain polymers. Next, we will discuss our band-structure results at the LDF and tight-binding levels which show that these tubules will exhibit properties ranging from good semiconductors to good metals depending on their radii and helical structure [81, 82, 98]. The metallic tubules are also predicted to be stable against a Peierls distortion at and far below room temperature, making them truly outstanding synthetic targets for novel lightweight metals [81]. Finally, we will conclude this section with a discussion of the energetics and elastic properties of these tubules [99]. We find that the strain energy per carbon relative to the graphite sheet depends sensitively only on the tubule radius. Altogether, the results discussed in this section yield a series of simple relations for estimating electronic and structural properties of these tubules as a function of radius and helicity.

In Section 6.6, we will provide concluding remarks. We will begin with a brief summary. Next, we will close by touching upon some of the current challenges to theory provided by the fullerenes and their derivatives and relatives. These challenges include issues involving the very early stages of fullerene formation, the isomers of Bf, the perplexing absence of certain higher fullerenes, and the promise of lower (C_{2N}; $10 \leq N < 30$, $N \neq 11$) fullerenes.

6.1   Theoretical Methods

Because of their large size, pure fullerenes represent an impossible challenge to current ab initio, all-electron, MCSCF, quantum-chemical methods that have been so successful in treating small molecules containing five or fewer heavy atoms. (Here heavy refers to any atom other than hydrogen.) Indeed, (D_{5h}) C_{70} is currently close to the upper limit that can be studied accurately even at the simpler but less exact ab initio, all-electron, Hartree-Fock (HF) level. In addition, comparable calculations for endohedral fullerene derivatives can become significantly more difficult. The interior atom often prefers not to sit at the cage center with a concomitant loss in symmetry [101, 102, 114], and this atom can also be so heavy that relativistic effects [115] must be included in any accurate all-electron study.

For larger fullerene derivatives such as micrometer-length graphitic tubules [80], the sheer number of atoms in a single tubule makes any cluster-based approach inappropriate. For long tubules, however, end effects can often be neglected so that translational symmetry can be used to reduce the problem to the point that ab initio HF and first-principles LDF band-structure calculations become feasible. LDF band-structure calculations are often preferable because, as for spheroidal fullerenes, they are computationally more tractable, typically predict geometries as well as or better than HF, and directly yield effective one-electron valence levels that, in contrast to HF calculations, compare favorably in *both* ordering and spacing to measured x-ray and ultraviolet valence photoelectron spectra (XPS and UPS).

Although the assumption of infinite length greatly reduces the complexity of tubule electronic structure calculations, these systems are still formidable, if not intractable, using typical band-structure methods relying on three-dimensional plane-wave basis sets and translational symmetry. An array of parallel packed tubules will have appreciable electron dispersion only along their axes and hence nearly flat bands perpendicular to that direction. The highly anisotropic nature of these tubules makes three-dimensional plane waves not an optimal choice for a basis set. Although this problem can be avoided by using a local Gaussian basis set [110], accurate calculations often remain intractable because the primitive translational unit cell can easily contain several hundred atoms for single tubules with diameters as small as 0.2 nm [98]. To avoid directly treating such a large number of atoms requires band-structure methods especially designed to take advantage of the helical symmetry of these tubules.

Computer simulations of the dynamic chemical reactivity of pure fullerenes and their derivatives also represent a challenge to current state-of-the-art theory and techniques. The traditional valence-force-field classical potentials for carbon do not allow for the dynamic making and breaking of bonds [116]. Also, fullerene carbon atoms have intermediate valences corresponding to neither purely sp^2 nor sp^3 hybridizations. In addition, to allow for the dynamic making and breaking of bonds, any realistic classical carbon potential used to model all but the simplest fullerene statics and any fullerene reactions should be many-body in nature to account for the hidden electronic degrees of freedom that can smoothly change the carbon valence with changes in the local carbon environment.

Over more than a decade, we have developed an array of computational methods that are well suited to the study of fullerene structures and dynamics [106–113]. As outlined in the introduction, these methods are reviewed in this section.

6.1.1 Electronic Structure Methods

We calculate the LDF electronic structure and total energy using molecular orbitals constructed from linear combinations of Gaussian-type functions [106]. The total energy and one-electron molecular orbitals are determined by variational minimization of the LDF expression for the total energy, which for a spin-restricted case is given by [117]:

$$E \equiv -\frac{1}{2} \sum_i \int d^3\mathbf{r}\, n_i\, \phi_i(\mathbf{r}) \nabla^2 \phi_i(\mathbf{r}) - \sum_j \int d^3\mathbf{r}\, \frac{Z_j}{|\mathbf{r} - \mathbf{R}_j|}\, \rho(\mathbf{r}) \quad (6.1)$$

$$+ \frac{1}{2} \int d^3\mathbf{r} \int d^3\mathbf{r}'\, \frac{\rho(\mathbf{r})\rho(\mathbf{r}')}{|\mathbf{r} - \mathbf{r}'|} + \int d^3\mathbf{r}\, \rho(\mathbf{r})\, \epsilon_{xc}\,[\rho(\mathbf{r})]$$

$$+ \frac{1}{2} \sum_j \sum_{j'}{}'\, \frac{Z_j Z_{j'}}{|\mathbf{R}_j - \mathbf{R}_{j'}|},$$

where Z_j denotes the nuclear charge of an atom located at $\mathbf{R}_j$, ϕ_i the one-electron molecular orbitals, n_i the corresponding occupation numbers, $\rho(\mathbf{r})$ the charge density, and ϵ_{xc} the local exchange-correlation potential. Extension to a local-spin-density formalism is straightforward and incorporated in the current computational scheme. Within this approach the electron repulsion and exchange-correlation potentials are evaluated using a variational fitting of the charge density [106] with orbital and potential fitting basis sets constructed from products of solid-spherical harmonics and Gaussian functions [107]. This variational fitting of the charge density results in a large reduction in computational overhead for systems such as the fullerenes, without introducing any first-order errors in the LDF total energy. In addition, efficient use of symmetry has been incorporated into a computer code [108] capable of calculating both the total energy as well as the one-electron energies and wave functions. The LDF used is the Perdew-Zunger [118] fit of the free-electron gas results of Ceperley and Alder [119].

A band-structure version of this approach has also been developed to calculate the LDF electronic structure and total energy of infinite-length chain polymers [110]. In addition, this approach has recently been generalized to treat helical polymers [111]. For a system that is symmetric under a screw operation $\mathcal{S}(h, \alpha)$, where h is the translation and α is the rotation angle associated with the screw operation $\mathcal{S}$, the one-electron wave functions are given as linear combinations of generalized Bloch functions $\varphi_i(\mathbf{r};\kappa)$ constructed from local functions $\chi_j(\mathbf{r})$:

$$\varphi_i(\mathbf{r}; \kappa) = \sum_m e^{-i\kappa m}\, \mathcal{S}^m\, \chi_j(\mathbf{r}). \quad (6.2)$$

The one-electron wave functions thus belong to irreducible representations of the screw symmetry group with a dimensionless analog of the wave vector, κ. Starting from Eq. (6.2), extension of the molecular methods and translationally periodic band-structure techniques to the helical case is described in detail else-where [111]. These LCAO-LDF molecular and extended-state methods have been shown to yield accurate values of equilibrium bond distances, relative binding energies, harmonic force constants, ionization potentials, and electron affinities, in a range of molecular and polymeric systems [85, 86, 106–110, 120]. Additional electronic properties including hyperfine coupling constants, UPS and XPS, and absorption spectra can be directly calculated from the one-electron wave functions and eigenvalues [88, 91, 110, 111, 121–124].

On occasion we also use an all-valence, nearest-neighbor, empirical, tight-bindng model to calculate the electronic structure of larger structures. This approach is particularly useful in scanning the properties of a range of tubular or related fullerene structures [82, 98, 100] that would be computationally too expensive to study using the first-principles methods. The carbon Slater-Koster [125] parameters characterizing this model are determined [112] by fitting to earlier LDF band-structure calculations on polyacetylene [110]. Within the no-tation of Ref. [125], these tight-binding parameters are given by $V_{ss\sigma} = -4.76$ eV, $V_{sp\sigma} = 4.33$ eV, $V_{pp\sigma} = 4.37$ eV, and $V_{pp\pi} = -2.77$ eV. Without loss of generality, we fix the diagonal term for the carbon p orbital by requiring $\varepsilon_p = 0$, which results in the s diagonal term being given by $\varepsilon_s = -6.0$ eV.

6.1.2   Molecular-Dynamics Methods

The empirical hydrocarbon potential energy function used in our studies was originally developed to model the chemical vapor deposition of diamond films [113]. However, because both diamond and fullerenes are carbon-based sys-tems, this empirical function has proved useful in a number of fullerene studies. This formalism, which was initially derived from chemical pseudopotential the-ory [126], assumes that the binding energy can be written in a form similar to a pair potential [127],

$$E_b = \sum_i \sum_{i>j} [V_r(r_{ij}) - B_{ij}(f) V_a(r_{ij})]. \qquad (6.3)$$

The functions V_r and V_a represent pair additive core-core repulsions and at-tractions due to valence electrons, respectively; $r_{ij} \equiv |\mathbf{R}_i - \mathbf{R}_j|$ is the scalar distance between atoms i and j; and the functional B_{ij} is a many-body empirical bond order that depends through the function f on quantities such as atomic coordination and bond angles. The pair additive terms are assumed to be trans-ferable among various bonding environments, and consequently many-body ef-fects are incorporated into the bond-order function.

For the potential used in this study, the pair terms and an analytic expression for the bond order have been fitted to the lattice constants and cohesive energies of graphite, diamond, face-centered cubic and simple-cubic lattices, the va-cancy formation energies of diamond and graphite, carbon-carbon and carbon-hydrogen bond energies in hydrocarbon molecules, the barrier for rotation about the carbon-carbon double bond in ethylene, and the barrier for the H + H$_2$

exchange reaction [113]. Parameterized in this way, this many-body potential yields reasonable predictions for a wide range of properties of hydrocarbon systems. For example, atomization energies for hydrocarbon molecules ranging from highly strained alkanes to large aromatics to radicals are predicted to 1% or better for over 80% of the molecules tested [113]. This potential also predicts surface reconstructions and relaxations on various diamond surfaces that are in agreement with more sophisticated total-energy methods [128–130]. For Bf, the potential predicts bond lengths of 1.42 and 1.45 Å and a strain energy of 0.3 eV relative to graphite. These numbers compare favorably to experimentally derived bond lengths of 1.391 and 1.444 Å [51] and a strain energy of approximately 0.3 eV estimated by comparing our total-energy LDF results for Bf [85, 86] to LDF band-structure results for 3D graphite [131].

In addition to providing reasonable first estimates for static energies and structures, our empirical potential function allows for bond breaking and making with associated changes in hybridization. Therefore, unlike more traditional valence-force-field-based potentials [116], this many-body potential can be used to model dynamic chemical reactivity involving changes in covalent bonding. Thus this potential allows us not only to study static properties, but also to simulate possible chemical processes involving fullerenes. MD simulations using this potential are implemented by integrating Hamilton's equations of motion using a high-order predictor-corrector method [132].

6.2   Pure Fullerenes and Their Derivatives

Krätschmer, Huffman, and co-workers in their original paper presented convincing experimental evidence, based on time-of-flight mass spectrometric, infrared absorption, and x-ray diffraction studies, that they had indeed prepared a new allotrope and the first molecular solid of carbon [38]. Their data implied that this molecular solid was composed of C_{60} molecules with the anticipated Bf shape.

With the wide availability of macroscopic quantities of C_{60}, the detailed characterization of this new building block for new materials has proceeded at a remarkable pace. Taylor et al. [39], Johnson et al. [40], and Ajie et al. [41] all obtained single-line ^{13}C NMR spectra consistent with the icosahedral symmetry of Bf. In other experiments, Bethune et al. [42] measured the vibrational Raman spectra of purified solid films of C_{60}, leading to results in good agreement with early theoretical predictions for Bf. Soon thereafter, Wragg et al. [43] reported STM images of the spherically shaped molecules making up their fullerite samples, again consistent with the Bf hypothesis. Also, Lichtenberger et al. [44, 45] later obtained both gas- and solid-phase valence UPS spectra for C_{60} exhibiting only a relatively few distinct peaks consistent with theoretical predictions for the electronic spectrum of Bf [33–36]. Subsequently, Weaver et al. [46] supported and extended Lichtenberger et al.'s results through a detailed synchrotron-radiation photoemission study. Wang et al. [47] also measured the electron affinity of C_{60}. At about the same time, Hawkins et al. [48] succeeded in breaking the symmetry of C_{60} through an osmylation reaction allowing for the growth of good-quality derivatized crystals. Their x-ray crystallographic studies proved that C_{60} in this new material was a carbon cage with the anticipated Bf structure. (The NMR data, although providing compelling evidence in favor of the Bf hypothesis, did not strictly prove that the molecule was Bf. A truncated dodecahedron will also yield a C_{60} carbon cage with I_h symmetry [20, 28] and hence only one symmetry-inequivalent carbon atom and two symmetry-inequivalent C-C bonds. Each carbon atom in this structure is also threefold coordinated, but the bonding pattern requires 20 three-membered and 12 ten-membered rings. The presence of these chemically bizarre rings results in highly strained C-C-C bond angles, which effectively exclude this C_{60} isomer. Although most soccer balls are based on a truncated icosahedron, there are some that are based on a truncated dodecahedron.) These experiments also yielded the first measurements of the two different carbon-carbon bond lengths in Bf. Subsequently, Yannoni et al. [49] determined these bond distances from ^{13}C NMR experiments on Bf alone, while Hedberg et al. [50] later derived these distances from electron diffraction studies of gas-phase Bfs. More recently, David et al. [51] determined the crystal structure and bonding of C_{60} from neutron powder diffraction data. There is no doubt that the structure of C_{60} is as depicted at the top of Figure 6.1, that is, Bf.

Experimental studies have not only focused on C_{60} but also on C_{70} and the larger fullerenes that can be extracted from the Krätschmer-Huffman material [39, 41, 42, 52–56, 58]. In their original work on C_{60}, both Taylor et al. [39] and Ajie et al. [41] reported ^{13}C NMR data consistent with the proposed D_{5h} structure for C_{70}. Bethune et al. [42] also assigned some of the Raman lines they observed to C_{70}. More recently, Jost et al. [52] reported synchrotron-radiation photoemission and inverse photoemission studies of C_{70} and compared and contrasted their results to those of Weaver et al. for Bf. Subsequently, McKenzie et al. [53] confirmed the D_{5h} structure for C_{70} through electron diffraction studies. In other developments, Diederich et al. [54] succeeded in isolating and partially characterizing the higher fullerenes C_{76}, C_{78}, C_{90}, and C_{94}. Their NMR studies of C_{78} provided the first convincing evidence for fullerene isomers [55], while Ettl et al. [56] found that C_{76} is chiral. Very recently, Lamb et al. [58] extracted and obtained STM images of giant spherical fullerenes up to 2 nm in diameter.

There has also been encouraging progress in preparing new chemicals from fullerenes. In addition to the work of Hawkins et al. [48], leading to the first structurally characterized fullerene derivative, Haufler et al. [72] were the first to report a chemical reaction on the surface of Bf leading to the new chemical, $C_{60}H_{36}$. Subsequently, Fagan et al. [73] prepared and characterized a rich series of Bf metal complexes, while Wudl et al. [74] reported a series of functionalized C_{60} fullerenes—the fulleroids. This growing wealth of experimental data on the pure fullerenes and their chemical derivatives provides a continually expanding challenge to chemical theory.

Prior to the production of macroscopic quantities of C_{60} in 1990, many theoretical studies of C_{60} had already appeared in support of the Bf hypothesis [8, 9, 16–36]. These studies showed that Bf was a conjugated closed-shell molecule with a sizable HOMO-LUMO gap [16–22] and significant resonance stabilization [21, 22], which was favored over a range of other C_{60} isomers including: chains [8, 23] and rings [8, 29]; graphite [8, 23] and diamond fragments [8]; and the same mass but different geometry carbon cages [8, 9, 20, 28, 29]. Many of these early theoretical studies were inspired by Kroto, Curl, Smalley, and co-workers' landmark experimental results [1]. However, unknown to the authors of Ref. [1], Bochvar and Gal'pern [16] and Davidson [17] had already concluded years before that Bf was a closed-shell system with a sizable HOMO-LUMO gap—that is, a good synthetic target. Bochvar and Gal'pern and Davidson reported Hückel molecular-orbital (HMO) electronic structures for Bf. After publication of Ref. [1], a barrage of additional HMO-based studies [18–22] confirmed and extended these early HMO results. Among those theoretical papers were those of Haymet [18] reporting work concurrent with Ref. [1].

Newton and Stanton [23] were the first to publish a full geometry optimization of Bf using a semiempirical quantum-chemical method. Their modified-neglect-of-diatomic-overlap (MNDO) calculations led to the earliest predicted optimized values for the two different bond lengths in Bf and hence the geometry of this molecule. Soon thereafter, a number of other optimized geometries of Bf appeared based on a range of semiempirical quantum-chemical methods [24–29]. Disch and Schulman [30] were the first to predict a geometry-optimized structure for Bf using ab initio HF methods. Subsequently Lüthi and Almlöf [31] reported HF results obtained using a larger basis set. Although several fixed-geometry LDF studies of Bf had already appeared [33, 34], the earliest LDF optimized geometry of this molecule was that of Ref. [35]. Summarized in Table 6.1 are pre-1990 semiempirical, quantum-chemical, ab initio HF, and LDF optimized values of the two different carbon-carbon bond lengths in Bf. All methods yielded predictions close to the later experimental results [51].

Early theoretical work also focused on the relative stability of Bf [8, 9, 20, 23, 28–31, 133]. The isolated pentagon and related rules were used to explain the unusual relative abundance of C_{60}, C_{70}, and some smaller C_N clusters observed in the cluster-beam experiments [8, 9]. In addition to yielding results that suggested that Bf was more stable than other C_{60} isomers [8, 9, 20, 23, 28, 29], early theoretical studies also suggested that Bf was more stable than lower fullerenes [8, 23, 30]. However, early theoretical results also indicated that Bf was not as stable as some larger fullerenes [8, 20, 23, 133], even down to the size of (D_{5h}) C_{70} [20, 23], suggesting that Bf had a kinetic rather than thermodynamic origin. A good deal of theoretical attention was also given to enumerating possible giant, higher, and lower fullerenes [8, 134]. These early studies of possible fullerene structures proved important years later in identifying specific isomers of higher fullerenes extracted from the Krätschmer-Huffman material [54–56, 135].

Table 6.1 Pre-1990 semiempirical quantum-chemical, ab initio HF, and LDF predictions for the two different carbon-carbon bond lengths in Bf in Å compared to the recent experimental results of David et al [51]. The basis sets used in the HF and LDF calculations are indicated parenthetically. The bond shared by a pentagon and hexagon is labeled 5-6, while the bond shared only by hexagons is labeled 6-6. The acronyms MNDO, HF, and LDF are defined in the text. INDO Stands for intermediate neglect of differential overlap; CGHM stands for Coulson-Goiebienski (self-consistent) Hückel method; PRDDO stands of partial retention of diatomic differential overlap; and SMOM/CI stands for self-consistent molecular orbital (with) configuration interaction

Method	5-6	6-6	Ref.
MNDO	1.474	1.400	[23]
GCHM	1.434	1.403	[24]
PRDDO	1.436	1.360	[25]
SMOM/CI	1.443	1.396	[26]
INDO	1.449	1.397	[28]
HF(STO-3G)	1.465	1.376	[30]
HF($7s3p/4s2p$)	1.453	1.369	[31]
LDF($11s6p$)	1.43	1.39	[35]
Experiment	1.444	1.391	[51]

In addition to studies of the geometry and relative stability of Bf, theoretical work prior to 1990 also addressed other properties of this special fullerene [23, 25–27, 32, 34]. A fine example was the fixed-geometry, complete-neglect-of-differential-overlap (CNDO/S) calculations of Larsson, Volosov, and Rosén [32]. These CNDO calculations, especially parameterized for spectroscopy (/S), predicted that the first ionization potential of Bf was ~7.55 eV, consistent with a range of experimental values (6.42–7.87 eV) that could be inferred from earlier cluster-beam studies [136]. These calculations also predicted that the electron affinity of neutral Bf was 2.4 eV, close to subsequent results (2.6–2.8 eV) from early UPS studies of mass-selected C_{60} negative ions [14]. At the time, this level of agreement between theory and experiment was cited as providing good indirect evidence for the Bf hypothesis [14]. Another excellent example of a successful early prediction was Stanton and Newton's MNDO [137] results, which concluded that Bf should have only four infrared-active T_{1u} modes at about 577, 719, 1353, and 1628 cm^{-1}. This prediction, subsequently supported by several force constant model calculations [138–140], played a key role in Krätschmer, Huffman, and co-workers' initial identification of C_{60} in their carbon smoke samples [37, 38].

Following the production of macroscopic amounts of C_{60} in 1990, MNDO, ab initio HF, and first-principles LDF theoretical studies of gas-phase fullerenes and their simpler chemical derivatives have proliferated [85–91, 141–159]. The earlier gas-phase LDF Bf calculations [35] were improved using larger basis sets [85,

86] and averaging over sets of grid points [87]. Additional HF studies of this molecule were also implemented using larger basis sets [141] and these HF results further improved using Möller-Plesset perturbation theory [142]. At least 1,812 C_{60} fullerene isomers were enumerated [143, 144], and a favorable few generated from Bf using the Stone-Wales mechanism [145] were found in geometry-optimized MNDO and HF calculations to be close to 2 eV higher in energy than Bf [146]. The geometries of (D_{5h}) C_{70} [147–152] and selected higher fullerenes [151–154] were also optimized using MNDO [147, 148], HF [149, 150], and LDF [151, 152, 154] methods, and MNDO [148] and LDF [86] studies of several giant fullerenes have appeared. In addition, MNDO [155], HF [141, 149, 156], and LDF [85, 90, 91] studies of hydrogenated and fluorinated Bf derivates and selected endohedral fullerene complexes [101, 102, 114] have been reported. In the condensed phase, geometry-optimized LDF band-structure results for C_{60} fullerites [157–159] and fixed-geometry LDF band-structure studies of several alkali fullerides have appeared [160, 161]. Important new computational results continue to be reported weekly [162].

For our part, using the theoretical methods outlined in Section 6.1, we have carried out not only increasingly detailed studies of Bf [85–87], but also begun to address some of the higher [88, 89] and giant fullerenes [86, 87] as well. Our work has also extended to some of the simpler hydrogenated and fluorinated fullerene derivatives [85, 90, 91]. These studies and results as outlined in the introduction are reviewed in the remainder of this section.

6.2.1 Buckminsterfullerene

The latest experimental carbon-carbon bond distances and their uncertainties for solid C_{60} [49, 51] are compared in Table 6.2 to the more accurately determined theoretical bond distances [85, 87, 141, 142, 158]. The HF and second-order Möller-Plesset perturbation (MP2) corrected HF bond distances were computed using an all-electron triple-zeta plus polarization basis—19 contracted Gaussian

Table 6.2 The predicted and measured two symmetry-inequivalent equilibrium Bf bond distances in Å. (The uncertainties are given in parentheses.) NDD refers to experimental results from neutron diffraction data

Method	5-6	6-6	Ref.
HF	1.448	1.370	[141]
MP2	1.446	1.406	[142]
LDF(solid phase)	1.444	1.382	[158]
LDF(gas phase)	1.445(3)	1.387(3)	[85, 87]
NMR	1.450(15)	1.400(15)	[49]
NDD	1.444(9)	1.391(10)	[51]

basis functions per carbon atom [141, 142]. Our LDF gas-phase calculation, like all the molecular LDF calculations discussed herein, also used 19 contracted Gaussian basis functions per carbon atom, albeit obtained in a different fashion [85, 86]. These LDF (gas) bond distances are those of Ref. [85] expanded to four significant digits with a corresponding uncertainty added [87].

In contrast to ab initio methods, analytic basis sets cannot be used directly to evaluate the density-functional total energy, because the exchange and correlation energy expressions invariably depend nonanalytically on the density. Thus an additional three-dimensional numerical integration, or something roughly equivalent to it, such as fitting, is required. The LDF (gas-phase) uncertainty in Table 6.2 is the standard deviation of ten optimizations of the 6-6 bond distance (the carbon-carbon bond shared by two hexagons) using ten different sets of sampling points [109]. For the price of this added numerical integration, Table 6.2 shows, as is typically the case, that LDF calculations agree better with experiment than HF and roughly as well as MP2, provided good basis sets are used. The difference in the 6-6 bond distance between our LDF gas- and the solid-phase entries in Table 6.2 could arise from solid-state effects. Our gas-phase results agree with those from another molecular gas-phase calculation [163] that is similar to this solid-state LDF calculation; that is, that calculation also uses the pseudopotential approximation and a plane-wave basis, but with an artificially large unit cell to simulate the isolated molecule. Because of improvements in basis sets, particularly the ability to use diffuse functions to fit the exchange and correlation potential, these LDF (gas-phase) bond distances are slightly different from the values 1.43 and 1.39 Å we reported in Ref. [35]. Our predicted LDF and the experimentally measured bond distances—especially the results of David et al. [51]—are in striking agreement.

Figure 6.2 depicts the bound one-electron eigenvalue spectrum obtained from our Bf geometry-optimized LDF calculations [85, 86]. In agreement with theoretical studies starting at least with the work Bochvar and Gal'pern [16], we found that Bf is a closed-shell molecule, with a sizable electric-dipole-forbidden HOMO-LUMO gap. The LDF occupied electronic configuration is given by $2t_{1g}4a_g6t_{1u}2t_{2g}6t_{2u}6g_u10h_g6g_g6h_u$, where the highest occupied orbital of each I_h symmetry is ordered according to increasing one-electron eigenvalues. This occupied electronic configuration agrees with the HF result [141]. The fact that the $6h_u$ HOMO and the $7t_{1u}$ LUMO states are both antisymmetric under inversion has led us to suggest that there might exist a novel, retarded, symmetric-charge, transfer-mediated component to the superconducting pairing interaction acting across the C_{60} molecules in the alkali fullerides [103].

Deep electron affinities (EAs) and shallow ionization potentials (IPs) predicted from our LDF geometry-optimized Bf calculations are given in Table 6.3 [86]. Rough first estimates are given in column 1, where the entries were computed by taking the negative of the corresponding LDF one-electron eigenvalues, while much better estimates are given in column 2, where the entries were computed by applying LDF transition-state theory [164]. The best

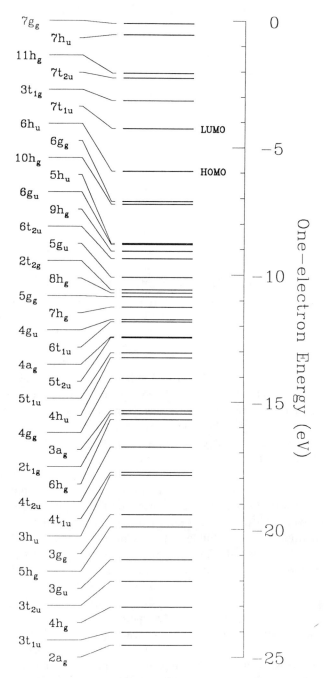

Figure 6.2 Bound (<0 eV) LDF one-electron levels together with their symmetry labels for Bf that has a large electric-dipole-forbidden HOMO-LUMO gap.

Table 6.3 LDF IPs computed using the one-electron eigenvalues (ε), the transition-state (TS) method, and computed directly (LDF-ΔSCF). The EAs, IPs, and parenthetically the separation from the next higher IP or EA are given in eV. EAs lie above the vertical break in the table, while IPs lie below

State	ε	TS	LDF-ΔSCF
$3t_{1g}$	3.15 (1.11)	1.61 (1.11)	1.71 (1.11)
$7t_{1u}$	4.26 (0.00)	2.72 (0.00)	2.82 (0.00)
$6h_u$	5.94 (0.00)	7.49 (0.00)	7.60 (0.00)
$6g_g$	7.12 (1.18)	8.68 (1.19)	8.78 (1.18)
$10h_g$	7.24 (1.30)	8.79 (1.30)	8.88 (1.28)
$5h_u$	8.78 (2.84)	10.40 (2.91)	10.48 (2.88)
$6g_u$	8.83 (2.89)	10.38 (2.89)	10.47 (2.87)

estimates are given in column 3, where the entries were computed from LDF self-consistent-field calculations before and after an electron was either added to or removed from the appropriate one-electron orbitals, that is, LDF-ΔSCF calculations. Our predicted LDF-ΔSCF first ionization potential of C_{60} (7.60 eV) agrees with experiment [44–46, 63], while our predicted LDF-ΔSCF first EA of C_{60} (2.8 eV) is only slightly higher than the recent experimental estimate [47] of 2.65 ± 0.05 eV. This large C_{60} electron affinity suggests that this molecule will have a rich chemistry, as Fagan et al. [73] and Wudl et al. [74] have found.

The parenthetical entries in Table 6.3 show that LDF deep Bf EAs can be accurately approximated by a rigid 1.44 eV upward shift of the negative of the LDF unoccupied one-electron eigenvalues. Similarly, the LDF shallow Bf IPs are seen to be accurately approximated by a rigid 1.67 eV upward shift of the negative of the occupied eigenvalues. All other shifts including the reversal of the $5h_u$ and $6g_u$ IPs are more than an order of magnitude smaller (0.03 eV). These results implied that the C_{60} valence photoelectron spectra of Lichtenberger et al. [44, 45] and Weaver et al. [46] could be compared in detail to our LDF one-electron occupied states, provided the varying cross sections for the various excitation processes—known to be important for carbon systems at least since the work of Gelius on benzene nearly 20 years ago [165]—were properly taken into account. The corresponding matrix elements pick out rapid spatial variations (e.g., near the core) in the molecular wave functions and hence are difficult, if not impossible, to evaluate readily in LDF approaches employing plane-wave basis sets and/or pseudopotentials. However, when combined with first-order time-dependent perturbation theory and a semiclassical description of the radiation-matter interaction, our LCAO-LDF-based approach can be used to calculate easily these matrix elements and hence the photoelectron cross sections. This approximate procedure has been shown to lead to results in good agreement with experiment for a range of large molecular and polymeric systems [123].

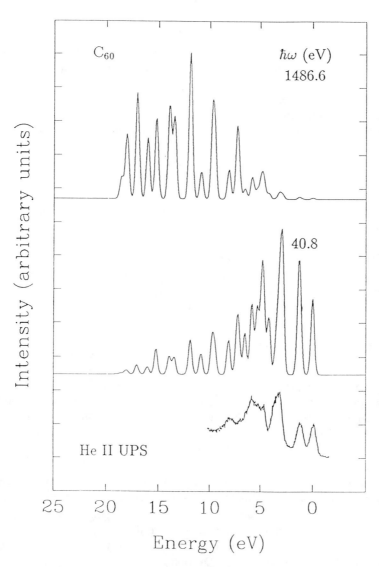

Figure 6.3 Predicted photoelectron line shapes from LDf results for Bf at incident photon energies of 40.8 and 1486.6 eV. Note that the zero of energy coincides with the calculated and measured threshold (−7.6 eV) and relative heights of the two curves are adjusted to have equal maximum values for the valence spectra displayed. The measured He II UPS (40.8 eV) is from Ref. [45].

Our predicted x-ray and ultraviolet-photoelectron (XPS and UPS) line shapes for Bf [88] calculated according to Ref. [123] are shown in Figure 6.3. Also shown in the lower portion of this figure is the He(II) UPS (40.8 eV) C_{60} spectrum obtained by Lichtenberger et al. [45]. There is overall good agreement between theory and experiment both in positions and relative intensities of the peaks. Our calculated line shapes, while confirming the assignments of Lichtenberger et al. for the two highest-energy peaks, also allowed assignment of the three lower-energy features. Weaver et al. [46] also reported experimental C_{60} photoelectron spectra obtained over a wider range of energies using a synchrotron source. They compared their results to one-electron densities of states obtained from pseudopotential-based LDF calculations. With higher incident photon energies, and the resolution typical of synchronton sources, they discerned features in addition to those seen earlier by Lichtenberger et al. Through an analysis of the line shapes obtained by Weaver et al., including how these line shapes change with incident photon energy, we have identified the origin of these additional features [88]. We have also found excellent agreement between their measured [46] and our predicted XPS (1486.6 eV) results [88]. This fine level of agreement between measured (solid-phase) and calculated (gas-phase) photoemission line shapes for C_{60} provided further firm additional evidence in favor of the Bf hypothesis, while also showing that Bf retains much of its molecular identity in the condensed phase.

6.2.2   Giant and Higher Fullerenes

Calculations on much larger carbon molecules using our variational LCAO-LDF approach are easily performed because of this approach's high efficiency. For example, we have studied the giant icosahedral fullerenes C_{180} and C_{240} depicted in Figure 6.4 using this method and compared our results to C_{60} [86]. Optimized geometries for these giant fullerenes were found using our empirical hydrocarbon potential [113]. This procedure was checked by comparing our LDF results obtained at the LDF optimized geometry for Bf to the corresponding LDF results obtained at the slightly different Bf geometry predicted by the empirical potential [86]. This check showed that the LDF electronic configuration was unchanged at the empirical potential geometry with none of the associated one-electron states altered by more than ± 0.1 eV. In addition, the LDF total energy at the empirical geometry was only increased by ≈ 0.01 eV per carbon from its value at the LDF optimized geometry.

Consistent with earlier Hückel calculations [7, 22, 133], we found that C_{180} and C_{240} are both closed-shell fullerenes. The LDF valence electronic structure is $1a_{1u}9a_g9t_{1g}17t_{1u}17t_{2u}9t_{2g}18g_g18g_u27h_g19h_u$ for C_{180} and $10a_g22t_{2u}22t_{1u}24g_u14t_{1g}14t_{2g}24g_g34h_g26h_u2a_u$ for C_{240}. The HOMO and LUMO levels of these molecules were found to have the same symmetry as the HOMO and LUMO levels of Bf. Hence the HOMO-LUMO gap is electric-dipole forbidden in both species. For C_{180} (C_{240}) this gap is $\approx 80\%$ ($\approx 70\%$) that of Bf. The deep LDF EAs and shallow LDF electron IPs for C_{180} and C_{240} are compared to those of Bf in

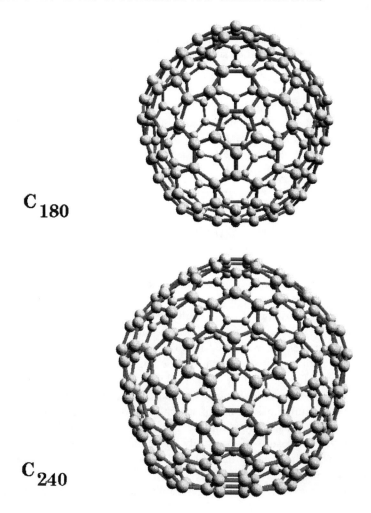

Figure 6.4 The giant closed-shell icosahedral fullerenes C_{180} and C_{240}.

Table 6.4. A side benefit of these calculations is that they can be used to show (as with our Bf results before) that the use of one-electron LDF eigenvalue differences in practical calculations of photoelectron line shape is accurate to within 0.1 eV within the LDF model. The predicted XPS and UPS line shapes for these giant fullerenes are given in Ref. [86] Our studies of these giant fullerenes also predicted that the binding energy per carbon in C_{180} (C_{240}) is 0.22 eV (0.26 eV) greater than the binding energy per carbon in Bf [86]. Hence, in comparison to Bf, these giant fullerenes are significantly more stable, suggesting a kinetic rather than thermodynamic origin for the copious production of Bf.

In addition to these giant fullerene studies, we also used our LCAO-LDF methods to compute the XPS and UPS spectra of a range of higher fullerenes

Table 6.4 LDF-ΔSCF EAs for C_{60}, C_{180}, and C_{240} lie above the vertical break in the table, while IPs lie below. Under each molecule the orbital to which an electron is added or subtracted is followed by the EA or IP

C_{60}		C_{180}		C_{240}	
$3t_{1g}$	1.71	$10t_{1g}$	3.21	$15t_{1g}$	3.31
$7t_{1u}$	2.82	$18t_{1u}$	3.56	$23t_{1u}$	3.61
$6h_u$	7.60	$19h_u$	6.91	$26h_u$	6.52
$6g_g$	8.78	$27h_g$	7.19	$34h_g$	6.93
$10h_g$	8.88	$18g_u$	7.24	$24g_g$	7.03
$6g_u$	10.47	$18g_g$	7.43	$14t_{2g}$	7.55

at the optimized geometries predicted by our empirical potential [88]. These results showed that the loss of I_h symmetry in these higher fullerenes should lead to a detectable broadening in their photoelectron line shapes relative to the corresponding line shapes for C_{60}. For example, rather than the sharp peak at threshold arising from a single transition observed in C_{60}, we predicted a broad transition arising from twenty valence electrons in six states ($19e_1''$, $14a_2''$, $7a_2'$, $19e_2''$, $23e_2'$, and $23e_1'$) as the first UPS peak for C_{70}—our one-electron eigenvalues are contained in a 0.7 eV interval; the experimental line should have additional instrumental broadening. This prediction was later experimentally confirmed by Jost et al. [52].

We have also computed a lower bound on the LDF binding energies for a range of higher fullerenes (some of which have been isolated in macroscopic quantities [54–56, 135]) by using the geometries predicted by our empirical potential [89]. The results of that study, which are compared to our geometry-optimized Bf binding energy in Table 6.5, again suggest a kinetic rather than a thermodynamic preference for Bf. Indeed, Table 6.5 shows that Bf is not energetically favored over graphite, giant fullerenes, higher fullerenes, and even (D_{5h}) C_{70}.

6.2.3 Hydrogenated and Fluorinated Bf Derivatives

Haufler et al. [72] were the first to report chemically reacting Bf through a Birch reduction to yield $C_{60}H_{36}$. To characterize this new molecule further, we examined the tetrahedral structure suggested by Haufler et al. [72] with 12 isolated double bonds and all hydrogen atoms bonded exohedrally, as depicted at the top of Figure 6.5 [85, 90, 91]. We first determined an optimized geometry for this isomer of $C_{60}H_{36}$ using our empirical hydrocarbon potential and then used this geometry in LDF electronic structure calculations. The basis sets used in these LDF calculations are described in detail in Ref. [85]. The resultant LDF electronic structure was found to be closed shell, $6a_u 6e_u 11a_g 11e_g 22t_g 27t_u$, with an electric-dipole-forbidden HOMO-LUMO excitation of 3.75 eV [85, 90]. This one-electron gap will only increase in an

Table 6.5   LDF binding energy (eV) per carbon, ΔE, for a series of higher fullerenes, giant fullerenes, and graphite measured with respect to Bf. The graphite estimate is obtained by comparing optimized LDF results for Bf [85, 86] to LDF band structure results for 3D graphite [131]

Fullerene	Symmetry	ΔE
C_{60}^a	I_h	0.0
C_{70}^b	D_{5h}	0.01
C_{76}^b	D_2	0.06
C_{84}^b	D_{6h}	0.08
C_{180}^a	I_h	0.22
C_{240}^a	I_h	0.26
Graphite	—	0.30

[a]Ref. [86].
[b]Ref. [89].

LSD-ΔSCF calculation. The predicted XPS and UPS line shapes for this isomer of $C_{60}H_{36}$ are given in Ref. [91]. Comparison of the minimum energy for this structure with that given by the empirical potential for the same molecule but constrained to have a combination of ideal sp^2 (planar) and sp^3 (tetrahedral) bonding suggested that there would be little strain associated with this structure. This result is consistent with our calculated energetics for substracting (adding) two hydrogen atoms from (to) this isomer, where we found that subtracting two hydrogens from $C_{60}H_{36}$ was endothermic by 6.86 eV, while adding two hydrogens to form $C_{60}H_{38}$ was exothermic by 5.38 eV [85].

We also predicted the geometry, electronic structure, and photoelectron line shapes of the so far hypothetical, fully saturated species $C_{60}H_{60}$ and $C_{60}F_{60}$ [85, 91]. The basis sets used in the LDF calculations are given in Ref. [85]. Table 6.6 compares optimized $C_{60}H_{60}$ and $C_{60}F_{60}$ geometries predicted by HF calculations [141] to optimized geometries predicted by our empirical hydrocarbon potential ($C_{60}H_{60}$) and by our first-principles LDF method ($C_{60}H_{60}$ and $C_{60}F_{60}$). Note that these results all assume that all monovalent atoms are bonded to the exterior of the cage (Fig. 6.5, middle). The largest difference in bond lengths between the predicted HF and the LDF geometries is 0.02 Å. The agreement is also good between our empirical-potential and LDF results. In fact, the total LDF energies at the LDF and empirical potential minima differ by only 0.54 eV. The slight bending away from the radial direction of the origin-carbon-hydrogen bond angle in the empirical-potential geometry is in the plane of the origin and the 6-6 bonds and shortens the 6-6 H-H third-neighbor distance. The LDF electronic structure of $C_{60}H_{60}$ was found to be closed shell with an electronic configuration $4a_g7t_{1u}7t_{2u}3t_{1g}11h_g7g_u3t_{2g}7g_g7h_u$, which agrees with the HF result [141]. The fully fluorinated species, $C_{60}F_{60}$, was found to have an LDF electronic structure given by $7a_g14t_{1u}14t_{2u}15g_g8t_{1g}1a_u15g_u8t_{2g}22h_g16h_u$, which also agrees with the HF result [141]. The one-electron LDF HOMO-LUMO

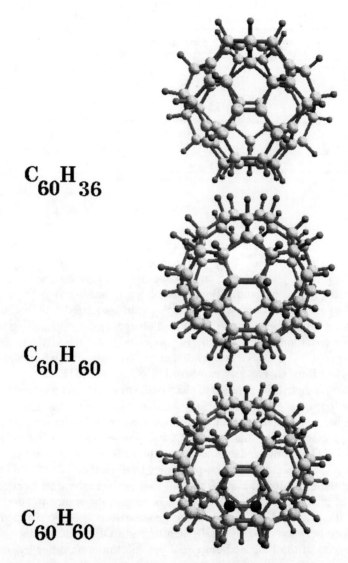

$C_{60}H_{36}$

$C_{60}H_{60}$

$C_{60}H_{60}$

Figure 6.5   A series of hydrogenated Bfs. From top to bottom: suggested tetrahedral structure of $C_{60}H_{36}$ with the 36 hydrogen atoms (smaller spheres) bonded to the exterior of the C_{60} cage; hypothetical $C_{60}H_{60}$ isomer with all hydrogens bonded to the exterior of the cage; and hypothetical lower-energy $C_{60}H_{60}$ isomer with two of the sixty hydrogens bonded to the interior of the cage at the carbon sites depicted in black.

Table 6.6 HF [141] and LDF [85] $C_{60}F_{60}$ and $C_{60}H_{60}$ and empirical potential (EP) [85] $C_{60}H_{60}$ optimized structural parameters

Molecule	5-6 C-C bonds (Å)	6-6 C-C bonds (Å)	$R_{c\text{-}x}$ bonds (Å)	Center-C-X angle
$C_{60}H_{60}$ (EP)	1.55	1.54	1.09	178°
$C_{60}H_{60}$ (HF)	1.57	1.55	1.08	
$C_{60}H_{60}$ (LDF)	1.55	1.54	1.10	180°
$C_{60}H_{60}$ (HF)	1.60	1.59	1.34	
$C_{60}F_{60}$ (LDF)	1.61	1.60	1.35	180°

gaps for $C_{60}H_{60}$ and $C_{60}F_{60}$ were found to be 3.72 and 3.48 eV, respectively. Again, LDF-ΔSCF calculations will only increase these gaps. The LUMOs for $C_{60}H_{60}$ and $C_{60}F_{60}$ were found to be $5a_g$, and $8a_g$, respectively. Thus, HOMO-LUMO excitations are electric-dipole forbidden in both molecules. The predicted XPS and UPS spectra for $C_{60}H_{60}$ are given in Ref. [91].

In contrast to $C_{60}H_{36}$, we found that $C_{60}H_{60}$ is significantly strained compared to ideal (for this case completely tetrahedral) bonding. We recognized early that such strain could be partially alleviated by bonding some of the hydrogen atoms to the interior cage wall [85]. Indeed, using our empirical potential, we found that the strain energy could be reduced, if between two and twelve of the hydrogen atoms were bonded inside Bf (Fig. 6.5, bottom) [85]. The stability gained by bonding hydrogen atoms inside the $C_{60}H_{60}$ cluster was later confirmed by other calculations using a more traditional valence-force-field-based potential [166]. Consistent with a small strain energy for the $C_{60}H_{36}$, we found that for this cluster the total energy went up for bonding hydrogen inside the Bf cage [85, 90].

6.3 Bf During High-Energy Collisions and Under Compression

An established method for analyzing the structure and stability of gas-phase clusters is to break them apart either by uv irradiation or through collision with surfaces or target gases. The mass spectra of the daughter ions can then be used to infer thresholds and channels for dissociation and hence gain information regarding the structure of the parent. Such probes when brought to bear on the fullerenes have yielded a series of remarkable results [13, 62–69].

O'Brien et al. [13] found in early laser-induced fragmentation experiments of positive carbon cluster ions that clusters smaller than C_{31}^+ readily fragment by the loss of C_3; however, starting at C_{31}^+ the photodissociation process dramatically changes. Beyond C_{31}^+ all even-membered clusters become very difficult to dissociate, and when dissociation begins it occurs through the loss of the high-energy product C_2. Also, any odd-membered clusters larger than C_{31}^+ now begin to fragment through the loss of C_1. These observations, although unanticipated, were fully consistent with the proposed highly stable fullerene structures for the even-membered clusters [13, 15].

In subsequent experiments Whetten and co-workers [62] and McElvany, Ross, and Callahan [63] found that C_{60} rather than dissociating—even by the loss of C_2—often simply rebounded when collided with surfaces at velocities far above those needed to fragment small aromatic hydrocarbons and alkali halide clusters. Additional experiments confirmed this remarkable resilience of C_{60} during hypervelocity (>2 km/sec) surface impacts [64]. The results of these three separate surface scattering experiments were surprising, because they suggested that C_{60} not only could survive violent short-time surface impacts without chemically reacting, but also could remain intact for comparably long times (at least a typical radiation time) following the impact—this despite probable levels of internal excitation well above the already high levels needed for photodissociation.

More recently, Schwarz and co-workers observed the formation of $C_N He^+$ fragments ($N = 46, 48, 50, 52, 54, 56, 58$) during high-energy collisions of C_{60}^+ with He [65]. They postulated that these fragments were endohedral complexes (He is trapped inside the fullerene cage). In subsequent experiments Ross and Callahan [66] not only confirmed these results but also detected the presence of $C_{60}He^+$. Ross and Callahan's experiments also convincingly demonstrated that the $C_{60}He^+$ complexes they observed were $(He@C_{60})^+$. Since then other rare gas atoms and small molecules have been found to form collision-induced endohedral complexes with fullerenes [67, 68]. The ability to trap a target atom during a high-energy collision with a projectile is unprecedented in mass spectrometric experiments and represents yet another novel feature of the fullerenes.

Bf has not only been subjected to high-energy collisions with surfaces and target gases, but also rapidly crushed in diamond-anvil cells [59] and shock compressed using hypervelocity copper flyer plates [60]. Regueiro et al. [59] crushed C_{60} fullerites to diamond while Yoo and Nellis [60] shocked fullerites to graphite and amorphous carbon. The results of such experiments are of fundamental interest not only because they clarify how C_{60} responds under extreme conditions and yield a novel synthetic route to diamond [59], but also because they might yield new phases of carbon. Just as lonsdaleite, originally discovered in granular form in meteorites, was subsequently prepared through the shock compression of graphite, new allotropes of carbon could result from the shock compression of fullerites with their unusual open structure. Theory suggests that carbon could have many quite different, potentially important allotropes at ambient conditions [167–173]—if only a way could be found to prepare these materials.

To gain insight into the behavior of C_{60} during hypervelocity impacts with surfaces, accompanying the entrapment of He, and under anisotropic compression we have carried out a series of MD simulations of these processes [92–

96]. These studies and results as outlined in the introduction are reviewed in this section.

6.3.1 High-Energy Collisions of Bf with Diamond

Our simulations of hypervelocity impacts of Bf [92, 93] with surfaces were carried out at the same time as the early surface scattering experiments were begun [62, 63]. In these simulations we chose to model the hypervelocity impact of Bf with a hydrogen-terminated diamond (111) surface, allowing all elements of the simulation (Bf, the surface, and the Bf-surface interaction) to be modeled using our single hydrocarbon potential described in Section 6.1.2. The hydrogen-terminated diamond (111) surface was modeled by ten layers of carbon atoms with 64 atoms per layer topped by a single layer of 64 hydrogen atoms. Periodic boundary conditions were applied in the directions parallel to the surface to approximate the semi-infinite crystal. The bottom two carbon layers were fixed during the simulations, while frictional forces were used to maintain the next seven layers of atoms at room temperature, 300 K [174]. The Bf molecule was given an initial temperature of 300 K, and its center-of-mass kinetic energy varied to give impact energies of 150, 200, and 250 eV with respect to the surface normal. At each energy, fifty trajectories were calculated with the surface impact point chosen randomly. Each trajectory was followed until the molecule either rebounded from the surface or became chemisorbed.

Figure 6.6 shows a series of snapshots of a typical nonreactive collision begun with an impact energy of 250 eV. The bottom of the molecule flattens, and the upper layers of the crystal are compressed as the collision begins (Fig. 6.6). Near the midpoint of the collision (Fig. 6.6), the molecule is approximately compressed to one half its equilibrium diameter. The compression of the top few layers of the diamond lattice is also evident. Desorption occurs as C_{60} regains its initial shape, and the surface atoms are driven back to their initial positions by the crystal (Fig. 6.6). Although containing excess vibrational energy, on average, Bf resumes an approximately spherical shape as it rebounds from the diamond surface. These results imply that Bf exhibits a remarkable ability to regain its original shape after severe dynamic compression.

Table 6.7 shows the probabilities for forming different product species at each of the three collision energies studied. At a collision energy of 150 eV, only nonreactive scattering occurs. For collisions with energies of 200 eV, 14% of the collisions result in chemical reactions. The reaction probability increases to 72% for a collision energy of 250 eV. The majority of the chemical reactions involve transfer of hydrogen atoms from the surface to the cluster. The predominance of this type of reaction results from the exposed position of the singly bonded hydrogen atoms that are easier to extract than the somewhat protected fourfold-coordinated surface carbons. Also, note that at the highest energy, chemisorption is the most probable event.

The amount of energy deposited into the cage vibrational modes during the

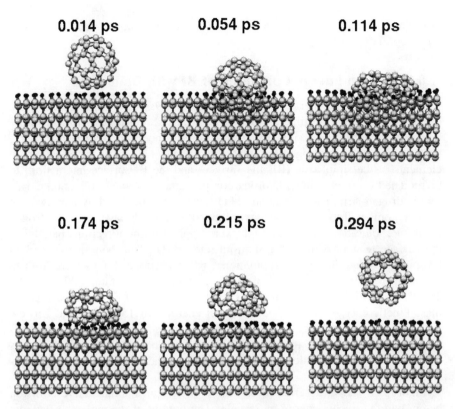

Figure 6.6 Snapshots of a simulated unreactive collision of Bf with a hydrogen-terminated (111) diamond surface. The initial collision energy is 250 eV.

scattering process is especially important because this heating might lead to dissociation at time scales that are orders of magnitude beyond those that can be simulated. The final internal and center-of-mass kinetic energies for unreactive molecules were 43.4 and 45.5, 51.5 and 53.9, and 60.7 and 57.4 eV for collision energies of 150, 200, and 250 eV, respectively. The internal and center-of-mass kinetic energies are almost equal for each set of trajectories. Approximately 20–30% of the initial center-of-mass energy is converted to internal energy, which is consistent with the conversion rate observed in large clusters [175]. These internal energies correspond to temperatures of 3000–4000 K so that dissociation might occur at longer times. High internal and low center-of-mass energies are found for many reactive collisions involving transfer of carbon atoms. During these collisions an ephemeral bond forms between the molecule and the surface. As the molecule rebounds, this bond exerts a retarding force, producing an elongation of the cage that leads to additional intramolecular vibrational excitation.

The predicted high stability of Bf during high-energy collisions with surfaces

Table 6.7 Composition of scattered molecules and probability of their formation as a function of collision energy

Scattered molecule	Collision energy		
	150 eV	200 eV	250 eV
C_{57}	0	0	2
C_{59}	0	2	4
C_{60}	100	86	28
$C_{58}H$	0	0	2
$C_{59}H$	0	4	2
$C_{60}H$	0	4	16
$C_{61}H$	0	0	6
$C_{60}H_2$	0	0	6
$C_{61}H_2$	0	0	2
Chemisorption	0	4	32
Total Reaction Probability	0%	14%	72%

is consistent with experimental observations. The simulations predict that the threshold for reaction is between 150 and 200 eV. In collisions of C_{60}^+ with stainless steel surfaces at a collision energy of 60 eV, only nonreactive scattering was observed [63]. Similarly, only nonreactive scattering was observed in experiments studying collisions of C_{60}^+ with silicon and graphite surfaces at energies up to 250 and 170 eV, respectively [62]. In collisions with graphite, fragmentation was not observed until the collision energy was raised above 250 eV [64].

6.3.2 High-Energy Collisions of Bf with He

In addition to simulating the hypervelocity impacts of Bf with surfaces, we have also simulated the formation of endohedral complexes during high-energy collisions of Bf with He [94, 95] and made direct comparisons with experimental observations [95]. Again Bf was modeled using our empirical hydrocarbon potential augmented by a screened Coulomb potential used to describe the He-carbon interaction [176, 177]. These combined potentials give barrier heights for passing through the centers of five- and six-membered rings of the Bf cage of 13.1 and 9.35 eV, respectively. For comparison, the barrier height for moving a He to the center of a benzene ring is calculated to be 10.7 eV using ab initio second-order Møller-Plesset perturbation theory [178], while first-principles LDF theory predicts that 10.4 eV is needed to move a He to the center of an unrelaxed six-membered ring in Bf [102]. At each collision energy, multiple trajectories with different Bf impact points were followed.

A comparison of the predicted and experimental probabilities for forming He@C_{60} is shown in Figure 6.7 [94, 95]. The overall agreement between the two sets of data is good. However, the maximum observed abundance of He@C_{60}

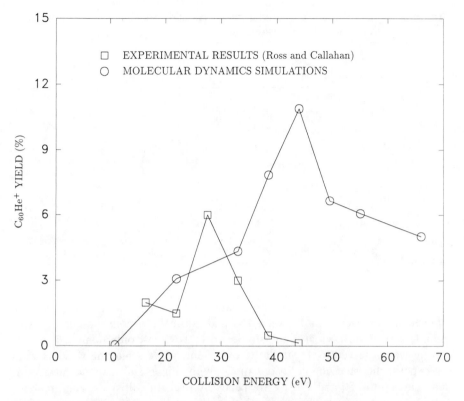

Figure 6.7 Effect of center-of-mass collisional energy on $(He@C_{60})^+$ formation efficiency predicted from our MD simulations and measured by Ross and Callahan [95].

is less than predicted and occurs at lower collision energies. Both differences are consistent with fragmentation of a portion of the $He@C_{60}$ formed initially in the experiments. $He@C_{60}$ formed at collision energies of either 27.6 or 44 eV will have internal energies well above the dissociation energy of the complex. However, the complex formed at the higher energy will have an additional 16.4 eV of internal energy and hence should have a higher dissociation rate [94]. The microseconds between collision and detection in the experimental system provide ample time for many of the initially formed complexes to dissociate. Therefore, as the collision energy increases, the measured energy dependence of the $He@C_{60}$ abundance results from the combined effects of an increasing probability of initially forming the complex and a decreasing probability that complex will survive to detection. In contrast, the dynamics of the collision are followed for less than a picosecond in the simulations so that any fragmentation processes occurring at a longer time cannot be observed.

The simulations predict an increase in the trapping efficiency to a maximum followed by a decrease as the collision energy increases. An analysis of the trajectories leading to trapping provides a qualitative explanation for this behavior. At low collision energies most trajectories leading to trapping occur for low-impact collisions in which the He passes through the center of one of the carbon rings. Low-energy collisions with high impact parameters tend to scatter from the exterior of the molecule because the tilted orientation of the rings of carbon atoms results in a higher barrier for penetration. High-energy collisions with low-impact parameters have low trapping probabilities because the He atom frequently passes completely through the cage. However, the contribution to the trapping probability from collisions with large impact parameters increases for high-energy collisions because more energy is available to cross the higher barrier. There is a higher probability for collisions with large impact parameters so the trapping efficiency is weighted toward higher energies.

A typical trajectory leading to trapping is shown in Figure 6.8. After passing into the cage's interior the He atom loses energy through multiple collisions

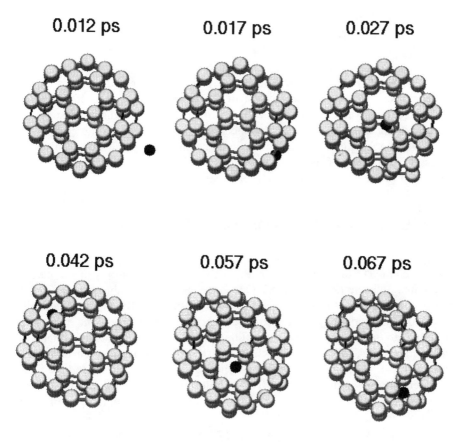

Figure 6.8 Snapshots of a simulated He entrapment by Bf. The initial He kinetic energy is 22 eV.

with carbon atoms. The He atom's kinetic energy eventually drops below the amount needed to escape, and it is endohedrally trapped.

6.3.3 Bf under Anisotropic Compression

Inspired by the similarities between graphite and Bf, early suggestions were made that Bf might yield a solid-state lubricant with unique properties. To explore whether the fullerenes could withstand the relatively high pressures that such an application would require, we performed several simulations of the anisotropic compression of a single Bf molecule between both graphite layers and hydrogen-terminated diamond surfaces [96]. These simulations should also provide insight into the behavior of fullerites both during nonhydrostatic crushing and extremely rapid shock compression [59, 60].

The starting point of these simulations was again our hydrocarbon potential outlined in Section 6.1.2. At relatively low pressures, we found that an individual C_{60} molecule was able to roll between graphite sheets. As the cluster was further compressed, it flattened to the point that its motion changed from rolling to sliding. At the highest pressures studied, the diameter of the cluster was one-third its original value. Upon removing the pressure, the cluster was able to recover its original spherical shape. Similar behavior was observed when a Bf molecule was compressed between hydrogen-terminated surfaces of diamond. These results are reminiscent of the observed resilience of Bf during simulated hypervelocity impacts with hydrogen-terminated diamond surfaces. So far, there are no reports of the successful application of fullerenes as lubricants, suggesting that other factors could be inhibiting this application. However, the flattening of the Bf cage we observe in these simulations suggests a first step in the shock compression of fullerites to graphite [60].

6.4 On the Way to Buckminsterfullerene

Despite the tremendous pace of fullerene research, the fundamental mechanisms leading to the high yield of Bf in the chaotic conditions of the carbon arc remain elusive [97, 179–188]. One key issue is size. Table 6.5 shows that despite its spherical shape and lack of abutting pentagons Bf is not energetically favored over graphite, giant fullerenes, some higher fullerenes, and even (D_{5h}) C_{70}. Indeed the strained cages of the fullerites are reminiscent of some energetic molecular solids that, when shocked, can detonate. Roughly, the closer the fullerene is to graphite, the greater the binding energy per carbon and the more energetically favored the cluster. Thus, if cluster growth in the cooling carbon vapor yields fullerenes larger than C_{70}, then these clusters would seem unlikely to shrink to C_{60} [179].

Under proper conditions, however, the yield of Bf can exceed 20% in the Krätschmer-Huffman process. Therefore, there likely exists an efficient kinetic mechanism that intervenes, pointing carbon clusters down the road to Bf before they have an opportunity to grow to a larger size. One clue to this process has come from MD studies we have made of high-temperature graphitic fragments [97] suggested to condense in the cooling carbon vapor prior to fullerene formation [180]. While consistent with the postulated importance of pentagons in the early stages of Bf formation [5, 11, 179–183], these studies also point to the central role of high temperatures. The results of these simulations are also consistent with Heath's recent model of the early stages of Bf formation [183]. Our studies and results as outlined in the introduction are reviewed in this section.

Graphitic fragments were modeled using our empirical hydrocarbon potential outlined in Section 6.1.2. The reactive aspect of this potential was important both to model the reactive edges of these fragments and to allow for any possible fragmentation at high temperatures. The fragments studied were treated as ribbons as an idealization of their probable low-symmetry, far-from-circular shapes. The series of initially planar ribbons used in these studies are depicted in Figure 6.9. These ribbons, with sizes ranging from 32 to 108 atoms, have varying lengths and widths. In addition, onefold-coordinated atoms were included at the edges of most of these ribbons to allow for the dynamic formation of edge pentagons. The starting conditions are generated by assuming the velocities for the individual atoms in the relaxed carbon ribbon are initially distributed with arbitrary directions in 3D space according to a Boltzmann distribution. The dynamics of these isolated ribbons are then followed at constant energy to simulate their motion in the carbon vapor between collisions. The effective temperature of an isolated ribbon was defined as the average kinetic energy per degree of freedom in units of temperature K. Assuming the conditions in the carbon arc reported by Haufler et al. [181] (a temperature of 1000 to 2000°C and a pressure between 100 and 200 torr), the average time between collisions of He, C, or C_{60} will be on the order of nanoseconds [189]. The 250 ps length of our simulations is an order of magnitude less than this collisional time.

At 300 K, all ribbons remained close to planar, showing only minor out-of-plane fluctuations. In addition, no pentagons formed at the ribbon edges. However, at higher temperatures pentagonal rings did form, as illustrated in Figure 6.10 for a 108-atom ribbon trajectory started at an initial temperature of 1150 K. During the first 15 to 20 ps, the internal temperature increased to near 2500 K because of the heat produced by bond formation along the edges of the ribbon, yielding five-membered rings. During this time, the amplitude of the fluctuations from planarity also increased dramatically both from this internal heating and from the curvature induced by the edge pentagons. These large fluctuations are illustrated in Figure 6.10(b) and (c). Although the ribbon started in a planar configuration, by 25 ps, it can be far from planar, as shown by the 28.5 or 39.0 ps snapshots. Finally, by ∼70 ps the low-frequency, large-amplitude oscillatory behavior was quenched as bonds formed across the ribbon's ends, preventing it from uncurling, as shown in Figure 6.10(d). This fusion of the ribbon's ends also resulted in an additional heating of the cluster.

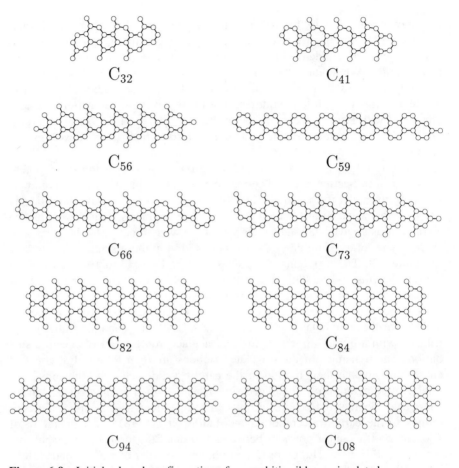

Figure 6.9   Initial relaxed configurations for graphitic ribbons simulated.

The hollow open-ended structure shown in Figure 6.10(d) should represent an excellent fullerene precursor.

Summarized in Table 6.8 are results for all the ribbons shown in Figure 6.9. Each of these simulations was started with Boltzmann-distributed atom velocities to generate a temperature of 1000 K, yielding actual initial temperatures in the range of 800 to 1200 K. With the exception of the two smaller strips, these results show that those ribbons with onefold-coordinated edge atoms all had high closure probabilities (20–80%) within times from 37.5 to 212.5 ps. In contrast, the two ribbons without any onefold-coordinated edge atoms, C_{59} and C_{94}, did not form any open-ended hollow structures. These ribbons also did not form such structures even when the starting temperatures were increased to 2500 K to compensate for their lack of self-heating. Therefore, the presence of pentagonal rings formed along the edges of these ribbons significantly re-

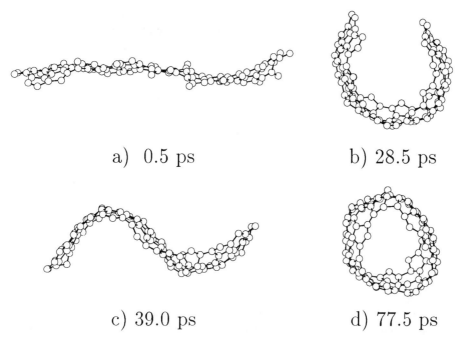

a) 0.5 ps   b) 28.5 ps

c) 39.0 ps   d) 77.5 ps

Figure 6.10   Snapshots of the curling and closure of the 108-atom ribbons started at an initial temperature of 1150 K at times of (a) 0.50, (b) 28.5, (c) 39.0, and (d) 77.5 ps.

duced the barrier to closure and consequently increased the probability that they will form good fullerene precursors.

However, our results also indicated that the existence of pentagons is not the only factor important in the formation of such precursors. First, as Table 6.8 shows, if the ribbon was too short, then it was unlikely to close even if a large number of pentagons formed at its edges. In addition, we found that even when these ribbons were of a size that the initial closure was exothermic and the barrier to this process was easily overcome at higher temperatures, this barrier could still effectively prevent closure at lower temperatures. Thus, graphitic ribbons growing at these lower temperatures should be less likely to yield good fullerene precursors. Rather, at these lower temperatures, such fragments might continue to spiral, eventually forming larger spheroidal particles with several interior layers such as envisioned by Zang et al. [5], Kroto and McKay [182], and Kroto [11]. This picture then suggests that the size distribution of any fullerenes formed in this manner should be dependent on the temperature of the carbon vapor, with lower temperatures leading to larger fullerenes or soot rather than C_{60}.

It might seem counterintuitive that so many of the ribbons simulated spon-

Table 6.8   Summary of 250 ps MD ribbon trajectories. Tabulated are the ribbon size, number of trajectories exhibiting closure out of the total number of trajectories for that ribbon size, and the times at which this closure occurred

Size	Ratio closed	Time of closure (ps)
C_{32}	0/4	
C_{41}	0/4	
C_{56}	1/5	91.5
C_{59}	0/5	
C_{66}	1/4	212.5
C_{73}	4/8	65.5, 67.25, 86.75, 230.5
C_{82}	1/5	76.5
C_{84}	4/5	37.5, 156.0, 157.0, 211.0
C_{94}	0/5	
C_{108}	3/6	70.5, 182.0, 186.5

taneously generate open-ended hollow structures. After all, these structures form while the ribbon is isolated, and hence statistically this process must be driven by an increase in entropy. The entropy of the *higher-temperature* open-ended hollow structure, however, is arguably higher than the entropy of the *lower-temperature* ribbon, because the hollow structure can be closer to quasiequilibrium for the longer fragments [97]. Hence, to prepare such fragments as ribbons was to prepare them further from equilibrium, in which case they can spontaneously close accompanied by an increase in entropy. If the preparation was viewed as a growth step, then the MD simulations showed that high-temperature ribbons with edge pentagons exceeding a critical length were likely to relax quickly to hollow structures that did not dissociate. The MD simulations also showed that, in addition to pentagons, high temperature was a key ingredient in readily surmounting the barrier to closure prior to additional growth.

Once a hollow open-ended structure is formed, additional growth should favor the closure of the open ends to form an actual fullerene. Because even relatively small ribbons form such hollow fullerene precursors, this suggests that lower fullerenes often first form and then grow up to C_{60}. A similar conclusion was reached by Heath [183]. But such behavior requires that C_{60} represent a special stopping point in fullerene growth, which is indeed implied by the isolated pentagon rule [179, 181]. Although our starting assumption of graphitic ribbons is an idealization of the probable low-symmetry, far-from-circular shapes of graphitic fragments in the carbon vapor, key indicators of size such as the presence of pentagons, the role of low-frequency, large-amplitude oscillations at high temperatures, and the ability of large enough clusters to close spontaneously between growth steps should survive in more detailed treatments of fullerene formation.

6.5   Graphitic Tubules

Following the synthetic breakthrough of Krätschmer, Huffman, and co-workers yielding macroscopic amounts of C_{60} and C_{70}, the possibility of related forms of carbon with potentially unique properties has been widely recognized. Recently, Iijima has observed needlelike carbon fibers consisting of 2 to 50 concentric tubules with diameters as small as 2 nm and lengths up to 1 μm growing at the cathode in an apparatus similar to that used for the mass production of C_{60} [80]. If the synthesis and processing of these fibers can be precisely controlled, then they might yield new materials with important structural and electronic properties [80–84, 190].

To gain further insight into the properties of such tubules, we have carried out an extensive set of studies of the electronic structure [81, 82, 98] and energetics [99] of the set of tubules constructed by rolling up a single graphite sheet into a cylindrical tube with a constant radius. The results of these studies as outlined in the introduction are reviewed in this section.

6.5.1   Tubule Symmetries

Each tubule can be visualized as a conformal mapping of a two-dimensional honeycomb lattice (depicted in Fig. 6.11) onto the surface of a cylinder subject to periodic boundary conditions both around the cylinder and along its axis [80, 82–84, 99]. The proper boundary condition around the cylinder can only be satisfied if the cylinder circumference maps to one of the Bravais lattice vectors of the graphite sheet. Thus each real lattice vector of the two-dimensional hexagonal lattice (the Bravais lattice for the honeycomb) defines a different way of rolling up the sheet into a tubule. Each such lattice vector, $\mathbf{R}$, can be defined in terms of the two primitive lattice vectors $\mathbf{R}_1$ and $\mathbf{R}_2$ and a pair of integer indices $[n_1, n_2]$, such that $\mathbf{R} = n_1\mathbf{R}_1 + n_2\mathbf{R}_2$. The point-group symmetry of the honeycomb lattice will make many of these equivalent, however, so that truly unique tubules are only generated using a one-twelfth irreducible wedge of the Bravais lattice. Within this wedge only a finite number of tubules can be constructed with a circumference below any given value such as that shown by the dashed line in Figure 6.11.

The construction of the tubule from a conformal mapping of the graphite sheet shows that each tubule can have up to three inequivalent (by point group symmetry) helical operations derived from the primitive lattice vectors of the graphite sheet. Thus while *all* tubules have a helical structure, tubules constructed by mapping directions equivalent to $\Theta = 0$ or 30° in Figure 6.11 (which correspond to lattice translation indices of the form $[n,0]$ and $[n,n]$, respectively) to the circumference of the tubule will possess a reflection plane. These high-symmetry tubules will therefore be achiral. For convenience we have labeled these special structures based on the shapes made by the most direct continuous path of bonds around the circumference of the tubule [99]. Specif-

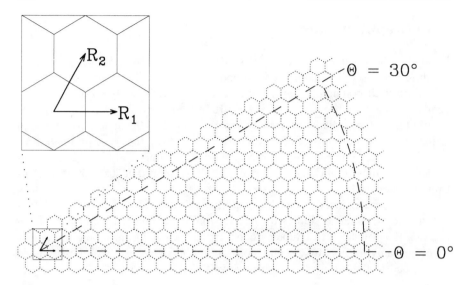

Figure 6.11 Irreducible wedge of the graphite lattice. Primitive lattice vectors R_1 and R_2 are depicted in inset. Θ defines the angle that the circumference vector makes with the primitive lattice vector. Dashed line depicts a 9 Å diameter cutoff for the tubule structures.

ically, the $[n,0]$-type structures were labeled as sawtooth and the $[n,n]$-type structures as serpentine. For other values of Θ, the tubules are chiral and have three inequivalent helical operations. The [5,5] serpentine and [9,0] sawtooth tubules that can be capped with a hemisphere of C_{60} are shown in Figure 6.12.

Because the primitive reciprocal lattice vectors of the hexagonal lattice (the Bravais lattice of the honeycomb lattice) are scalar multiples of real lattice vectors, each tubule generated by the conformal mapping can be shown to be translationally periodic down the tubule axis [82, 98, 99]. However, even for relatively small-diameter tubules, the minimum number of atoms in a translational unit cell can be quite large. For example, if $n_1 = 10$ and $n_2 = 9$, then the radius of the tubule is less than 0.7 nm, but the translational unit cell contains 1084 carbon atoms. The rapid growth in the number of atoms that can occur in the minimum translational unit cell makes recourse to the helical and higher point-group symmetry of these tubules practically mandatory in any comprehensive study of their properties as a function of radius and helicity. These symmetries can be used to reduce to two the number of atoms necessary to generate any tubule [98], for example, reducing the matrices that have to be diagonalized in a calculation of the tubule's electronic structure to a size no larger than that encountered in a corresponding electronic structure calculation of two-dimensional graphite.

The rotational and helical symmetry of a tubule defined by **R** can be seen by using these symmetry operators to generate the tubule [98]. This is done

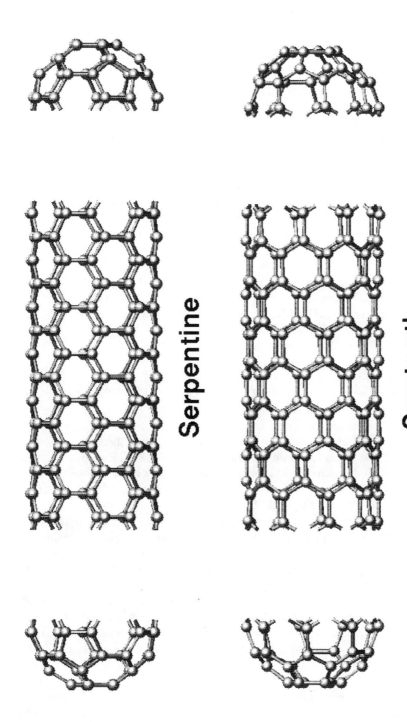

Figure 6.12   The [5,5] serpentine and [9,0] sawtooth fullerene tubules. These tubules can be capped with a hemisphere of Bf as shown.

by first introducing a cylinder of radius $|\mathbf{R}|/2\pi$ with an axis chosen along the $\hat{\mathbf{z}}$ axis in a right-handed coordinate system with Cartesian unit vectors $\hat{\mathbf{x}}$, $\hat{\mathbf{y}}$, $\hat{\mathbf{z}}$. The two carbon atoms located at $\mathbf{d} \equiv (\mathbf{R}_1 + \mathbf{R}_2)/3$ and $2\mathbf{d}$ in the [1,1] primitive unit cell of Figure 6.11 are then mapped to the surface of this cylinder. The first atom is mapped to $(|\mathbf{R}|/2\pi)\hat{\mathbf{x}}$, which requires that the position of the second be found by rotating $(|\mathbf{R}|/2\pi)\hat{\mathbf{x}}$ by $2\pi(\mathbf{d}\cdot\mathbf{R})/|\mathbf{R}|^2$ radians about the $\hat{\mathbf{z}}$ axis in conjunction with a translation $|\mathbf{d} \times \mathbf{R}|/|\mathbf{R}|$ units along this axis. Next, note that the $\hat{\mathbf{z}}$ axis must coincide with a C_N axis for the tubule, where N is the largest common divisor of n_1 and n_2. Thus, the positions of these first atoms can be used to locate $2(N - 1)$ additional atoms on the cylinder surface by $(N - 1)$ successive $2\pi/N$ rotations about the $\hat{\mathbf{z}}$ axis. Altogether, these $2N$ atoms complete the specification of the primitive helical motif, which corresponds to an area on the cylinder surface given by: $A_M = N|\mathbf{R}_1 \times \mathbf{R}_2|$. This primitive helical motif can then be used to tile completely the remainder of the tubule by repeated operation of a single screw operation $\mathcal{S}(h, \alpha)$ representing a translation h units along the $\hat{\mathbf{z}}$ axis in conjunction with a rotation α radians about this axis. To find h and α and hence determine $\mathcal{S}(h, \alpha)$, first note that there must exist a real lattice vector $\mathbf{H} = p_1\mathbf{R}_1 + p_2\mathbf{R}_2$ in the honeycomb lattice such that $h = |\mathbf{H}| \sin \theta$ and $\alpha = (2\pi|\mathbf{H}|/|\mathbf{R}|) \cos \theta$, where θ is the angle between $\mathbf{H}$ and $\mathbf{R}$. In terms of $\mathbf{H}$ and $\mathbf{R}$, the area of the primitive motif on the cylinder surface, A_M, equals $|\mathbf{H} \times \mathbf{R}|$. However, A_M also equals $N|\mathbf{R}_1 \times \mathbf{R}_2|$. Therefore, $|\mathbf{H} \times \mathbf{R}| = N|\mathbf{R}_1 \times \mathbf{R}_2|$ or equivalently,

$$p_2 n_1 - p_1 n_2 = \pm N. \tag{6.4}$$

There are no other constraints on $\mathbf{H}$ and hence on $\mathcal{S}(h,\alpha)$. If a set of integers $[p_1, p_2]$ satisfy Eq. (6.4), then so too will the set $[(p_1 \pm n_1), (p_2 \pm n_2)]$. For uniqueness, we restrict $\mathbf{R}$ to the irreducible wedge of Figure 6.11 $[n_1 \geq n_2 \geq 0]$, take $p_1 \geq 0$, choose the plus sign in Eq. (6.1), and then find the single solution set that yields the minimum value of $|\mathbf{H}|$. These choices restrict $\mathcal{S}(h,\alpha)$ to the right-handed screw operation along the $\hat{\mathbf{z}}$ axis that yields the minimum twist angle α around this axis. Note that h does not depend on $\mathbf{H}$, because $h = |\mathbf{H}| \sin \theta = |\mathbf{H} \times \mathbf{R}|/|\mathbf{R}| = N|\mathbf{R}_1 \times \mathbf{R}_2|/|\mathbf{R}|$.

These results prove by construction that every tubule defined by $\mathbf{R}$ can be generated by first mapping only two atoms onto the surface of a cylinder of radius $|\mathbf{R}|/2\pi$ and then using the rotational and helical symmetry operators to determine the remainder of the tubule. As an example consider the [6,3] tubule shown in Figure 6.13, which is defined by $\mathbf{R} = 6\mathbf{R}_1 + 3\mathbf{R}_2$. Then according to our prescription the first atom of this tubule is mapped to $(3\sqrt{21}/2\pi)|\mathbf{d}|\hat{\mathbf{x}}$ and the position of the second then found by rotating this vector $\pi/7$ radians around $\hat{\mathbf{z}}$ in conjunction with a translation $|\mathbf{d}|/(2\sqrt{7})$ units along this axis. Because N equals 3, $\hat{\mathbf{z}}$ must coincide with a C_3 axis. Thus, the positions of these first two atoms can be used to locate four additional atoms on the cylinder surface by two successive $2\pi/3$ rotations around $\hat{\mathbf{z}}$. Altogether, these 6 atoms complete the specification of the primitive helical motif for this tubule. To determine $\mathcal{S}(h,\alpha)$ used to generate the remainder of the tubule from this motif,

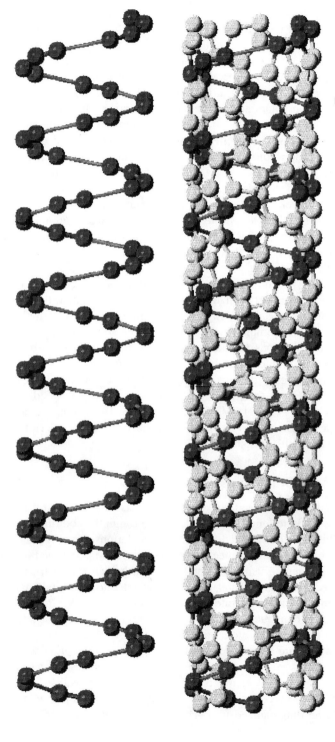

Figure 6.13 Top: the one-third of the [6,3] tubule generated by applying $\mathcal{G}(\alpha, h)$ with $\alpha = 3\pi/7$ and $h = 3|\mathbf{d}|/(2\sqrt{7})$ to only the first two atoms mapped to the cylinder. The gray lines are included as a guide to the eye. Bottom: this same one-third (dark) plus the remaining two-thirds (light) of the [6,3] tubule generated by applying $\mathcal{G}$ to the full six-atom primitive helical motif.

we then solve Eq. (6.4) subject to our constraints to find the solution set [1,1]. Hence, $\mathbf{H} = \mathbf{R}_1 + \mathbf{R}_2$, which in turn implies that $h = 3|\mathbf{d}|/(2\sqrt{7})$ and $\alpha = 3\pi/7$. If this resultant $\mathcal{S}(h,\alpha)$ is applied to only the first two atoms mapped to the cylinder surface, then one-third of the tubule is generated, as illustrated at the top of Figure 6.13. However, if the full helical motif is used, the entire structure is generated, as shown at the bottom of Figure 6.13.

In addition to their rotational and helical symmetries, all tubules defined by $|\mathbf{R}|$ in the hexagonal lattice have a translational symmetry along the tubule axis with a repeat length given by $\sqrt{3}|\mathbf{R}|$. This result can be verified by inscribing a hexagon with a side coinciding with $\mathbf{R}$ in the honeycomb lattice prior to mapping to the cylinder. Hence, $M \equiv \sqrt{3}|\mathbf{R}|/h$ and $T \equiv M\alpha/2\pi$ give the number of motifs and complete (2π radian) helical turns in this translational repeat unit, respectively. These results allow every tubule to be uniquely labeled by $2N * M/T$ and h [98]. It might happen that M and T are not relatively prime, which implies that $\sqrt{3}|\mathbf{R}|$ is not the minimum translational repeat length. Indeed, we have shown [98] that the minimum translational repeat length is given by $\sqrt{3}|\mathbf{R}|/(N\hat{N})$, where $\hat{N} = 1$, unless there exists an integer m such that $(n_1 - n_2)/N = 3m$, in which case $\hat{N} = 3$. Because the symbol "/" indicates that any common factors between M and T should be eliminated, the $2N * M/T$ notation accounts for the possibility that the minimum repeat length (corresponding to a least one complete helical turn) can be shorter than $\sqrt{3}|\mathbf{R}|$, without changing the definitions M and T. In this notation both the [n,0] sawtooth and [n,n] serpentine tubules are labeled as $2n * 2n/1$ helices, but $h = 3|\mathbf{d}|/2$ for the sawtooth tubules while $h = \sqrt{3}|\mathbf{d}|/2$ for the serpentine tubules. As an additional example, both the [6,5] and [9,1] tubules have $h = 3|\mathbf{d}|/(2\sqrt{91})$, but the [6,5] tubules is a 2*182/33 helix while the [9,1] tubule is a 2*182/163 helix. As a last example, the [6,3] tubule shown in Figure 6.13 is labeled as a 6*14/3 helix with $h = 3|\mathbf{d}|/(2\sqrt{7})$.

6.5.2 Tubule Band Gaps

We have carried out a detailed study of the effects of helicity and radius on the band gaps of infinite-length tubules defined through the conformal mapping from the graphite sheet [81, 82, 98]. These calculations (employing both our first-principles LDF and all-valence, tight-binding methods outlined in Section 6.1.1) were implemented using the rotational and helical symmetries of these tubules described previously. Thus, symmetry-adapted local orbitals, which are eigenstates of the rotational operator C_N with eigenvalues $e^{(-2\pi i l/N)}$, $l = 0, \ldots,$ $(N - 1)$, were first constructed for each of the atomic-centered basis functions associated with the first two atoms mapped to the cylinder. Next, generalized Bloch sums, which are eigenstates of the screw operator $\mathcal{S}(h,\alpha)$ with eigenvalues $e^{-i\kappa}$, $-\pi < \kappa \leq \pi$, were constructed from each of these symmetry-adapted orbitals. The matrix elements of the one-electron effective Hamiltonian between these Bloch functions then vanish, unless $l = l'$ and $\kappa = \kappa'$, reducing the matrices to be diagonalized to a size no larger than that encountered in a corresponding band-structure calculation of two-dimensional graphite. Note that if $\alpha = 2\pi m$ with m an integer, then κ corresponds to a normalized quasimomentum; that is, $\kappa = kh$, where k is the traditional one-dimensional wave vector from Bloch's theorem.

Our initial band-gap studies [81], completed prior to Iijima's announcement [80], focused on the electronic structure of the [n,n] serpentine tubules with special emphasis on the [5,5] tubule. This tubule can be capped by a hemisphere of Bf oriented along a fivefold axis as depicted in Figure 6.12. These early studies revealed that all serpentine tubules are metals with two bands of different symmetry crossing at $\kappa_F \approx \pm 2\pi/3$, with the Fermi level, ϵ_F, coinciding with this crossing. These features are illustrated by our first-principles LDF band-structure results for the [5,5] tubule shown in Figure 6.14, where the a_1 and a_2 bands (obtained using a $7s3p$ Gaussian basis set and labeled by the irreducible representation of the C_{5v} point group) cross at $\kappa_F \approx \pm 0.69\pi$. In that work we also showed that the four states of the serpentine tubules at ϵ_F (a_1 and a_2 states at $\pm\kappa_F$ for Fig. 6.14) are closely related to the four inequivalent states at the Fermi level of graphite. These four graphite states correspond to the two inequivalent (by translational symmetry) well-known K points in the first Brillouin zone of two-dimensional graphite where the graphite bonding and antibonding π bands are degenerate. Although our LDF calculations demonstrated that the tubule curvature slightly shifts the crossing at ϵ_F away from the K point of graphite, we found that these bands continue to cross, because they have different symmetries.

We also found that the tubule states in the neighborhood of ϵ_F are analogous to those in the neighborhood of the Fermi level of a single sheet of graphite provided that $\mathbf{k}$ in the Brillouin zone of graphite is further quantized by imposing Born-von Karmann boundary conditions over a width equivalent to the tubule circumference, $|\mathbf{R}|$ [81]; that is, $\mathbf{k} \cdot \mathbf{R} = 2\pi m$, where m is an integer and $\mathbf{k}$ is a reciprocal lattice vector in the first Brillouin zone of the honeycomb lattice. Then combining this result with a mean-field analysis, we demonstrated that the [5,5] and serpentine tubules of larger diameter should be stable against a Peierls distortion at room temperature and far below [81]. In addition, using the density of states at ϵ_F that follows from this analysis, we found that an array of close-packed [5,5] and other serpentine tubules of somewhat larger diameter should have a carrier density comparable to a good metal [81]. Furthermore, the carrier mobility along the axis of these tubules should be comparable to the excellent intraplanar carrier mobility of graphite. Therefore, we found that all nanoscale diameter serpentine tubules are rigid enough to avoid a Peierls distortion at and far below room temperature—which would turn them into semiconductors—while enjoying the high intraplanar carrier mobility of pure graphite, without the low carrier density that makes pure two-dimensional graphite a semimetal [81]. *This remarkable combination of properties established these small-diameter serpentine tubules as truly outstanding synthetic targets for novel, lightweight metals.*

Following Iijima's suggestion that many of the nanoscale graphite tubules

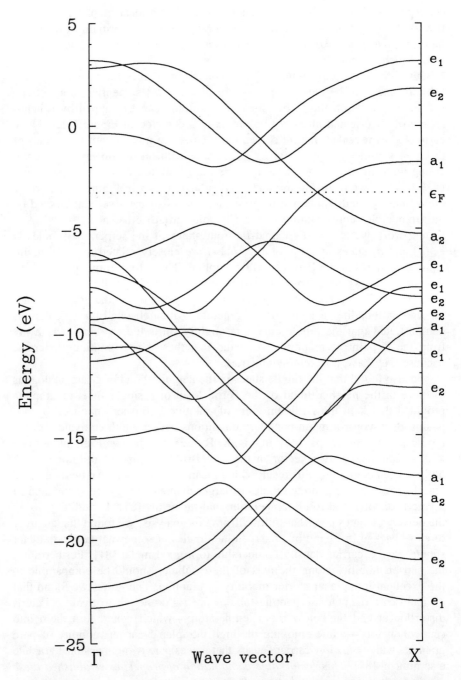

Figure 6.14 Local-density-functional valence-band structure of the [5,5] serpentine fullerene tubule. The Fermi level, ϵ_F, is depicted with a dotted line. Γ and X correspond to the dimensionless ''wave vector'' coordinate κ of 0 and π, respectively.

he observed were chiral, we undertook a comprehensive study of the effects of helicity and radius on the band gaps of infinite length tubules defined through the conformal mapping from the graphite sheet [82, 98]. Our tight-binding results for the band gaps of achiral and chiral tubules with a radius less than 15 Å are summarized in Figure 6.15. These results show that no tubules other than the serpentine ones discussed have a zero band gap at the Fermi level. These results also show that among those tubules with nonzero band gaps those with **R** such that $n_1 - n_2 = 3m$ with m an integer have very small band gaps.

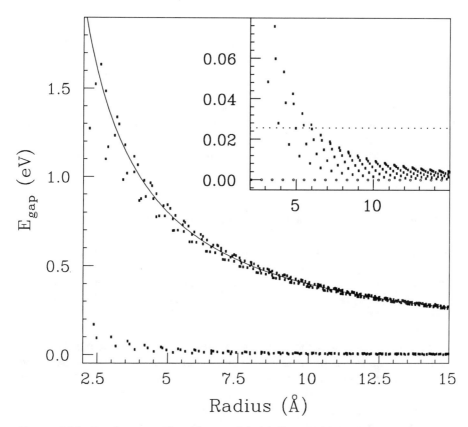

Figure 6.15 Band gaps predicted by our tight-binding model for sawtooth and chiral tubules (i.e., $[n_1,n_2]$ tubules with $n_1 \neq n_2$) up to 15 Å in radius. All tubules with $n_1 \neq n_2$ are predicted to have nonzero band gaps, but tubules with $n_1 - n_2 = 3m$, where m is an integer not equal to zero (solid squares), exhibit much smaller band gaps than similar radii tubules with $n_1 - n_2 \neq 3m$ (dots). Data for the small band-gap sawtooth and chiral tubules ($n_1 - n_2 = 3m$, $m \neq 0$) are presented on an expanded scale in the insert. Also included in the insert are points corresponding to the zero-band-gap serpentine tubules ($n_1 = n_2$). The horizontal dotted line intersecting the insert vertical axis at 0.026 eV corresponds to room temperature. The solid curve in the main portion of the figure is given by $E_g = -V_{pp\pi}|\mathbf{d}|/r_T$ approximately valid when $n_1 - n_2 \neq 3m$.

We also found that within this set of tubules those with larger Θ have smaller band gaps than those with similar radii and a smaller value of Θ. Thus, the achiral sawtooth ($\Theta = 0°$) and the serpentine ($\Theta = 30°$) tubules yield the upper and lower limits for the tubule band gap for a given radius with the serpentine tubules always having zero band gaps. In addition, these tight-binding studies imply that all the narrow-band-gap sawtooth and chiral tubules ($n_1 - n_2 = 3m$, $m \neq 0$) have band gaps less than room temperature if their radii, $r_T \equiv |\mathbf{R}|/(2\pi)$, exceed $r_T \approx 6.0$ Å. Hence these larger-diameter narrow-band-gap sawtooth and chiral tubules can be effectively considered metals.

The occurrence of the subset of small-band-gap achiral sawtooth and chiral tubules is readily understood by generalizing the analysis of the tubule states in the neighborhood of the Fermi level in terms of the band structure of planar graphite as Hamada et al. [83] and Saito et al. [84] have shown. Specifically, the graphite analogy suggests immediately that the tubule could be metallic if $\mathbf{k} \cdot \mathbf{R} = 2\pi m$, with m an integer and $\mathbf{k}$ evaluated at a K point of graphite [84]; for example, $\mathbf{k} = (4\pi/9|\mathbf{d}|^2)(\mathbf{R}_1 - \mathbf{R}_2)$. But this implies at once that $n_1 - n_2 = 3m$, where m is an integer. The graphite analogy, however, predicts that these tubules are true metals [84], while the full calculation shows they are only narrow-band-gap semiconductors.

This discrepancy arises because of the neglect of curvature effects in the graphite model [82–84, 98]. As with the serpentine tubules [81], curvature shifts the attempted point of crossing away from the K point of graphite. However, in contrast to the corresponding bands for the serpentine tubules, these bands have the same symmetry so that they couple, leading to a small gap at ϵ_F. Furthermore, for tubules with moderate band gaps ($n_1 - n_2 \neq 3m$), we have shown [98] by additional analysis of the graphite states in the neighborhood of a K point that $E_g \approx -V_{pp\pi}|\mathbf{d}|/r_T$, where $V_{pp\pi} = -2.77$ eV and $|\mathbf{d}| = 1.44$ Å. The solid line in Figure 6.15 demonstrates that this expression provides an excellent approximation to the calculated band gap as the tubule radius increases. Hence these tubules do not have band gaps approaching room temperature until their radii exceed $r_T \approx 150.0$ Å.

We conclude that only the serpentine tubules ($n_1 = n_2$) are true metals. However, there is a subset of the sawtooth and chiral tubules ($n_1 - n_2 = 3m$, $m \neq 0$) that are narrow-gap semiconductors that effectively become metallic if their radii exceed $r \approx 6.0$ Å. Within the set of small- or zero-band-gap tubules ($n_2 - n_2 = 3m$), the achiral sawtooth ($\Theta = 0°$) and the serpentine ($\Theta = 30°$) tubules yield the upper and lower limits for the tubule band gap for a given radius, with the serpentine tubules always having zero band gaps. The remaining tubules ($n_1 - n_2 \neq 3m$) are semiconductors with a band gap given approximately by $E_g = -V_{pp\pi}|\mathbf{d}|/r_T$. These tubules do not have band gaps approaching room temperature until their radii exceed $r_T \approx 150.0$ Å.

6.5.3 Tubule Energetics

In addition to a study of their band gaps as a function of radius and helicity, we have examined the energetics and elastic properties of all tubules with radii less than 9 Å [99]. These studies were implemented using not only our empirical hydrocarbon potential sketched in Section 6.1.2 (referred to as EP2), but also Tersoff's empirical carbon potential [127] (referred as EP1). Where feasible, the empirical potential results were also checked using our first-principles total-energy LDF band-structure methods. In these studies, we first generated an initial tubule structure with periodic boundary conditions matching the minimum translational periodicity along the tubule axis using the conformal mapping from the graphite sheet described in Section 6.5.1. Once these tubules were generated, we relaxed the constraint of conformal mapping, and minimized the energy with respect to their configuration and periodic boundary along the tube axis for both these empirical potentials.

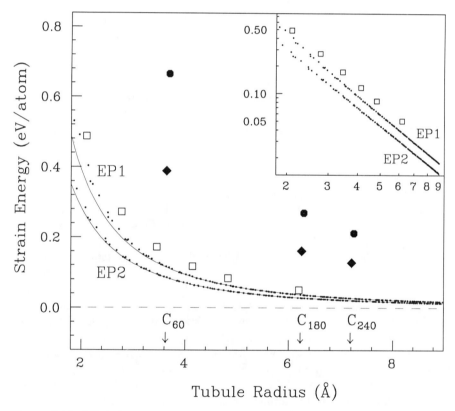

Figure 6.16 Minimized strain energy relative to graphite (eV per carbon atom) as a function of tubule radius for potentials EP1 and EP2. Zero energy corresponds to the equilibrium graphite energies of -7.3995 and -7.3756 eV per atom for EP1 and EP2, respectively. The solid lines are the r_T^{-2} approximation resulting from best linear fit to log-log data given in the inset. Open squares give LDF strain energies for unoptimized serpentine structures relative to extrapolated limit. Isolated symbols give corresponding strain energies per carbon atom using EP1 (circles) and EP2 (diamonds) for fullerene cage structures C_{60}, C_{180}, and C_{240} at the radii indicated.

Figure 6.16 depicts the strain energy per carbon (relative to that of the graphite sheet) for these optimized tubule structures as a function of tubule radius for both empirical potentials. As expected, the strain energy decreases with increasing radius for both potentials, with the energy per carbon approaching the limiting graphite value shown as dashed lines in Figure 6.16. The results using EP1, however, show a larger dependence of the strain energy on tubule radii compared to the results using EP2. Although the results depicted in Figure 6.16 are for tubules with Θ values ranging from 0 to 30 degrees, the strain energy depends sensitively only on the radius and thus is practically independent of the helicity of the tubule.

For comparison, we also calculated total energies for a series of high-symmetry serpentine tubules with $n = 3, \ldots, 7$ and $n = 9$ using our first-principles LDF methods. The results of these total-energy studies are plotted as open circles in Figure 6.16. The open circle corresponding to the [5,5] tubule shown in Figure 6.12 was obtained by direct minimization of the LDF total energy, which yields a minimum energy structure with a radius of 3.47 Å with both types of carbon-carbon bonds having essentially equal lengths of 1.44 Å. The remaining open circles were obtained by using unoptimized tubule structures generated from the conformal mapping from graphite assuming a 1.44 Å carbon-carbon bond length. The LDF strain energies are all slightly larger than the corresponding energies predicted from either of the empirical potentials but show a similar monotonically decreasing trend with increasing radius. This increased strain energy in the LDF results compared to the empirical potential results might arise from the explicit treatment of π-bonding energy in the LDF approach that is not incorporated in the empirical potentials.

The inset of Figure 6.16 presents a log-log plot of the same tubule data presented on a linear scale in the main portion of the figure. A linear regression using the natural logarithms of the data yields a slope of -2.0 ± 0.06 for both empirical potentials and the LDF results, with a high correlation coefficient. Using the results of this fit, we have drawn solid lines in the main portion of Figure 6.16, demonstrating how well the r_T^{-2} behavior fits the results for the empirical potentials. Thus, these results show that the r_T^{-2} dependence in the strain energy predicted from continuum elastic theory [191] persists to very small-radius tubules. Also shown in Figure 6.16 are the energies per atom with respect to graphite for the icosahedral fullerene clusters C_{60}, C_{180}, and C_{240} calculated using the empirical potentials. These clusters represent high symmetric structures that have the strain energy well distributed around the cluster [86]. For both potentials the strain energy associated with these clusters is much larger than the infinite tubules with comparable radii, indicating the formation of fullerenes—including giant fullerenes—rather than tubules during condensation is probably controlled by growth kinetics rather than energetics.

We have also examined the energetics of stretching and compressing a tubule [99]. Figure 6.17 depicts total-energy results versus strain along the tubule axis for the [5,5] serpentine tubule, in which we compare results of fully optimized structures for a fixed repeat length along the tubule axis using both the

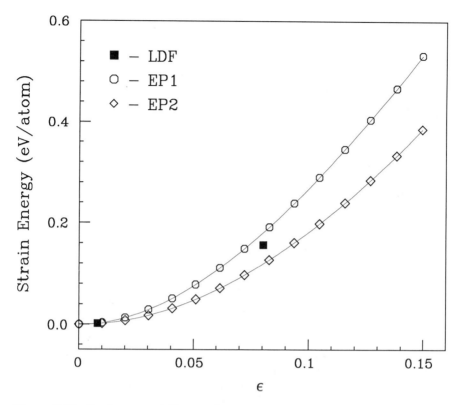

Figure 6.17 Strain energy (eV per carbon atom) versus uniform tensile strain in the tubule axis direction for [5,5] serpentine tubule using empirical potentials EP1 (open circles), EP2 (open diamonds), and the LDF method (solid squares). Solid lines for empirical potentials are used as a guide to the eye.

empirical potentials and the LDF method. We see that the empirical methods are in good agreement with each other, and with the first-principles LDF results.

After this successful check on the reliability of the empirical potentials for calculating this effective elastic modulus of the tubule, we extended our empirical potential calculations on the same set of 169 tubules used previously for strain energies to the numerical second derivatives of the total energy with respect to strain along the tubule axis. These results (in terms of strain energy per carbon) are depicted in Figure 6.18. Again, as the radii increase, these values approach a limiting value. In the limit of infinite radius, we can correlate these results with elastic constants of graphite if we neglect interactions between layers. In this case our results for the second derivative of the total energy per carbon with respect to linear strain should equal the product of the graphite c_{11} elastic constant and the specific volume per carbon, V_0. Using experimentally determined lattice constants $a_0 = 2.462$ Å and $c_0 = 6.707$ Å,

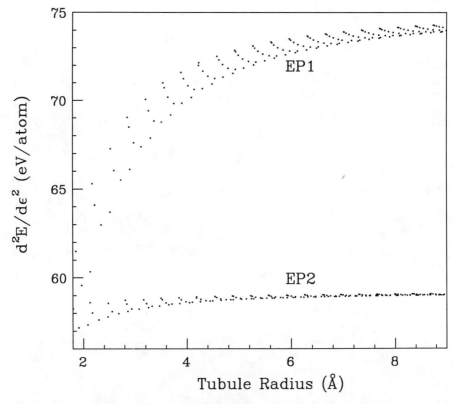

Figure 6.18 Numerical second derivatives of energy per carbon with respect to uniform strain along the tubule axis direction for the potentials EP1 and EP2.

and the elastic constant $c_{11} = 1.06$ TPa [192, 193], we found that $V_0 \approx 8.80$ Å^3 and $V_0 c_{11} \approx 58.2$ eV/atom. This close agreement of EP2 with experiment and excess stiffness using the EP1 potential has been noted in other calculations on graphitic systems [127, 194].

For both potentials, the tubules tend to get softer with smaller radii, with EP1 showing almost an order of magnitude greater dependence of the stiffness of the tubule as a function of radius than EP2. Unlike the energy, however, the stiffness of the tubules is dependent on Θ as well as tubule radius, and this dependence is maximized for the smaller, more strained tubules. We found that tubules with smaller Θ are softer than those with a similar radius and larger value of Θ. Thus the achiral sawtooth ($\Theta = 0°$) and serpentine ($\Theta = 30°$) tubules yield the lower and upper limits of the stiffness along the tubular axis, respectively, for a given radius.

6.6 Summary and Concluding Remarks

In this chapter we have reviewed some of our recent theoretical studies of Bf and its derivatives [35, 81, 82, 85–105]. After outlining our theoretical methods [106–113] in Section 6.1, we began by showing in Section 6.2 that our calculated results for the geometry [35, 85–87], ionization potentials [86, 87], electron affinity [86, 87], and photoelectron line shapes [88] for Bf were all in good agreement with the corresponding experimental data [44–51, 63]. Further, in Section 6.2 we discussed a series of total-energy calculations that showed that carbon was less bound in Bf than in many not-much-larger higher fullerenes [86, 87, 89], suggesting the importance of kinetics in understanding the remarkably high yield of Bf in the Krätschmer-Huffman process. We also showed in Section 6.2 that a suggested structure for $C_{60}H_{36}$ [72] yielded a large electric-dipole-forbidden HOMO-LUMO gap [85, 90]. In addition, in Section 6.2 we discussed results that suggested that if $C_{60}H_{60}$ can be made, then isomers with some of the hydrogens bonded to the interior of the cage were likely to be lower in energy [85].

Next, in Section 6.3 we reviewed MD studies of the Bf during high-energy collisions with diamond surfaces [92, 93] and He target atoms [94, 95]. These MD studies led to results in good agreement with experiment [62–64, 95], while providing further insight into the scattering processes involved. We also discussed in Section 6.3 studies of Bf under anisotropic compression [96], which suggested a first step in the shock compression of fullerites to graphite [60].

Then, in Section 6.4 we discussed MD studies of the dynamic curling and closure of graphitic fragments as an early step to Bf formation [97]. The results of these simulations, while consistent with the importance of pentagons in these early processes [5, 11, 179–183] also suggested the key roles of high temperatures, size, and low-frequency, large-amplitude oscillations in readily surmounting the barrier to initial closure prior to growth beyond the size of C_{60}.

Finally, in Section 6.5 we reviewed extensive studies of graphitic tubules, which have yielded a series of simple relations estimating tubule electronic and structural properties as a function of radius and helicity [81, 82, 98, 99]. In general, we found that the elastic properties of these tubules were what would be expected from extrapolating the properties of graphite systems to nanometer dimensions [99]. The small-radii serpentine tubules were identified as truly outstanding candidates for excellent high-strength, lightweight metals [81].

The discovery of fullerenes [1] and fullerites [38] has opened vast new areas for important research. Even within the very restricted confines of the fullerene-related problems addressed in this chapter, there are many key issues that deserve immediate attention. Why fullerenes first form remains a central question. Further theoretical [183–185] and experimental [183, 188] studies of the chain-to-ring transition thought to occur prior to initial fullerene formation should be undertaken using both model potentials and first-principles methods. In addition, the free energies of other likely fullerene precursors should be calculated and compared. A better understanding of the remarkable process of fullerene condensation could ultimately allow synthesis of specific fullerenes tailored to particular uses. Also, such knowledge could assist in uncovering new synthetic routes based on selected chemical precursors designed to leapfrog Bf to yield higher fullerenes in larger amounts. This synthetic goal may seem unattainable, but experimental evidence from laser vaporization studies of $C_{36}O_6$ ring compounds suggests that specific precursors can alter the resulting fullerene distribution so that C_{70} predominates over C_{60} as a final product [195]. Unfortunately, these laser vaporization techniques do not provide a route to producing macroscopic amounts of either fullerene.

Bf itself remains a fascinating problem partly because there are potentially 1811 other fullerene isomers of C_{60} [143, 144]. Although these C_{60} fullerenes all have abutting pentagons and hence even the most favorable are probably several eV higher in energy than Bf, many should nevertheless persist indefinitely at room temperature in a vacuum. However, with one exception [196], Bf is the only isomer of C_{60} so far observed. Because of the high yield of Bf, it is not sufficient to argue that these other isomers are so reactive that they quickly grow beyond the size of C_{60}. Rather, a facile mechanism probably exists for rearranging pentagons on the surface of C_{60} and the lower fullerenes, which ultimately leads to Bf in the Krätschmer-Huffman process. The possibility should also not be discounted that there exists a very selective condensation process that yields Bf without generating these other isomers [186]— even as intermediate species. However, if such isomerization or condensation mechanisms exist, their details are unknown. The Stone-Wales transformation [145] provides a convenient method for rearranging pentagons on the surface of fullerene models, but whether nature chooses to follow this Woodward-Hoffmann forbidden path requiring the breaking of at least two single carbon-carbon fullerene bonds remains a mystery. A challenge for theory is to devise physically plausible fullerene isomerization and/or direct condensation mechanisms and then test these mechanisms computationally. A challenge for experiment is to generate other C_{60} fullerenes starting from Bf and then study these other C_{60} fullerenes as they return to the energetically favored Bf state.

Another continuing challenge to theory is provided by the higher fullerenes. For example, why are C_{76}, C_{78}, C_{84}, C_{90}, and C_{94} produced in trace amounts in the Krätschmer-Huffman process but not C_{72} [197]? All fullerenes starting at C_{70} have at least one isolated-pentagon isomer [198]. Also, the D_{6h} isolated-pentagon isomer of C_{72} is a closed-shell species with a binding energy per carbon comparable to (D_{5h}) C_{70} [89]. Why then has (D_{6h}) C_{72} not been separated even in trace amounts from the Krätschmer-Huffman material, while (D_{5h}) C_{70} can be produced in high yield [39, 41]? The HOMO-LUMO gap of C_{72} is about 13% less than C_{70} [89], which might account for a higher reactivity. However, contrary to experiment [55, 135], a related test predicts that the isolated-pentagon, closed-shell D_{3h} isomer of C_{78} should be more prevalent than the isolated-pentagon, pseudo-closed-shell, C_{2v} and D_3 isomers of C_{78} [199]. Although all higher fullerenes so far characterized obey the isolated-pentagon rule [135], the number of such preferred structures rapidly increases with increasing fullerene size [197, 199]: C_{76} has 2 isomers with isolated pentagons; C_{78} has 5; C_{84} has 24; C_{90} has 46; and C_{94} has 134 [197]. A convincing resolution of why only certain higher fullerenes are observed will probably go beyond relative energy criteria to provide insight into fullerene growth and isomerization mechanisms.

Lower fullerenes have not been synthesized in gram quantities or even separated in trace amounts, but they too are deserving of more attention. Structural [182] and HMO-based [201] surveys of the relative stability of lower fullerenes have appeared, and a range of experimental data from laser vaporization studies is available [1, 3, 9, 10, 136, 202, 203], but there have been no systematic studies of lower fullerenes using either MNDO, HF, or LDF methods. However, many lower fullerenes contain few enough atoms that systematic studies using these techniques are feasible. For example, C_{44} has only 87 distinct fullerene isomers [105]. The results of a series of such studies could yield additional insight into fullerene growth and production mechanisms. At the very least these studies will allow additional testing of proposed fullerene stability criteria [8, 182, 200, 201]. They also will allow contact to be made with studies of similar-size silicon clusters [104].

All lower fullerenes contain abutting pentagons and hence are rather reactive. A challenge then is to devise techniques to compensate the reactive sites on the surfaces of these cages with foreign chemical species to yield robust new molecules akin to $C_{20}H_{20}$—dodecahedrane, $C_{20}H_{20}$, was synthesized directly in 1976 [204]. "Shrink-wrapping" experiments vividly demonstrate that lower fullerene cages are small enough to interact strongly with single endohedral species [15]. Systematic algorithms exist for generating all possible lower fullerenes [8, 201]. A current challenge for theory is to use state-of-the-art computational methods to determine which of these structures might represent likely candidates for either exohedral or endohedral stabilization using other chemical species. This area seems ripe for a fruitful interplay between theory and experiment, as the work of Guo et al. demonstrates [115].

Just as we were completing this review, Ebbesen and Ajayan [205] announced the first successful synthesis of gram quantities of graphitic nanotubes using a variant of the Krätschmer-Huffman technique. This key advance will undoubtedly stimulate many additional studies aimed at determining the structural, electronic, and mechanical properties of these tubules and tailoring their growth conditions. Many questions immediately come to mind. What are the preferred structures of these concentric tubules? Are they all chiral, or are there some that are serpentine and sawtooth [99]? How can they be capped [206], and can they be connected [100]? What are their electrical properties [81–84]? Are they all semiconductors, or are there some that are excellent metals or even good insulators? For those that are semiconductors, can they be successfully doped? What are the mechanical properties of these tubules? How will they respond under compression and stress? Do they have the high strengths and rigidity that their graphitic and tubular structure implies [99]? Can they be used as precursors to yet more exotic forms of carbon? Will they provide new solid-state lubricants with unique properties? How are these tubules formed? Can techniques be devised that optimize the growth and allow the extraction of macroscopic amounts of selected tubules? These questions all deserve immediate attention.

Although the vast opportunities for important research made possible by the discovery of the fullerenes and fullerites are only beginning to be explored, this work has already led to new chemicals and materials with potentially important technological applications. The growing wealth of experimental data on the pure fullerenes and their chemical derivatives provides a continually expanding challenge to theory. Buckminsterfullerenes transcend the boundaries between applied mathematics, chemistry, solid-state physics, and materials science to the benefit of all. Just as Krätschmer, Huffman, and co-workers' initial identification of Bf in their carbon soot samples benefitted from a fruitful dialogue between theory and experiment, so too should continuing advances in this emerging field of fullerenes and their relatives.

Acknowledgments

This work was supported by the U.S. Office of Naval Rsearch (ONR) through the Naval Research Laboratory (NRL) and directly through the ONR Physics, Chemistry, and Materials Divisions through contract numbers N00014-92-WX-24183 and N00014-92-WX-24007. Computational support for the work described herein was provided in part by grants of computer resources from NRL. D.H.R. acknowledges an NRL-National Research Council Resident Research Associateship. J.A.H. acknowledges an NRL-Office of Naval Technology Resident Research Associateship.

References

1. H. W. Kroto, J. R. Heath, S. C. O'Brien, R. F. Curl, and R. E. Smalley, *Nature (London)* **318**, 162 (1985).

2. J. R. Heath, S. C. O'Brien, Q. Zhang, Y. Liu, R. F. Curl, H. W. Kroto, F. K. Tittel, and R. E. Smalley, *J. Am. Chem. Soc.* **107**, 7779 (1985).

3. E. A. Rohlfing, D. M. Cox, and A. Kaldor, *J. Chem. Phys.* **81**, 3322 (1984).

4. D. W. Thompson, *On Growth and Form*, Cambridge University Press, London, 1963, Chapter IX, p. 737.

5. Q. L. Zang, S. C. O'Brien, J. R. Heath, Y. Liu, R. F. Curl, H. W. Kroto, and R. E. Smalley, *J. Phys. Chem.* **90**, 525 (1986).

6. B. Grumbaum and T. S. Motzkin, *Can. J. Math.* **15**, 744 (1963).

7. P. W. Fowler, *Chem. Phys. Lett.* **131**, 444 (1986).

8. T. G. Schmalz, W. A. Seitz, D. J. Klein, and G. E. Hite, *J. Am. Chem. Soc.* **110**, 1113 (1988).

9. H. W. Kroto, *Nature (London)* **329**, 529 (1987).

10. R. F. Curl, and R. E. Smalley, *Science* **242**, 1017 (1988).

11. H. W. Kroto, *Science* **242**, 1139 (1988).

12. Y. Liu, S. C. O'Brien, Q. Zhang, J. R. Heath, F. K. Tittel, R. F. Curl, H. W. Kroto, and R. E. Smalley, *Chem. Phys. Lett.* **126**, 215 (1986).

13. S. C. O'Brien, J. R. Heath, R. F. Curl, and R. E. Smalley, *J. Chem. Phys.* **88**, 220 (1988).

14. S. H. Yang, C. L. Pettiette, J. Conceicao, O. Cheshnovsky, and R. E. Smalley, *Chem. Phys. Lett.* **139**, 233 (1987).

15. F. D. Weiss, J. L. Elkind, S. C. O'Brien, R. F. Curl, and R. E. Smalley, *J. Am. Chem. Soc.* **110**, 4464 (1988).

16. D. A. Bochvar and E. G. Gal'pern, *Dokl. Alkad. Nauk. SSSR* **209**, 610 (1973).

17. R. A. Davidson, *Theoret. Chim. Acta (Berl.)* **58**, 193 (1981).

18. A. D. J. Haymet, *Chem. Phys. Lett.* **122**, 421 (1985); *J. Am. Chem. Soc.* **108**, 319 (1986).

19. R. C. Haddon, L. E. Brus, and K. Raghavachari, *Chem. Phys. Lett.* **125**, 459 (1986).

20. P. W. Fowler and J. Woolrich, *Chem. Phys. Lett.* **127**, 78 (1986).

21. D. J. Klein, T. G. Schmalz, G. E. Hite, and W. A. Seitz, *J. Am. Chem. Soc.* **108**, 1301 (1986).

22. D. J. Klein, W. A. Seitz, and T. G. Schmalz, *Nature (London)* **323**, 703 (1986).

23. M. D. Newton and R. E. Stanton, *J. Am. Chem. Soc.* **108**, 2469 (1986).

24. M. Ozaki and A. Takahasi, *Chem. Phys. Lett.* **127**, 242 (1986).

25. D. S. Marynick and S. Estreicher, *Chem. Phys. Lett.* **132**, 383 (1986).

26. M. Kataoka and T. Nakajima, *Tetrahedron* **42**, 6437 (1986).

27. I. László and L. Udvardi, *Chem. Phys. Lett.* **136**, 418 (1987).

28. T. Shibuya and Y. Masaaki, *Chem. Phys. Lett.* **137**, 13 (1987).

29. M. L. McKee and W. C. Herndon, *J. Mol. Struc. (Theochem.)* **153**, 75 (1987).

30. R. L. Disch and J. M. Schulman, *Chem. Phys. Lett.* **125**, 465 (1986); M. Schulman, R. L. Disch, M. A. Miller, and R. C. Peck, *Chem. Phys. Lett.* **141**, 45 (1987).

31. H. P. Lüthi and J. Almöf, *Chem. Phys. Lett.* **135**, 357 (1987).

32. S. Larsson, A. Volosov, and A. Rosén, *Chem. Phys. Lett.* **137**, 501 (1987).

33. S. Satpathy, *Chem. Phys. Lett.* **130**, 545 (1986).

34. P. D. Hale, *J. Am. Chem Soc.* **108**, 6087 (1986).

35. B. I. Dunlap, *Int. J. Quantum Chem. Symp.* **22**, 257 (1988).

36. A. Rosén and B. Wästberg, *J. Chem. Phys.* **90**, 2525 (1989).

37. W. Krätschmer, K. Fostiropoulos, and D. R. Huffman, *Chem. Phys. Lett.* **170**, 167 (1990).

38. W. Krätschmer, L. D. Lamb, K. Fostiropoulos, and D. R. Huffman, *Nature (London)* **347**, 354 (1990).

39. R. Taylor, J. P. Hare, A. K. Abdul-Sada, and H. W. Kroto, *J. Chem. Soc. Chem. Commun.* 1423 (1990).

40. R. D. Johnson, G. Meijer, and D. S. Bethune, *J. Am. Chem. Soc.* **112**, 8983 (1990).

41. H. Ajie, M. M. Alvarez, S. J. Anz, R. D. Beck, F. Diederich, K. Fostiropoulos, D. R. Huffman, W. Krätschmer, Y. Rubin, K. E. Schriver, D. Sensharma, and R. L. Whetten, *J. Phys. Chem.* **94**, 8630 (1990).

42. D. S. Bethune, G. Meijer, W. C. Tang, H. J. Rosen, *Chem. Phys. Lett.* **174**, 219 (1990).

43. J. L. Wragg, J. E. Chamberlain, H. W. White, W. Krätschmer, and D. R. Huffman, *Nature (London)* **348**, 623 (1990).

44. D. L. Lichtenberger, M. E. Jatcko, K. W. Nebesny, C. D. Ray, D. R. Huffman, and L. D. Lamb, *Mat. Res. Soc. Symp. Proc.* **206**, 673 (1991).

45. D. L. Lichtenberger, K. W. Nebesny, C. D. Ray, D. R. Huffman, and L. D. Lamb, *Chem. Phys. Lett.* **176**, 203 (1991).

46. J. H. Weaver, J. L. Martins, T. Komeda, Y. Chen, T. R. Ohno, G. H. Kroll, N. Troullier, R. E. Haufler, and R. E. Smalley, *Phys. Rev. Lett.* **66**, 1741 (1991).

47. L.-S. Wang, J. Conceicao, J. Changming, and R. E. Smalley, *Chem. Phys. Lett.* **182**, 5 (1991).

48. J. M. Hawkins, A. Meyer, T. A. Lewis, S. Loren, F. J. Hollander, *Science* **252**, 312 (1991); J. M. Hawkins, T. A. Lewis, S. D. Loren, A. Meyer, J. R. Heath, Y. Shibato, and R. J. Saykully, *J. Org. Chem.* **55**, 6250 (1990).

49. C. S. Yannoni, P. P. Bernier, D. S. Bethune, G. Meijer, and J. R. Salem, *J. Am. Chem. Soc.* **113**, 3190 (1991).

50. K. Hedberg, L. Hedberg, D. S. Bethune, C. A. Brown, H. C. Dorn, R. D. Johnson, and M. de Vries, *Science* **254**, 410 (1991).

51. W. I. F. David, R. M. Ibberson, J. C. Matthewman, K. Prassides, T. J. S. Dennis, J. P. Hare, H. W. Kroto, R. Taylor, and D. R. M. Walton, *Nature (London)* **353**, 147 (1991).

52. M. B. Jost, P. J. Benning, D. M. H. Poirier, J. H. Weaver, L. P. F. Chibante, and R. E. Smalley, *Chem. Phys. Lett.* **184**, 423 (1991).

53. D. R. McKenzie, C. A. Davis, D. J. Cockayne, D. A. Muller, and A. M. Vassallo, *Nature (London)* **355**, 622 (1992).

54. F. Diederich, R. Ettl, Y. Rubin, R. L. Whetten, R. Beck, M. Alvarez, S. Anz, D. Sensharma, F. Wudl, K. C. Khemani, and A. Koch, *Science* **252**, 548 (1991).

55. F. Diederich, R. L. Whetten, C. Thilgen, R. Ettl, I. Chao, and M. M. Alvarez, *Science* **254**, 1768 (1991).

56. R. Ettl, I. Chao, F. Diederich, and R. L. Whetten, *Nature (London)* **353**, 149 (1991).

57. Z. Li, M. Chander, J. C. Patrin, J. H. Weaver, P. F. Chibante, and R. E. Smalley, *Science* **253**, 429 (1991).

58. L. D. Lamb, D. R. Huffman, R. K. Workman, S. Howells, T. Chen, D. Sarid, and R. F. Ziodo, *Science* **255**, 1413 (1992).

59. M. N. Regueiro, P. Monceau, and J-L. Hodeau, *Nature (London)* **355**, 237 (1992).

60. C. S. Yoo and W. J. Nellis, *Science* **254**, 1489 (1991).

61. J. Milliken, T. M. Keller, A. P. Baronavski, S. W. McElvany, J. H. Callahan, and H. H. Nelson, *Chem. Mater.* May/June, 386 (1991).

62. R. D. Beck, P. St. John, M. M. Alvarez, F. Diedrich, and R. L. Whetten, *J. Phys. Chem.* **95**, 8402 (1991).

63. S. W. McElvany, M. M. Ross, and J. H. Callahan, *Acc. Chem. Res.* **25**, 162 (1992).

64. H.-G. Busmann, Th. Lill, and I. V. Hertel, *Chem. Phys. Lett.* **187**, 459 (1991).

65. T. Weiske, D. K. Böhme, J. Hrušák, W. Krätschmer, and H. Schwarz, *Angew. Chem., Int. Ed. Engl.* **30**, 884 (1991).

66. M. M. Ross and J. H. Callahan, *J. Phys. Chem.* **95**, 5720 (1991).

67. K. A. Caldwell, D. E. Giblin, C. S. Hsu, D. Cox, and M. L. Gross, *J. Am. Chem. Soc.* **113**, 8519 (1991).

68. T. Weiske, D. Böhme, and H. Schwarz, *J. Phys. Chem.* **95**, 8451 (1991); T. Weiske, J. Hrušák, D. Böhme, and H. Schwarz, *Helv. Chim. Acta* **75**, 79 (1992); *Chem. Phys. Lett.* **186**, 459 (1991).

69. Z. Wan, J. F. Christian, and S. L. Anderson, *J. Chem. Phys.* **96**, 3344 (1992).

70. R. D. Johnson, M. S. de Vries, J. Salem, D. S. Bethune, and C. S. Yannoni, *Nature (London)* **355**, 239 (1992).

71. Y. Chai, T. Guo, C. Jin, R. E. Hautler, L. P. Chibante, J. Fure, L. Wang, J. M. Alford, and R. E. Smalley, *J. Phys. Chem.* **95**, 7564 (1991).

72. R. E. Haufler, J. Conceicao, L. P. F. Chibante, Y. Chai, N. E. Byrne, S. Flanagan, M. M. Haley, S. C. O'Brien, C. Pan, Z. Xiao, W. E. Billups, M. A. Ciufolini, R. H. Hauge, J. L. Margrave, L. J. Wilson, R. F. Curl, and R. E. Smalley, *J. Phys. Chem.* **94**, 8634 (1990).

73. P. J. Fagan, J. C. Calabrese, and B. Malone, *Science* **252**, 1160 (1991); *J. Am. Chem. Soc.* **113**, 9408 (1991); The Chemical Nature of C_{60} as Revealed by the Synthesis of Metal Complexes. In *Fullerenes*, G. S. Hammond and J. K. Valerie, Eds., American Chemical Society, Washington, DC, 1992, Chap. 12.

74. F. Wudl, *Acc. Chem. Res.* **25**, 157 (1992); F. Wudl, A. Hirsch, K. C. Khemani, T. Suzuke, P.-M. Allemand, A. Koch, H. Echert, G. Srdanov, and H. M. Webb, Survey of Chemical Reactivity of C_{60}, Electrophile and Dieno-polarophile Par Excellence. In *Fullerenes*, G. S. Hammond and J. K. Valerie, Eds., American Chemical Society, Washington, DC 1992, Chap. 11.

75. A. F. Hebard, M. J. Rosseinsky, R. C. Haddon, D. W. Murphy, S. H. Glarum, T. T. M. Palstra, A. P. Ramirez, and A. R. Kortan, *Nature (London)* **350**, 600 (1991).

76. M. J. Rosseinsky, A. P. Ramirez, S. H. Glarum, D. W. Murphy, R. C. Haddon, A. F. Hebard, T. T. M. Palstra, A. R. Kortan, S. M. Zahurak, and A. V. Makhija, *Phys. Rev. Lett.* **66**, 2830 (1991).

77. K. Holczer, O. Klein, S.-M. Huang, R. B. Kaner, K.-J. Fu, R. L. Whetten, and F. Diederich, *Science* **252**, 1154 (1991).

78. S. P. Kelty, C.-C. Chen, and C. M. Lieber, *Nature (London)* **352**, 223 (1991).

79. K. Tanigaki, T. W. Ebbesen, S. Saito, J. Mizuki, J. S. Tsai, Y. Kubo, and S. Kuroshima, *Nature* **352**, 222 (1991).

80. S. Iijima, *Nature (London)* **354**, 56 (1991).

81. J. W. Mintmire, B. I. Dunlap, and C. T. White, *Phys. Rev. Lett.* **68**, 631 (1992).

82. J. W. Mintmire, D. H. Robertson, B. I. Dunlap, R. C. Mowrey, D. W. Brenner, and C. T. White, *Mater. Res. Soc. Sym. Proc.* **247**, 339 (1992).

83. N. Hamada, S. Sawada, and A. Oshiyama, *Phys. Rev. Lett.* **68**, 1579 (1992).

84. R. Saito, M. Fujita, G. Dresselhaus, and M. S. Dresselhaus, *Appl. Phys. Lett.* **60**, 2204 (1992); *Phys. Rev. B* **46**, 1804 (1992).

85. B. I. Dunlap, D. W. Brenner, J. W. Mintmire, R. C. Mowrey, and C. T. White, *J. Phys. Chem.* **95**, 5763 (1991).

86. B. I. Dunlap, D. W. Brenner, J. W. Mintmire, R. C. Mowrey, and C. T. White, *J. Phys. Chem.* **95**, 8737 (1991).

87. B. I. Dunlap, Isomerization and Icosahedral Fullerenes. In *Physics and Chemistry of Finite Systems: From Cluster to Crystals*, NATO ASW Series, P. Jena, S. N. Khanna, and B. K. Rao, Eds., Kluwer Academic Publishers, Dordrecht, 1992, p. 1295.

88. J. W. Mintmire, B. I. Dunlap, D. W. Brenner, R. C. Mowrey, and C. T. White, *Phys. Rev. B* **43**, 14281 (1991).

89. B. I. Dunlap, *Energetics and Fullerene Fractionation, Phys. Rev.* **47**, 4018 (1993).

90. B. I. Dunlap, D. W. Brenner, R. C. Mowrey, J. W. Mintmire, D. H. Robertson, and C. T. White, *Mater. Res. Soc. Proc.* **206**, 687 (1991).

91. B. I. Dunlap, J. W. Mintmire, D. H. Robertson, D. W. Brenner, R. C. Mowrey, and C. T. White, *Mater. Res. Soc. Proc.* **247**, 351 (1992).

92. R. C. Mowrey, D. W. Brenner, B. I. Dunlap, J. W. Mintmire, and C. T. White, *Mater. Res. Soc. Symp. Proc.* **206**, 357 (1991).

93. R. C. Mowrey, D. W. Brenner, B. I. Dunlap, J. W. Mintmire, and C. T. White, *J. Phys. Chem.* **95**, 7138 (1991).

94. R. C. Mowrey, D. W. Brenner, B. I. Dunlap, J. W. Mintmire, and C. T. White. Molecular-Dynamics Simulations of C_{60}/He Collisions. In *Physics and Chemistry of Finite Systems: From Cluster to Crystals*, NATO ASW Series, P. Jena, S. N. Khanna, and B. K. Rao, Eds., Kluwer Academic Publishers, Dordrecht, 1992, p. 1353.

95. R. C. Mowrey, M. M. Ross, and J. H. Callahan, *J. Phys. Chem.* **96**, 4755 (1992).

96. D. W. Brenner, J. A. Harrison, C. T. White, and R. J. Colton, *Thin Solid Films* **206**, 220 (1991).

97. D. H. Robertson, D. W. Brenner, and C. T. White, *J. Phys. Chem.* **96**, 6133 (1992).

98. C. T. White, D. H. Robertson, and J. W. Mintmire, *Helical and Rotational Symmetries of Nanoscale Graphitic Tubules*, *Phys. Rev. B* **47**, 5485 (1993).

99. D. H. Robertson, D. W. Brenner, and J. W. Mintmire, *Phys. Rev. B* **45**, 12592 (1992).

100. B. I. Dunlap, *Phys. Rev. B* **46**, 1933 (1992).

101. P. P. Schmidt, B. I. Dunlap, and C. T. White, *J. Chem. Phys.* **95**, 10537 (1991).

102. J. L. Ballester and B. I. Dunlap, *Phys. Rev. A* **45**, 7985 (1992).

103. C. T. White, M. Cook, B. I. Dunlap, R. C. Mowrey, D. W. Brenner, P. P. Schmidt, and J. W. Mintmire, Virtual Symmetric Charge Transfer Superconducting Pairing Excitations in C_{60}. In *Physics and Chemistry of Finite Systems: From Cluster to Crystals*, NATO ASW Series, P. Jena, S. N. Khanna, and B. K. Rao, Eds., Kluwer Academic Publishers, Dordrecht, 1992, p. 1397.

104. D. W. Brenner, B. I. Dunlap, J. A. Harrison, J. W. Mintmire, R. C. Mowrey, D. H. Robertson, and C. T. White, *Phys. Rev. B* **44**, 3479 (1991).

105. M. Lyons, B. I. Dunlap, D. W. Brenner, D. H. Robertson, R. C. Mowrey, J. W. Mintmire, and C. T. White, Relative Energetics of C_{44} Fullerene Isomers. In *Physics and Chemistry of Finite Systems: From Cluster to Crystals*, NATO ASW Series; P. Jena, S. N. Khanna, and B. K. Rao, Eds., Kluwer Academic Publishers, Dordrecht, 1992, p. 1347.

106. B. I. Dunlap, J. W. D. Connolly, and J. R. Sabin, *J. Chem. Phys.* **71**, 3396 (1979); ibid., **71**, 4993 (1979); J. W. Mintmire and B. I. Dunlap, *Phys. Rev. A* **25**, 88 (1982).

107. B. I. Dunlap *Phys. Rev. A* **42**, 1127 (1990).

108. B. I. Dunlap and N. Rösch, *Adv. Quantum Chem.* **21**, 317 (1990).

109. R. S. Jones, J. W. Mintmire, and B. I. Dunlap, *Int. J. Quantum Chem. Symp.* **22,** 77 (1988).

110. J. W. Mintmire and C. T. White, *Phys. Rev. Lett.* **50,** 101 (1983); *Phys. Rev. B* **28,** 3283 (1983).

111. J. W. Mintmire, Local-Density Functional Electronic Structure of Helical Chain Polymers. In *Density Functional Methods in Chemistry*, J. Labanowski and J. Andzelm, Eds., Springer-Verlag, New York, 1991, p. 125.

112. M. L. Elert, J. W. Mintmire, and C. T. White, *J. Phys. (Paris), Colloq.* **44,** C3-451 (1983); M. L. Elert, C. T. White, and J. W. Mintmire, *Mol. Cryst. Liq. Cryst.* **125,** 329 (1985); C. T. White, unpublished.

113. D. W. Brenner, *Phys. Rev. B* **42,** 9458 (1990).

114. J. Cioslowski, *J. Am. Chem. Soc.* **113,** 4139 (1991); J. Cioslowski and E. D. Fleischmann, *J. Chem. Phys.* **94,** 3730 (1991).

115. T. Guo, M. D. Diener, Y. Chai, M. J. Alford, R. E. Haufler, S. M. McClure, T. Ohno, J. H. Weaver, G. E. Scuseria, and R. E. Smalley, *Science* **257,** 1161 (1992).

116. U. Burkert and N. L. Allinger, *Molecular Mechanics*, American Chemical Society: Washington, 1982.

117. J. C. Slater, *Phys. Rev.* **81,** 385 (1951); R. Gáspar, *Acta Phys. Acad. Sci. Hung.* **3,** 263 (1954); P. Hohenberg and W. Kohn, *Phys. Rev.* **136,** B864 (1964).

118. J. P. Perdew and A. Zunger, *Phys. Rev. B* **23,** 5948 (1981).

119. D. M. Ceperley and B. J. Alder, *Phys. Rev. Lett.* **45,** 566 (1980).

120. J. W. Mintmire, *Inter. J. Quantum Chem. Symp.* **24,** 851 (1990).

121. C. T. White, F. W. Kutzler, and M. Cook, *Phys. Rev. Lett.* **56,** 252 (1986).

122. M. Cook and C. T. White, *Phys. Rev. Lett.* **59,** 1741 (1987); *Phys. Rev. B* **38,** 9674 (1988).

123. J. W. Mintmire, F. W. Kutzler, and C. T. White, *Phys. Rev. B* **36,** 3312 (1987); J. W. Mintmire and C. T. White, *Inter. J. Quantum Chem. Symp.* **17,** 609 (1983).

124. J. W. Mintmire, *Phys. Rev. B* **39,** 13350 (1989).

125. J. C. Slater and G. F. Koster, *Phys. Rev.* **94,** 1498 (1954).

126. G. C. Abell, *Phys. Rev. B* **31,** 6184 (1985).

127. J. Tersoff, *Phys. Rev. Lett.* **56,** 632 (1986); *Phys. Rev. B* **37,** 6991 (1988); ibid. **39,** 5566 (1989).

128. D. Vanderbilt and S. G. Louie, *Phys. Rev. B* **30,** 6118 (1981).

129. M. Page and D. W. Brenner, *J. Am. Chem. Soc.* **113,** 3270 (1991).

130. Y. L. Yang and M. D'Evelyn, *J. Am. Chem. Soc.* **114,** 2796 (1992).

131. S. B. Trickey, F. Müller-Plathe, G. H. F. Diercksen, and J. C. Boettger, *Phys. Rev. B* **45,** 4460 (1992) and references therein.

132. C. W. Gear, *Numerical Initial Value Problems in Ordinary Differential Equations*, Prentice-Hall, Englewood Cliffs, NJ, 1971, p. 148.

133. R. C. Haddon, L. E. Brus, and K. Raghavachari, *Chem. Phys. Lett.* **131,** 165 (1986).

134. P. W. Fowler, J. E. Cremona, and J. L. Steer, *Theor. Chim. Acta.* **73,** 1 (1988).

135. F. Diederich and R. L. Whetten, *Acc. Chem. Res.* **25,** 119 (1992).

136. D. A. Cox, D. J. Trevor, K. C. Reichmann, and A. Kaldor, *J. Am. Phys. Soc.* **108,** 2459 (1986).

137. R. E. Stanton and M. D. Newton, *J. Phys. Chem.* **92,** 2141 (1988).

138. Z. C. Wu, D. A. Jelski, and T. F. George, *Chem. Phys. Lett.* **137,** 291 (1987).

139. S. J. Cyvin, B. N. Brendsdal, B. N. Cyvin, and J. Brunvoll, *Chem. Phys. Lett.* **143,** 377 (1988).

140. D. E. Weeks and W. G. Harter, *Chem. Phys. Lett.* **144,** 366 (1988).

141. G. E. Scuseria, *Chem. Phys. Lett.* **176,** 423 (1991).

142. M. Häser, J. Almlöf, and G. E.. Scuseria, *Chem. Phys. Lett.* **181,** 497 (1991).

143. D. E. Manolopoulos, *Chem. Phys. Lett.* **192,** 330 (1992).

144. X. Liu, T. G. Schmalz, and D. J. Klein, *Chem. Phys. Lett.* **192,** 331 (1992).

145. A. J. Stone and D. J. Wales, *Chem. Phys. Lett.* **128,** 501 (1986).

146. K. Raghavachari and C. M. Rohlfing, *J. Phys. Chem.* **96,** 2463 (1992).

147. K. Raghavachari and C. M. Rohlfing, *J. Phys. Chem.* **95,** 5768 (1991).

148. D. Bakowies and W. Thiel, *J. Am. Chem. Soc.* **113,** 3704 (1991).

149. G. E. Scuseria, *Chem. Phys. Lett.* **180,** 451 (1991).

150. J. Baker, P. W. Fowler, P. Lazzeretti, M. Malagoli and R. Zanasi, *Chem. Phys. Lett.* **184,** 182 (1991).

151. S. Saito and A. Oshiyama, *Phys. Rev. B* **44,** 11532 (1991).

152. W. Andreoni, F. Gygi, and M. Parrinello, *Chem. Phys. Lett.* **189,** 241 (1992).

153. K. Raghavachari, *Chem. Phys. Lett.* **190,** 397 (1992).

154. X.-Q. Wang, C. Z. Wang, B. L. Zhang, and K. M. Ho, *Phys. Rev. Lett.* **69,** 69 (1992).

155. D. Bakowies and W. Thiel, *Chem. Phys. Lett.* **193,** 236 (1992).

156. J. Cioslowski, *Chem. Phys. Lett.* **181,** 68 (1991).

157. S. Saito and A. Oshiyama, *Phys. Rev. Lett.* **66,** 2637 (1991).

158. J. L. Martins, N. Troullier, and J. H. Weaver, *Chem. Phys. Lett.* **180,** 457 (1991).

159. Q.-M. Zang, J.-Y. Yi, and J. Bernholc, *Phys. Rev. Lett.* **66,** 2633 (1991).

160. S. C. Erwin and M. R. Pederson, *Phys. Rev. Lett.* **67,** 1610 (1991).

161. S. C. Erwin and W. E. Pickett, *Science* **254,** 842 (1991).

162. *The Almost (but never quite) Complete Buckminsterfullerene Bibliography.* Available via electronic mail at the internet address bucky@sol1.1rsm.upenn.edu.

163. B. P. Fueston, W. Andreoni, M. Parrinello, and E. Clementi, *Phys. Rev B* **44,** 4056 (1991).

164. J. C. Slater, Adv. Quan. Chem. **6,** 1, (1972).

165. U. Gelius, Molecular Orbitals and Line Intensities in ESCA Spectra. In *Proceedings of the International Conference on Electron Spectroscopy,* D. A. Shirley, ed., North-Holland, Amsterdam, 1972, p. 311.

166. M. Saunders, *Science* **253**, 330 (1991).

167. I. V. Stankevich, M. V. Nikerov, and D. A. Bochvar, *Russian Chemical Reviews* **53**, 640 (1984).

168. H. R. Karfunkel and T. Dressler, *J. Am. Chem. Soc.* **114**, 2285 (1992).

169. A. Y. Liu, M. L. Cohen, K. C. Hass, and M. A. Tamor, *Phys. Rev. B* **43**, 6742 (1991).

170. A. Y. Liu and M. L. Cohen, *Phys. Rev. B* **45**, 4579 (1992).

171. T. Lenosky, X. Gonze, M. Teter, and V. Elser, *Nature (London)* **355**, 333 (1992).

172. D. Vanderbilt and J. Tersoff, *Phys. Rev. Lett.* **68**, 511 (1992).

173. M. O'Keeffe, G. B. Adams, and O. F. Sankey, *Phys. Rev. Lett.* **68**, 2325 (1992).

174. H. J. C. Berendsen, J. P. M. Postma, W. F. van Gunsteren, A. DiNola, and J. R. Haak, *J. Chem. Phys.* **81**, 3684 (1984).

175. R. D. Beck, P. St. John, M. L. Homer, and R. L. Whetten, *Science* **253**, 879 (1991).

176. G. Z. Moliere, *Z. Natürforsch.* **2A**, 133 (1947).

177. D. J. O'Connor and R. J. MacDonald, *Radiat. Effects* **34**, 247 (1977).

178. J. Hrušák, D. K. Böhme, T. Weiske and H. Schwarz, *Chem. Phys. Lett.* **193**, 97 (1992).

179. R. E. Smalley, *Acc. Chem. Res.* **25**, 98 (1992).

180. R. F. Curl and R. E. Smalley, *Sci. Am.* **265**, 54 (1991).

181. R. E. Haufler, Y. Chai, L. P. F. Chibante, J. Conceicao, C. Jin, L.-S. Wang, S. Maruyama, and R. E. Smalley, *Mater. Res. Soc. Symp.* **206**, 627 (1991).

182. H. W. Kroto and K. McKay, *Nature (London)* **331**, 328 (1988).

183. J. R. Heath, Synthesis of C_{60} from Small Carbon Clusters. In *Fullerenes*, G. S. Hammond and J. K. Valerie, Eds., American Chemical Society: Washington, DC, 1992, Chap. 1.

184. J. R. Chelikowsky, *Phys. Rev. Lett.* **67**, 2970 (1991); ibid. **69**, 388 (1992).

185. C. Z. Wang, C. H. Xu, C. T. Chan, and K. M. Ho, *J. Phys. Chem.* **96**, 3563 (1992).

186. T. Wakabayashi and Y. Achiba, *Chem. Phys. Lett.* **190**, 465 (1992).

187. I. J. Ford and C. F. Clement, *Phys. Rev. Lett.* **69**, 387 (1992).

188. G. von Helden, M.-T. Hsu, P. R. Kemper, and M. T. Bowers, *J. Chem. Phys.* **95**, 3835 (1991).

189. See, for example, D. A. McQuarrie, *Statistical Mechanics*, Harper & Row, NY, 1975, pp. 358–73.

190. M. S. Dresselhaus, G. Dresselhaus, and R. Saito, *Phys. Rev. B*, **45**, 6234 (1992).

191. G. G. Tibbetts, *J. Cryst. Growth* **66**, 632 (1983).

192. M. S. Dresselhaus, G. Dresselhaus, K. Sugihara, I. L. Spain, and H. A. Goldberg, *Graphite Fibers and Filaments*, Springer-Verlag, Berlin, 1998.

193. B. T. Kelly, *Physics of Graphite*, Applied Sciences, London, 1988.

194. J. A. Harrison, R. J. Colton, C. T. White, and D. W. Brenner, *Mater. Res. Soc. Symp. Proc.* **239**, 573 (1992).

195. Y. Rubin, M. Kahr, C. B. Knobler, F. Diederich, and C. L. Wilkins, *J. Am. Chem. Soc.* **113**, 495 (1991).

196. J. F. Anacleto, H. Perreault, R. K. Boyd, S. Pleasance, M. A. Quilliam, P. G. Sim, J. B. Howard, Y. Makarovsky, and A. L. Laflure, *Rapid Commun. Mass Spec.* **6,** 214 (1992).

197. X. Liu, T. G. Schmalz, and D. J. Klein, *Chem. Phys. Lett.* **188,** 550 (1992).

198. D. J. Klein and X. Liu, as cited in Ref. [197].

199. P. W. Fowler, R. C. Batten, and D. E. Manolopoulos, *J. Chem. Soc. Faraday Trans.* **87,** 3103 (1991).

200. P. W. Fowler and D. E. Manolopoulos, *Nature (London)* **355,** 428 (1992).

201. D. E. Manolopoulos, J. C. May, and S. E. Down, *Chem. Phys. Lett.* **181,** 105 (1991).

202. Y. A. Yang, P. Xia, A. L. Junkin, and L. A. Bloomfield, *Phys. Rev. Lett.* **66,** 1205 (1991).

203. S. W. McElvany, H. H. Nelson, A. P. Baronavski, C. H. Watson, and J. R. Eyler, *Chem. Phys. Lett.* **134,** 214 (1987).

204. J. J. Pireaux, J. Riga, E. Thibaut, C. Tenret-Noel, R. Caudano, and J. J. Verbist, *Chem. Phys.* **22,** 113 (1977).

205. T. W. Ebbensen and P. M. Ajayan, *Nature (London)* **358,** 220 (1992).

206. M. Fujita, R. Saito, G. Dresselhaus, and M. S. Dresselhaus, *Phys. Rev. B* **45,** 13834 (1992).

7

Electronic Structure of the Fullerenes: Carbon Allotropes of Intermediate Hybridization

R. C. Haddon and K. Raghavachari

In order to close a sheet of carbon atoms into a fullerene, a finite curvature must be imposed. In addition to the benzenoid topology of graphite [six-membered rings (6MRs)], spheroid formation requires the presence of 12 5MRs. These simple principles have served to define a new class of carbon allotropes [1, 2]. The topological aspects of the fullerenes—placement of the 5MRs on the surface of the spheroid—is qualitatively accounted for in the simple HMO approach. In the present article we discuss the other feature of the fullerenes that distinguishes them from the familiar benzenoid hydrocarbons and graphite—*nonplanarity*. The geometrical aspects of fullerene formation as it relates to pyramidalization of the constituent carbon atoms has been recognized for some time [3–5]. Here we shall be concerned with the effect of nonplanarity on the electronic structure of the carbon atoms as it arises in the fullerenes [3, 4].

The primary response of a conjugated carbon atom to a deviation from planarity is a rehybridization of the sp^2 σ and p π orbitals, which are the rule in planar situations [6]. Our interest in the electronic structure of nonplanar conjugated organic molecules arose in connection with the stability of twisted π-electron systems, particularly the bridged annulenes, which show large skeletal dihedral angles together with compelling evidence for aromatic character [7, 8]. When examined with the π-orbital axis vector (POAV) analysis, it was found that these systems are able to maintain favorable π-orbital overlap by pyramidalization and rehybridization of the conjugated carbon atoms [9, 10]. The POAV analysis extends σ-π separability into three dimensions by use of the orbital orthogonality relationships that are the basis of standard hybridization theory [6]. In POAV2 theory, a π orbital is defined as that hybrid orbital that is locally orthogonal to the σ orbitals, whereas from the standpoint of the POAV1 analysis the π orbital is that hybrid orbital that makes equal angles with the σ orbitals (Fig. 7.1) [6].

The inhibition of conjugation as a result of π-orbital misalignment was modeled by studying the C_2 twisting of ethylene, and it was shown that the POAV and 3D-HMO analyses provide a good account of the electronic structure of nonplanar conjugated carbon atoms [9, 10]. In the present study the electronic structure of fullerene carbon atoms is modeled by the C_{2v} bending of ethylene, as this is the primary mode of structural distortion of the conjugated system that is necessary for spheroid formation. It should be noted that the pyramidalization and rehybridization in the fullerenes is driven by the geometrical constraints of the σ system, whereas in the torsionally distorted conjugated systems, pyramidalization and rehybridization occur so that the π system can maintain favorable π orbital alignment [7, 8]. As a result of their unique geometry the fullerenes maintain a high degree of π orbital alignment [4], and in the C_{2v} bending of ethylene considered in this study the alignment is perfect; thus the C_2 and C_{2v} distortions taken together model the two important disruptions to π bonding that occur in nonplanar conjugated organic molecules.

As discussed elsewhere, the use of ethylene to model extended π systems has limitations [8, 10, 11]. Apart from the introduction of C-H bonds in place of C-C σ bonds, the conjugation at the carbon atoms is *anisotropic* [11], as there is only one π bond to each carbon atom. The fullerenes have isotropic conjugation with π bonds in all three σ-bond directions. In this respect they are ideally suited for analysis with the POAV method, which equally weights each of the three σ bonds to a conjugated carbon in solving for the hybridization and orientation of the π orbital. For the same reason, ethylene provides the most difficult case for the POAV analysis [8, 10, 11]. The results obtained on ethylene therefore provide an upper bound on the errors that may be expected when the method is applied to the fullerenes.

The curvature imposed on the surface modifies the electronic structure of the sheet in two crucial ways, one of which relates to the electronic structure of the carbon atoms and the other to the bonding in the cluster [3, 4, 12]. The maximum curvature that can be accommodated at a conjugated carbon atom defines the lower limit for fullerene stability and is therefore of some interest.

The curvature of the surface at a carbon atom is most simply expressed by the pyramidalization angle $[(\theta_{\sigma\pi} - 90)°]$ [3, 4, 6, 7] shown in Figure 7.1. As the σ bonds at a conjugated carbon atom deviate from planarity, the primary effect is a change in hybridization—a rehybridization of the carbon atom so that the π orbital is no longer of purely p-orbital character and the σ orbitals no longer contain all of the s-orbital character. Thus, the fullerenes are of intermediate hybridization [3, 4, 12]. Using the standard nomenclature, the σ-bond hybridization falls between graphite (sp^2) and diamond (sp^3). The average POAV1 σ-bond hybridization for C_{60} is $sp^{2.278}$, and the π-orbital fractional s character is 0.085 (POAV1) and 0.081 (POAV2) [3, 4, 10]. An approximate

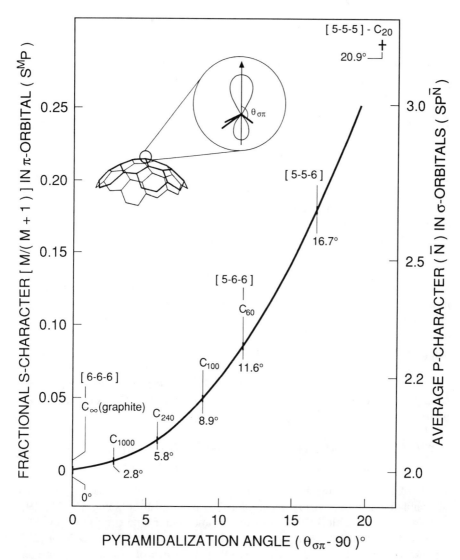

Figure 7.1 Rehybridization as a function of pyramidalization angle. The π-orbital axis vector (POAV1 approximation) is defined as that vector that makes equal angles to the three σ bonds at a conjugated carbon atom [6–8]. The common angle to the three σ bonds (which are assumed to <u>lie</u> along the internuclear axes) is denoted $\theta_{\sigma\pi}$. The average pyramidalization angle [$(\theta_{\sigma\pi} - 90)°$] shown for representative fullerenes (C_n) was obtained from Eq. (2) of Ref. [4] for $n > 60$. The fractional hybridization, [$m/(m + 1)$] = 2 $\cot^2 \theta_{\sigma\pi}$ (see Refs. [6–8]).

POAV1 treatment of the rehybridization required for closure of carbon spheroids of arbitrary size [4] is shown in Figure 7.1. A recent analysis of ^{13}C NMR coupling constants in a derivative of C_{60} led to an estimate of 0.03 for the fractional rehybridization in neutral C_{60} [13, 14]. Furthermore, other authors [15, 16] have adopted the POAV hybridization theory in analyzing C_{60}, so it seemed useful to subject the theory to a rigorous test along the lines of our previous work on twisted π systems.

7.1   Calculations

7.1.1   Generalized Valence Bond

GVB [17] calculations were carried out with the STO-3G basis set using a version of the GAUSSIAN 92 program [18] on a Stardent computer. The calculations were carried out in the perfect-pairing approximation, and the six electron pairs of highest energy were split [GVB(6/12)], so that fission of the four C-H and the σ C-C bonds were treated on an equal basis with the weakening of the π bond [9, 10]. The GVB wave function correctly treats bond-breaking processes, and the resulting orbitals may be clearly identified as σ and π bonds [9, 10]. In previous studies of torsionally (C_2) twisted ethylene, it was found that GVB/STO-3G calcuations of the geometries and energies provided a good account of the potential surface [9, 10]. The ethylene geometries were fully optimized at the GVB/STO-3G level within C_{2v} symmetry using the POAV1 pyramidalization angle $[(\theta_{\sigma\pi} - 90)°]$ [6] as the deformation coordinate.

The GVB wave functions were truncated at the carbon atoms, and the resulting orbital coefficients are plotted as the valence-state fractional hybridization $[m/(m + 1)]$, of the π orbital ($s^m p$) [10, 11].

7.1.2 π-Orbital Axis Vector Analysis and Three-Dimensional Hückle Molecular-Orbital Theory

The POAV and 3D-HMO analyses have been tested and applied to a number of nonplanar conjugated organic molecules and transition states [6, 11, 19], and are now available in the form of PC and mainframe programs (POAV3 Release 1.0) from the Quantum Chemistry Program Exchange (QCPE) [20]. Release 2.0, which incorporates most other measures of nonplanarity in conjugated organic molecules [11], is currently in beta testing and may be obtained from the author. The atomic coordinates of the molecule or molecular fragment of interest are the only data required for the execution of the POAV3 program.

7.2   Results

The structures and energies for C_{2v} distorted ethylene are given in Table 7.1 as a function of the POAV1 pyramidalization angle (as per the structure in the

Table 7.1 GVB/STO-3G calculations on C_{2v} distorted ethylene[a]

$\theta_{\sigma\pi}$ (deg)	Bond lengths (Å)		Bond angles (deg)		Energies		GVB π overlap
	C = C (Å)	C-H (Å)	C-C-H (deg)	H-C-H (deg)	Total (hartree)	Relative (kcal/mol)	
90.0	1.347	1.100	121.9	116.1	−77.18265	0.0	0.5843
92.5	1.348	1.100	121.7	116.0	−77.18147	0.7	0.5814
95.0	1.352	1.101	121.1	115.5	−77.17791	3.0	0.5726
97.5	1.358	1.101	120.1	114.8	−77.17194	6.7	0.5581
100.0	1.366	1.101	118.7	113.8	−77.16359	12.0	0.5387
102.5	1.378	1.102	116.9	112.5	−77.15286	18.7	0.5139
105.0	1.392	1.102	114.8	111.1	−77.13982	26.9	0.4854
107.5	1.409	1.103	112.4	109.3	−77.12448	36.5	0.4525
109.5	1.425	1.104	110.4	107.7	−77.11074	45.1	0.4241

[a]See Figures 7.2 and 7.3 for structural definition.

figures), for $0° < (\theta_{\sigma\pi} - 90)° < 19.5°$, which spans the range from planar to tetrahedral carbon.

Figure 7.2 shows the fractional s character in the GVB/STO-3G wave function as a function of $(\theta_{\sigma\pi} - 90)°$, compared with the values obtained from applying the POAV analysis to the same GVB-optimized geometries. The POAV methods apparently underestimate the rehybridization at small deformation angles, but provide a fairly good account of the carbon wave function for $(\theta_{\sigma\pi} - 90)° < 15°$, which encompasses all of the pyramidalization angles yet reported for free fullerenes and nonplanar conjugated organic molecules [11]. Beyond this range of pyramidalization, the POAV theory overestimates the s character of the π orbital.

The orientations of the π-orbital axis vectors from the GVB wave functions and POAV analysis at the GVB equilibrium geometries are collected in Figure 7.3 in the form of the angles made by the π-orbital axis vectors with the C-C internuclear axis. Particularly in the case of the POAV2 analysis, the agreement with the GVB values is excellent over the whole range of pyramidalization angles.

7.3 Discussion

7.3.1 Rehybridization

The C_{2v}-type distortion of ethylene and its derivatives has been considered by a number of authors, both experimentally and theoretically, and much of this work has recently been reviewed by Borden [21]. There is strong evidence for the effects of rehybridization from both standpoints. A recent analysis of ^{13}C

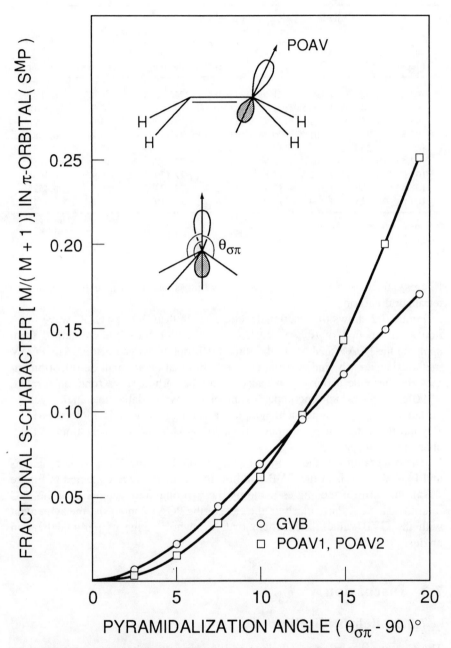

Figure 7.2 GVB/STO-3G and POAV fractional s character $[m/(m + 1)]$ in the π orbital $(s^m p)$ of ethylene as a function of the pyramidalization angle at the GVB/STO-3G equilibrium geometries (C_{2v} symmetry).

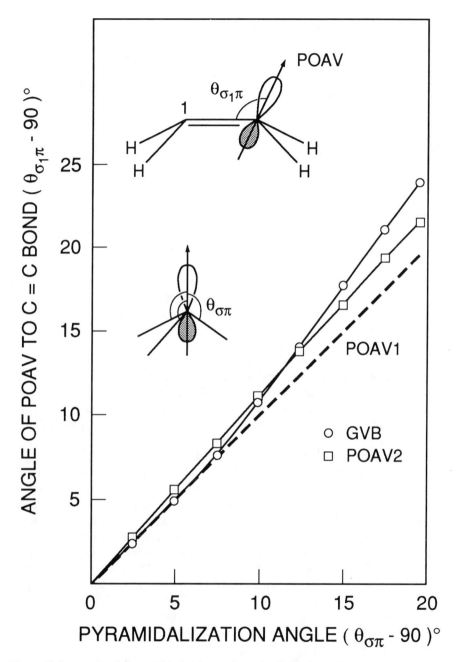

Figure 7.3 Angle of the π-orbital axis vector to the C-C internuclear axis in ethylene, as calculated by GVB and POAV theory as a function of the pyramidalization angle at the GVB/STO-3G equilibrium geometries (C_{2v} symmetry).

NMR coupling constants in a derivative of C_{60} led to an estimate of 0.03 for the fractional hybridization in neutral C_{60} [13, 14]. Reference to Figures 7.1 and 7.2 shows this to be much smaller than the values calculated by the POAV and GVB treatments. The π-orbital hybridization ($s^m p$) and fractional hybridization calculated for C_{60} are: $m = 0.09275$, $m/(m + 1) = 0.0849$ (POAV1) and $m = 0.08773$, $m/(m + 1) = 0.0807$ (POAV2) [3, 4, 10]. At the C_{60} pyramidalization angle of $(\theta_{\sigma\pi} - 90°) = 101.6°$, the POAV1, POAV2, and GVB treatments of the GVB-optimized structure of ethylene all give $m/(m + 1) \sim 0.086$. A POAV analysis has been carried through for the specific case of C_{60} by Troullier and Martins [16], who obtain a fractional hybridization of $[m/(m + 1)] = 0.284^2 = 0.0807$. These authors note that this carbon atom hybridization agrees qualitatively with their self-consistent local-density calculations of the electronic structure of C_{60}.

The GVB-calculated s character is higher than the POAV value for small values of the pyramidalization angle (Fig. 7.2); this is to be expected because there is little energetic cost to mixing in small amounts of s character. The same behavior was found in the analysis of torsionally twisted ethylene [9]. From the standpoint of extended Hückel theory, this defect in hybridization theory may be attributed to the neglect of valence-state ionization energies in obtaining the orbital compositions [22]. In solving for the carbon hybridizations, it is assumed that the $2s$ and three $2p$ orbitals will be equally represented in the ground-state wave function of the carbon atom. Given the fact that the carbon $2s$ atomic orbital lies so much lower in energy than the $2p$ orbitals, the standard hybridization theory underestimates the contribution of the $2s$ orbital. For example, Goddard and co-workers [23] have analyzed the GVB/STO-4G wave function for methane and found $sp^{2.1}$ hybridization (fractional s character: 0.321), whereas the standard hybridization for methane is sp^3 (fractional s character: 0.25). However, the s character in the σ-bond carbon hybrids in ethylene was slightly *lower* than expected from standard hybridization theory [23].

The high electron affinity of C_{60} [3, 24–33] supports the idea of s character in the π molecular orbitals, but it could be argued that in the neutral molecule this is manifested as σ strain, which is released on reduction [27, 32] (σ-bond bending) [8, 10]. In this way the s character would be progressively mixed into the π molecular orbitals as the molecule accepts electrons but would not be present in neutral C_{60} itself, and this would account for the small s character found in the NMR analysis of a C_{60} derivative [13, 14]. No doubt this occurs to some extent as the pyramidalization of carbanions is well known. This picture requires that the fullerenes are not only of intermediate hybridization, but also of *variable* hybridization. Variation in the carbon hybridization with electron addition would also relieve the electron repulsion due to the accumulation of electrons on the ball by extending the hybrids away from the surface. Car-Parrinello calculations on the $A_6 C_{60}$ phases [12] indicate that most of the electron density added to the C_{60} cluster ends up on the exterior of the molecule [34]. The small *decrease* in electron density discernible in the calculations [34] just beyond the surface of the molecule during the transformation $C_{60} \rightarrow C_{60}^{6-}$

may reflect the reorientation of the σ bonds accompanying rehybridization in the negative ion. Such an effect could lead to an increase in bandwidth due to improved overlap between the C_{60} molecules in alkali-metal-doped C_{60} phases [12], as opposed to neutral solid C_{60}.

It is likely that rehybridization in the fullerenes affects properties beyond the electron affinity [3, 5]. The larger than expected bulk modulus in C_{60} has been attributed to a greater electron density outside the molecule than inside as a result of rehybridization [35]. Rehybridization in the metallic A_3C_{60} compounds may affect the relaxation rate and Knight shift of the carbon atoms [36]. Current theories of superconductivity [37–40] in the A_3C_{60} phases [12] suggest that the enhancement in the superconducting transition temperature over that observed in the graphite intercalation compounds results from the curvature and rehybridization present in the C_{60} molecule.

7.3.2 Orientation of the π-Orbital Axis Vector

The orientation of the π-orbital axis vectors is well described by the POAV method (Fig. 7.3). The π-orbital axis vector makes equal angles of 11.64° with the three σ bonds in C_{60} in the POAV1 approximation, whereas in the POAV2 analysis the bond angles are 99.5°, 99.5°, and 105.5°. Troullier and Martins [16] obtain 105.45° for this latter angle.

7.3.3 The Limiting Pyramidalization Angle in the Fullerenes

Figure 7.1 provides an idealized view of the pyramidalization and rehybridization required for fullerene formation. It was constructed by defining the surface of the general fullerene, C_n, in terms of the area of $n/2$ unit cells of graphite sheet [4]. This leads to an ideal average behavior for the pyramidalization and rehybridization of the constiturent carbon atoms of the fullerenes. In fact there is bound to be a distribution of geometries and electronic structure of the carbon atoms in all fullerenes except C_{60} (and C_{20}), as a result of the presence of the 12 5MRs. Only C_{60} has all carbon atoms in an identical environment, and it is in this sense that C_{60} is *special*. For example, in calculations on C_{70}, Raghavachari and Rohlfing [41] have found pyramidalization angles ranging from 8.9° to 12.0° in the different regions of the molecule. Pyramidalization in the fullerenes is concentrated at the carbon atoms that are incorporated in a 5MR, and it is therefore the presence of the 12 5MRs and their distributions over the surface of the fullerene that are responsible for deviations from Figure 7.1.

We first consider the high-symmetry cases where the carbon atoms are all equivalent. In C_{60}, each carbon has bond angles of 108° (5MR), 120° (6MR), and 120° (6MR); this situation is denoted [5-6-6] in Figure 7.1. The three other possibilities are also shown in the figure. The [6-6-6] arrangement is well known in planar conjugated organic molecules, particularly the benzenoid hydrocarbons and graphite, where all the bond angles to the carbon atoms are 120°. The [5-5-5] structures correspond to C_{20}, which also has all carbon atoms equivalent, but seems to be too strained to exist.

In discussing the limits for fullerene pyramidalization, it is important to note that there are likely to be three different sets of experimental limits: (1) *Free fullerenes*—isolable fullerenes, (2) *transient fullerenes*—observable fullerenes, for example, in the gas phase or in matrix isolation, and (3) *complexed fullerenes*—fullerenes that are isolable or observable only in combination with some other chemical species such as a metal. Reports in the literature already suggest that these three sets of limits will be quite distinct, at least for small fullerenes.

In gas-phase experiments there is evidence that the smallest transient fullerene attainable falls in the vicinity of C_{28} and C_{32} [42–45], whereas the smallest complexed fullerene apparently corresponds to U@C_{28}, which is a fullerene of tetrahedral symmetry with the uranium atom inside the cluster [45]. The smallest free fullerene to be isolated so far is C_{60} [46], which has a unique pyramidalization angle of 11.6°. C_{60} represents an important threshold because any smaller fullerene requires a topology that no longer obeys the isolated pentagon rule [1, 2, 5, 42–45]; that is, the fullerene must contain carbon atoms that are at the vertices of *two* 5MRs—[5-5-6] in the notation adopted in Figure 7.1. If the bond angles to the carbon atom are ideal (108°, 108°, 120°), this requires a pyramidalization angle of 16.71°. In calculations on a C_{60} isomer that also violates the isolated pentagon rule, Raghavachari and Rohlfing [47] have found a pyramidalization angle of 15.0° at the [5-5-6] carbon atoms. There is some suggestion that such a species may be isolable [48]. This would be an important achievement as the most nonplanar conjugated organic molecule yet to be characterized has a pyramidalization angle of 13.04° [11].

Although Figure 7.1 suggests that the fullerenes will be increasingly strain-free at large size, this is only correct in the broadest sense. The 5MRs once again spoil this simple relationship between size and stability. Contrary to Figure 7.1, it is not always possible to distribute the strain fully over the whole fullerene, and, particularly in the case of large highly symmetric fullerenes, pyramidalization is concentrated at the carbon atoms contained in the 5MRs. This behavior is demonstrated in the calculations reported by Bakowies and Thiel on the icosahedral (I_h) fullerenes [49]. They find maximum pyramidalization angles of 21.54° (C_{20}), 11.64° (C_{60}), 10.42° (C_{80}), 9.70° (C_{180}), 9.67° (C_{240}), and 10.31° (C_{540}), which may be compared with the ideal values of 20.91° (C_{20}), 11.64° (C_{60}), 10.01° (C_{80}), 6.66° (C_{180}), 5.76° (C_{240}), and 3.84° (C_{540}), computed from the relationship given in Ref. [4] (for $n > 60$ in C_n). Whether this degree of pyramidalization or the large variation in the degree of pyramidalization over the surface is sufficient to make them unstable remains to be seen, although no highly symmetric higher fullerene has yet been isolated [50].

References

1. H. W. Kroto, J. R. Heath, S. C. O'Brien, R. F. Curl, and R. E. Smalley, *Nature* **318**, 162–64 (1985).

2. Q. L. Zhang, S. C. O'Brien, J. R. Heath, Y. Liu, R. F. Curl, H. W. Kroto, and R. E. Smalley, *J. Phys. Chem.* **90**, 525–28 (1986).

3. R. C. Haddon, L. E. Brus, and K. Raghavachari, *Chem. Phys. Lett.* **125**, 459–64 (1986).

4. R. C. Haddon, L. E. Brus, and K. Raghavachari, *Chem. Phys. Lett.* **131**, 165–69 (1986).

5. T. G. Schmaltz, W. A. Seitz, D. J. Klein, and G. E. Hite, *J. Am. Chem. Soc.* **110**, 1113–27 (1988).

6. R. C. Haddon, *Accts. Chem. Res.* **21**, 243–49 (1988).

7. R. C. Haddon and L. T. Scott, *Pure Appl. Chem.* **58**, 137–42 (1986).

8. R. C. Haddon, *J. Am. Chem. Soc.* **108**, 2837–42 (1986).

9. R. C. Haddon, *Chem. Phys. Lett.* **125**, 231–34 (1986).

10. R. C. Haddon, *J. Am. Chem. Soc.* **109**, 1686–85 (1987).

11. R. C. Haddon, *J. Am. Chem. Soc.* **112**, 3385–89 (1990).

12. R. C. Haddon, *Accts. Chem. Res.* **25**, 127–33 (1992).

13. J. M. Hawkins, S. Loren, A. Meyer, and R. Nunlist, *J. Am. Chem. Soc.* **113**, 7770–71 (1991).

14. J. M. Hawkins, *Accts. Chem. Res.* **25**, 150–56 (1992).

15. R. A. Jishi and M. S. Dresslehaus, *Phys. Rev B* **45**, 2597–600 (1992).

16. N. Troullier and J. L. Martins, *Phys. Rev. B* **46**, 1754–65 (1992).

17. W. A. Goddard III, T. H. Dunning, Jr., W. J. Hunt, and P. J. Hay, *Accts. Chem. Res.* **6**, 368–76 (1973).

18. GAUSSIAN 91, Development Version (Revision C), M. J. Frisch, G. W. Trucks, M. Head-Gordon, P. M. W. Gill, M. W. Wong, J. B. Foresman, B. Johnson, H. B. Schlegel, M. A. Robb, E. S. Replogle, R. Gomperts, J. L. Andres, K. Raghavachari, J. S. Binkley, C. Gonzalez, R. L. Martin, D. J. Fox, D. J. DeFrees, J. Baker, J. J. P. Stewart, and J. A. Pople, Gaussian, Inc., Pittsburgh, PA, 1991.

19. R. C. Haddon, *Tetrahedron* **24**, 7611–20 (1988).

20. R. C. Haddon, QCPE508, QCMP044, *QCPE Bull.* **8** (1988).

21. W. T. Borden, *Chem. Rev.* **89**, 1095–109 (1989).

22. R. Hoffmann, private communication.

23. J. H. Hay, W. J. Hunt, and W. A. Goddard III, *J. Am. Chem. Soc.* **94**, 8293–301 (1972).

24. R. F. Curl and R. E. Smalley, *Science* **242**, 1017–22 (1988).

25. L. S. Wang, J. Conceicao, C. Jin, and R. E. Smalley, *Chem. Phys. Lett.* **182**, 5–11 (1991).

26. P. A. Limbach, L. Schweikhard, K. A. Cowen, M. T. McDermott, A. G. Marshall, and J. V. Coe, *J. Am. Chem. Soc.* **113**, 6795–98 (1991).

27. P.-M. Allemand, A. Koch, F. Wudl, Y. Rubin, F. Diedrich, M. M. Alvarez, S. J. Anz, and R. L. Whetten, *J. Am. Chem. Soc.* **113,** 1050–51 (1991).

28. R. E. Hauffler, J. Conceicao, L. P. F. Chibante, Y. Chai, N. E. Byrne, S. Flanagan, M. M. Haley, S. C. O'Brien, C. Pan, Z. Xiao, W. E. Billups, M. A. Ciufolini, R. H. Hauge, J. L. Margrave, L. J. Wilson, R. F. Curl, and R. E. Smalley, *J. Phys. Chem.* **94,** 8634–36 (1990).

29. D. M. Cox, S. Behal, M. Disko, S. M. Gorun, M. Greaney, C. S. Hsu, E. B. Kollin, J. Millar, J. Robbins, W. Robbins, R. D. Sherwood, and P. Tindall, *J. Am. Chem. Soc.* **113,** 2940–44 (1991).

30. D. Dubois, K. M. Kadish, S. Flanagan, R. E. Haufler, L. P. F. Chibante, and L. J. Wilson, *J. Am. Chem. Soc.* **113** 4364–66 (1991).

31. B. Miller, J. M. Rosamilia, G. Dabbagh, A. J. Muller, and R. C. Haddon, *J. Electrochem. Soc.* **139,** 1941–45 (1992).

32. F. Wudl, *Accts. Chem. Res.* **25,** 157–61 (1992).

33. Q. Xie, E. Perez-Cordero, and L. Echegoyen, *J. Am. Chem. Soc.* **114,** 3978–80 (1992).

34. W. Andreoni, F. Gygi, and M. Parrinello, *Phys. Rev. Lett.* **68,** 823–26 (1992).

35. S. J. Duclos, K. Brister, R. C. Haddon, A. R. Kortan, and F. A. Thiel, *Nature* **351,** 380–82 (1991).

36. R. Tycko, G. Dabbagh, M. J. Rosseinsky, D. W. Murphy, R. Fleming, A. P. Ramirez, and J. C. Tully, *Science* **253,** 884–86 (1991).

37. K. H. Johnson, M. E. McHenry, and D. P. Clougherty, *Physica C* **183,** 319–23 (1991).

38. J. L. Martins, N. Troullier, and M. Schabel, submitted.

39. C. M. Varma, J. Zaanen, and K. Raghavachari, *Science* **254,** 989–92 (1991).

40. M. Lannoo, G. A. Baraff, M. Schlüter, and D. Tomanek, *Phys. Rev B* **44,** 12106–8; M. Schlüter, M. Lannoo, M. Needels, G. A. Baraff, and D. Tomanek, *Phys. Rev. Lett.* **68,** 526–29 (1992).

41. K. Raghavachari and C. M. Rohlfing, *J. Phys. Chem.* **95,** 5768–73 (1991).

42. H. W. Kroto, *Nature* **329,** 529–31 (1987).

43. R. F. Curl and R. E. Smalley, *Science* **242,** 1017–22 (1988).

44. H. W. Kroto, *Science* **242,** 1139–45 (1988).

45. T. Guo, M. D. Diener, Y. Chai, M. J. Alford, R. E. Hauffler, S. M. McClure, T. Ohno, J. H. Weaver, G. E. Scuseria, R. E. Smalley, Science **257,** 1661–64 (1992).

46. W. Krätschmer, L. D. Lamb, K. Fostiropoulos, and D. R. Huffman, *Nature* **347,** 354–58 (1990).

47. K. Raghavachari and C. M. Rohlfing, *J. Phys. Chem.* **96,** 2463–66 (1992).

48. J. F. Anacieto, H. Perreault, R. K. Boyd, S. Pleasance, M. A. Quilliam, P. G. Sim, J. B. Howard, Y. Makarovsky, and A. L. Lafleur, *Rapid Comm. Mass Spectrom.* **6,** 214–20 (1992).

49. D. Bakowies and W. Thiel, *J. Am. Chem. Soc.* **113,** 3704–14 (1991).

50. F. Diederich and R. L. Whetten, *Accts. Chem. Res.* **25,** 119–26 (1992).

CHAPTER

8

Theory of Electronic and Superconducting Properties of Fullerenes

Marvin L. Cohen and Vincent H. Crespi

Some definitive models of the electronic structure of solid C_{60} and M_xC_{60} ($M =$ alkali metal) have been proposed using state-of-the-art band-structure calculations. Much of the work relies on one-electron theory. The results appear to be plausible, although several researchers have questioned the applicability of theories of this kind in view of the possible importance of electron correlation in these systems. Here we describe the main conclusions of the theoretical results with the aim of using them as a starting point for understanding other properties such as optical and photoemission spectra and superconductivity.

After a review of the electronic structure of the C_{60} molecule and prototype systems, a discussion of the lattice vibrations is given. The electronic and vibrational properties are used to describe the electron–phonon coupling and its role in phonon-induced superconductive pairing. The ramifications of a BCS model are discussed along with an analysis of normal-state transport properties. Finally, some ideas for nonphonon mechanisms of superconductivity are discussed.

8.1 Electronic Structure

We begin with the electronic structure of an isolated C_{60} molecule. The bonding character of C_{60} is predominantly sp^2 with a small admixture of sp^3 character due to the nonzero curvature. For this reason, the electronic states can be decomposed into approximate π and σ states. The bonding σ states reside well below the highest occupied level, which is composed of orbitals having primarily π character [1, 2]. The nearly spherical structure of the C_{60} suggests a labeling of these electronic states in terms of spherical harmonics, with the σ and π electrons corresponding to different radial quantum numbers. The carbon-derived potential splits the degenerate energy levels of the spherical potential, producing the level structure shown in Figure 8.1. The nonspherical part of the potential is dominated by an $\ell = 10$ contribution, which splits electronic levels having $\ell \geq 5$ [3].

Fifty π electrons fill orbitals from $\ell = 0$ to 4, and ten electrons occupy part of the $\ell = 5$ shell. The carbon potential splits the eleven degenerate states of this $\ell = 5$ level into a fivefold-degenerate highest occupied molecular orbital (HOMO) of h_u symmetry, a threefold degenerate lowest unoccupied molecular orbital (LUMO) of t_{1u} symmetry, and an additional threefold-degenerate level at slightly higher energy. The ten $\ell = 5$ electrons fill the fivefold level, yielding a closed-shell configuration.

The C_{60}-based solids provide an interesting series of materials for electronic structure calculations. Electronic band theory provides a consistent explanation of semiconducting fcc C_{60} [3, 4], doped metallic systems such as K_3C_{60} [5, 6], and semiconducting K_6C_{60} [7]. We focus on fcc C_{60} and K_3C_{60} as prototype materials for examining the normal-state and superconducting properties of this class of materials.

Although there are still some open questions related to the effects of electron–electron correlation energy for these materials [8], it has been argued [3, 4] that one-electron band theory yields consistent results with bandwidths ~0.5 eV. The nature of the correlation effects and the limitations of the band theory are frontier areas of research at this point. With this caveat, we describe the two prototype systems using standard band theory.

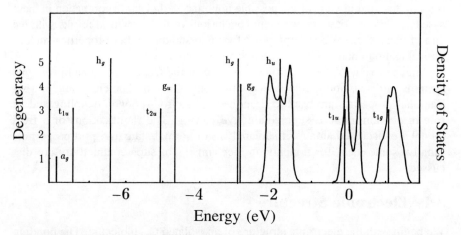

Figure 8.1 Electronic energy levels of π states in an isolated C_{60} molecule [1–3]. The highest σ states, which have been omitted for clarity, lie at roughly -4 eV. The density of states of the bands in K_3C_{60} that correspond to the h_u, t_{1u}, and t_{1g} states of the molecule has been superimposed [4, 5, 13–16]. The Fermi level of the solid is set at 0 eV.

Solid fcc C_{60} is bound by van der Waals attraction between the clusters, which have a nearest-neighbor center-to-center separation of 10 Å. This relatively large separation compared to the C–C separations within the clusters ($\sim$1.45 Å) results in flat bands formed from the predominantly sp^2 states. As in the isolated molecule, the resulting band states near the band gap have π character. If a spherical harmonics decomposition is used for the wave functions as in the molecular system, then the states near the band gap have angular momentum character of $\ell = 4$ and 5.

Two band-structure calculations by Saito and Oshiyama (SO) [5] and by Martins and Troullier (MT) [6] give similar results and differ only in the finer details. Both calculations employ pseudopotentials and the local-density approximation (LDA). SO employ a Gaussian-orbital basis that is considerably smaller than the plane-wave basis set used by MT, which has as many as $\sim$30,000 plane waves. Both the SO and MT calculations include total energy estimates of lattice constants, bulk moduli, and cohesive energies. It is encouraging that the results of these calculations are similar considering the differences in the wave-function expansion.

Some general features of the band structure can be discussed in terms of the relation of the bands to molecular orbitals. Since there are 240 electrons per primitive cell, one expects 120 (90 σ and 30 π) filled bands. The 90 filled σ bands are well below the Fermi energy. The top 5 bands are derived from the h_u fivefold-degenerate HOMO of C_{60}. The lowest 3 valence bands are related to the t_{1u} LUMO. The remaining 3 of the 11 $\ell = 5$ states are contained in an upper unoccupied t_{1g} state. Both the valence-band maximum and conduction-band minimum lie at the X point of the Brillouin zone, with a gap separation of $\sim$1.5 eV. The relative orientations of the C_{60} clusters can affect the band structure. Here the highest-symmetry configuration, T_h, is assumed for the fcc lattice of rigid molecules. The rotation of the molecules above $\sim$260 K [9] alters the band structure and broadens the bands.

Because the states near the band gap originate from states of u symmetry, optical transitions between the top of the valence band and the bottom of the conduction band are forbidden in lowest order. Although the LDA calculations are best suited for ground-state properties, photoemission [10] and inverse photoemission [11] data are in reasonably good agreement with the predictions of the pseudopotential-LDA calculations.

When alkali-metal atoms are placed in the two interstitial tetrahedral sites and the octahedral site, a rigid-band model would predict that the t_{1u}-derived bands are half full, implying that K_3C_{60} should be metallic. In the gas phase, the C_{60} molecule can accommodate two electrons and form a stable C_{60}^{2-} ion [12]. In the solid phase, analysis of K_3C_{60} and K_6C_{60} suggests that the potassium $4s$ electron is transferred to the C_{60} molecule, resulting in a strongly bonded ionic metal for K_3C_{60} and a semiconductor with a band gap of $\sim$1 eV [7] for K_6C_{60}. Hence, the charge state of C_{60} is strongly influenced by its local environment. In particular, the Madelung energy of the crystal, the electron affinity of the C_{60}, and the ionization energy of the alkali metal are major factors in the bonding and charge states.

Although evidence for the inadequacy of a strict rigid band model exists [10], the concept of alkali-metal doping appears to be representative of the situation for K_3C_{60}. Electrons are transferred, and the weakly bonded van der Waals crystal transforms into a more strongly bound ionic metal. The Fermi level crosses the lowest two t_{1u} bands with the electrons occupying almost all of the lowest band and more than half of the second of the three bands. The Fermi level lies on a steep downward slope in the density of states, as indicated in Figure 8.1. In addition, the density of states is a sensitive function of the lattice constant. For these reasons, theoretical results for the density of states at the Fermi level, $N(E_F)$, vary from 10 to 30 states per eV per C_{60}. Because of the band dispersion between Γ and X, holelike Fermi surfaces exist at Γ for both the first and second conduction bands. The first band also contains hole surfaces near the L point of the Brillouin zone. For the next highest band there are both hole and electron surfaces, yielding a Fermi surface with at least two holelike and two electronlike branches [13–18].

In contrast to the undoped material, K_3C_{60} exhibits low-temperature orientational disorder, with equal population of two orientations related by a $\pi/2$ rotation [19]. This disorder will smear out peaks in the density of states [20] and contribute to a large residual resistivity [21]. Electronic structure results based on a perfect crystal should be used with caution in light of this microscopic orientational disorder.

8.2 Phonons

The phonons of C_{60}-based solids can be conveniently divided into low-frequency interball and high-frequency intraball modes. The interball phonons are further divided into librational, vibrational, and alkali-metal-atom modes. Each of these modes could plausibly contribute to some degree to the electron–phonon coupling necessary for the BCS theory of superconductivity. The mechanisms of electron–phonon coupling will be described later; here we consider the various phonon modes and give a qualitative discussion of the possibilities for electron–phonon coupling for each class of modes.

The lowest-energy phonons of doped C_{60} are the rotational motions of the C_{60} molecules. In contrast to pristine C_{60}, the doped material does not exhibit quasifree molecular rotation [22]. Instead, the C_{60} molecules of the doped materials undergo small-amplitude librational motions. These librational modes have very low frequencies—on the order of 20 or 30 K [23], as expected due to the large oscillator mass and the rather weak van der Waals binding between the C_{60} molecules. Since the band dispersion of solid C_{60} is determined by the electronic overlap between molecules, which in turn is sensitive to the relative orientation of adjacent molecules, the librational modes could potentially exhibit significant electron–phonon coupling. However, the low frequency of these modes suggests that they will not make substantial contributions to the super-conducting transition temperature.

The fullerene systems also possess low-frequency modes of translational character. These are expected to have somewhat higher frequencies than the librational modes, since bulk translational motions should induce greater interaction between the molecules than rotations. A rough estimate of their frequencies based on the bulk modulus of the material [24] yields $\omega \sim 60$ K. Inelastic neutron scattering [25] and Raman scattering [26] reveal broad peaks near 50 and 150 K, which have been identified with these modes. Low to moderate $\mathbf{q}$ translational modes induce local changes in the lattice constant, which induce local changes in the bandwidth. A tight-binding picture of interball interaction yields small to moderate electron–phonon coupling for these distortions [27, 28].

The alkali-metal atoms reside in the tetrahedral and octahedral interstitial sites of the fcc C_{60} lattice. Electron energy loss spectroscopy [29] suggests that the vibrational modes of the alkali metals in these sites have frequencies around 100–200 K. Local-density calculations indicate that the sites are effectively "holes" in the C_{60} lattice, with little charge density other than that residing immediately atop the alkali-metal atom. This result suggests that the alkali-metal atoms reside in strongly anharmonic potentials with a significant quartic component and perhaps a weak multiple-well character. The large electron affinity of the C_{60} molecule guarantees that the alkali-metal atoms will be almost completely ionized. The positively charged ions will polarize the electrons on adjacent C_{60} molecules, thereby coupling the alkali-metal motions to the electronic system. An estimate of the potential electron–phonon coupling due to a static displacement of the alkali-metal atoms suggests extremely strong coupling [30]. However, a model of superconductivity based entirely on the alkali-metal atom vibrations encounters significant difficulties, as will be discussed in more detail later.

The high-frequency intraball phonons subdivide into two ill-defined classes, the lower-frequency radial modes and the higher-frequency tangential modes. The tangential modes involve stretching and compression of the carbon–carbon bonds, as opposed to the primarily radial motions of the carbon atoms for the lower-frequency modes. The situation is analogous to that in graphite, in which the modes with in-plane displacements have higher frequencies than those involving buckling motions out of the graphitic sheets. In contrast to doped C_{60}, alkali-metal-intercalated graphite systems have superconducting transition temperatures below 1 K [31]. The curvature of C_{60} produces hybridization between the radial p_z orbitals and the s orbitals, which allows new scattering channels for the electron–phonon coupling that are absent in planar graphite [32, 33].

We will concentrate on the C_{60} intraball phonon modes of A_g and H_g symmetry, since group therapy implies that only these modes will couple with the t_{1u} symmetry electronic states at the Fermi level. The relevant phonons are enumerated in Table 8.1 [34]. The lower-frequency A_g mode corresponds to a breathing motion, whereas the higher-frequency mode involves out-of-phase

TABLE 8.1 Experimental phonon frequencies[33] for intraball modes in *undoped* C_{eo} of the correct symmetry to mediate electron-phonon coupling. Upon doping, the energies shift by a few percent.

Mode	$\omega(K)$
$H_g(1)$	393
$H_g(2)$	629
$H_g(3)$	1022
$H_g(4)$	1071
$H_g(5)$	1581
$H_g(6)$	1799
$H_g(7)$	2055
$H_g(8)$	2266
$A_g(1)$	715
$A_g(2)$	2114

oscillations of the single and double bonds. The lowest-frequency H_g mode produces a radial distortion of the C_{60} molecule into an ellipsoid, whereas the higher-frequency H_g modes possess progressively more tangential character [35], as expected.

By fortunate coincidence, the selection rules for electron–phonon coupling and Raman scattering are the same, so that these same ten modes are also Raman active. Electron–phonon coupling will cause decay of phonons into electron–hole pairs, yielding a finite phonon lifetime, which may be evidenced by the nonzero widths of the Raman peaks [36]. In theory, $q = 0$ Raman-active modes will not couple to the electrons because of phase-space factors; however, the strong orientational disorder in the doped system will allow wave-vector-nonconserving processes.

A study of the broadening of the Raman lines upon doping indicates a significant coupling to the $H_g(2)$ mode at 560 K, and moderate coupling to the higher-frequency $H_g(7)$ mode [37]. Inelastic neutron scattering [25] also indicates substantial broadening of several of the H_g modes. If this broadening is interpreted in terms of electron–phonon coupling, then a reasonably consistent argument for electron–phonon superconductivity can be made.

8.3 Superconductivity

The 1986 discovery [38] of superconductivity in oxides above 30 K, and subsequent developments in this area have resulted in a renaissance in the field of superconductivity. Although there are copper oxides with transition temperatures $T_c > 125$ K, many of these ceramic materials are difficult to work with and progress is slow in taking advantage of their high T_c's and critical fields. Although at first, many researchers felt that coupling to phonons could not provide the high T_c's found in the oxides, this objection can be shown to be incorrect. Most current objections to phonon mechanisms for oxide superconductors are related to their unusual normal-state properties. One feature common to the Cu–O-based superconductors is planar geometries, which result in highly anisotropic properties. In contrast, the C_{60}-based superconductors [39] are three-dimensional systems and have T_c's exceeding 30 K. Normal-state and superconducting properties are generally isotropic, and there is hope that a thorough understanding of the underlying superconducting mechanism can lead to higher T_c's as well as practical applications. At first glance, the C_{60} systems look more amenable to theoretical analysis than the oxides. However, there is still not a general consensus on the origin of superconductivity or on the nature of various electron–electron interactions in these materials. We review the current situation with some emphasis on the conventional theory of electron–phonon interactions and superconductivity.

A popular point of view holds that the C_{60} systems are BCS superconductors. There are numerous differences between these materials and more conventional BCS metals such as Al, Sn, or Nb, but it has been proposed [30, 32, 40–43] that electron pairing caused by phonon exchange is the dominant feature causing superconductivity in C_{60} solids. To date, the primary objections to phonon-induced pairing in C_{60} systems have been related to the properties of the superconducting parameters and the attractiveness of alternate theories.

In BCS and Eliashberg [44, 45] models for T_c, the primary parameters determining the superconducting properties are the electron–phonon coupling constant λ, the Coulomb repulsion μ, and some average phonon energy E_D. For standard band-structure models of metals, μ is renormalized to μ^*

$$\frac{1}{\mu^*} = \frac{1}{\mu} + \ln \frac{E_F}{E_D} \tag{8.1}$$

and μ is typically reduced by a factor of 2 to 5. Strong-coupling effects arising from electron–phonon renormalization and quasiparticle damping change λ to $\lambda^* = \lambda/(1 + \lambda)$, yielding a McMillan [46] equation for T_c of the form

$$T_c = E_D e^{[-1/(\lambda^* - \mu^*)]}, \tag{8.2}$$

which is appropriate for $\lambda < 1.5$. Numerical constants can be included in Eq. (8.2), as was done by McMillan to fit the case of Nb.

The electron–phonon coupling parameter λ can be expressed [46] in terms of $\langle I^2 \rangle$, which is a Fermi surface average of the square of the matrix elements representing the scattering of electrons by an atomic displacement. We obtain

$$\lambda \sim \frac{N(E_F)\langle I^2 \rangle}{M\langle \omega^2 \rangle}, \tag{8.3}$$

where M is the atomic mass and $\langle \omega^2 \rangle$ represents an average of the square of the phonon frequencies [46]. Equation (8.3) for λ can be interpreted as the ratio of an average electronic spring constant and an average lattice vibrational or phonon spring constant. It can be argued [32, 33] that if the $\langle \omega^2 \rangle$ are related to the C–C bond oscillations, then one would expect similar phonon spring constants for C_{60} solids and graphite. Hence, the significantly higher T_c's for M_3C_{60}, where M is an alkali-metal atom, arises from the numerator in Eq. (8.3). As mentioned earlier, it is expected that the curvature of the C_{60} molecule can play an important role in the electron–phonon coupling. For example, in graphite, the mirror plane symmetry yields $\langle I^2 \rangle = 0$ for several displacement-induced electron scatterings when matrix elements are taken between s and p_z states. However, the curvature in the C_{60} case mixes these states and yields nonzero matrix elements, which can have large effects on λ when compared to graphite intercalated compounds.

The electron–phonon model has been refined by several researchers in an attempt to determine the importance of the various possible phonons in the electron pairing process. A widely held view is that the intraball vibrational modes dominate. In this model, the C_{60} molecules reside in a sea of electrons with sufficient charge transfer to create bands ≤ 1 eV in width. Electronic charge-density plots [5, 6] reveal spread-out charge distributions consistent with the itinerant-electron picture of this model. The diameters of the C_{60} molecules and their separations are both ~ 10 Å. Hence electrons pair via phonons localized on the molecules, but retain their bandlike itinerant nature. In most of these models, vibrations arising from coupling between molecules or from movement of the alkali-metal atoms are considered to be less important than intraball excitations because of the higher energies of the latter.

This model of pairing via intraball phonons has been justified to some extent by measurements of T_c as a function of pressure and as a function of different substitutions for M in M_3C_{60}. A negative dependence on pressure was found for $M = $ K and Rb [47], where $\partial T_c / \partial P \sim -0.8/$kbar. This result is interpreted in terms of a decrease in $N(E_F)$ arising from an increase in electron wavefunction overlap with decreasing volume. A similar effect arises when larger alkali metals are used [48]. It is argued that the increase in lattice constant or volume causes a decrease in bandwidth and an increase in $N(E_F)$. For example, T_c increases from 19.3 K for K_3C_{60} to 29.4 K for Rb_3C_{60} when the lattice constant increases from 14.253 to 14.436 Å.

Although the data for T_c versus lattice constant are monotonic, it is not exactly linear as is sometimes claimed. A striking linear relation has been found [49] between the measured T_c and the calculated $N(E_F)$. It is tempting to interpret this dependence as a signal of non-BCS behavior since the weak-coupling formula most commonly used for T_c is exponentially dependent on $N(E_F)$. However, T_c varies exponentially with λ at low λ and as $\sqrt{\lambda}$ at large λ. In fact, a good approximation to the solution of the Eliashberg equations for $\mu^* = 0$ is [50, 51]

$$T_c = 0.25 \frac{\sqrt{\langle \omega^2 \rangle}}{\sqrt{e^{2/\lambda} - 1}}. \tag{8.4}$$

This function has a linear dependence on λ in an intermediate range of λ. Hence, the linear dependence of T_c on $N(E_F)$ can be explained within conventional BCS–Eliashberg theory if the coupling parameters are in the proper range.

One assumption inherent in this model is that the variation of $N(E_F)$ does not significantly change λ or μ apart from a multiplicative factor. Screening effects could reduce the electron–phonon matrix elements for the larger values of $N(E_F)$, yielding a nonlinear dependence of λ on $N(E_F)$. This is an open problem at this point, and it bears on the appropriateness of the model.

The various theoretical phonon-induced pairing proposals differ primarily in the phonons considered to be important for the pairing. Because of the wide spectrum of phonon energies, the choice of the parameters λ and μ depend sensitively on the choice of phonon frequencies. In the models proposed, the average phonon energies and coupling strengths are $\omega \sim 1100$ K with $\lambda \sim 0.6$ [32], $\omega \sim 1300$ K with $\lambda \sim 0.6$ [41], $\omega \sim 2000$ K with $\lambda \sim 0.5$ [40], and $\omega \sim 500$ K with $\lambda \sim 1.0$ [42]. Since weakly and strongly coupled superconductors have different experimental "signatures" for various properties, these could be used to constrain the values of λ. Two experiments that bear on this distinction are the superconducting-gap–to–T_c ratio $2\Delta/kT_c$ and the discontinuity in the heat capacity at T_c. We are unaware of data for the latter, but tunneling [52] and infrared measurements [53] of Δ have been interpreted in terms of a gap ratio $2\Delta/kT_c \sim 5$, which suggests strong coupling and hence fairly large values of λ.

Raman scattering provides another approach for estimating λ [36]. By examining broadening effects on various phonon contributions to the Raman spectrum, an estimate of the electron–phonon coupling can be obtained. Some theoretical distillation of the data is necessary, and extrapolations to obtain contributions of phonons having nonzero q are needed. In general, the results yield couplings that are consistent with the theoretical models of the superconductivity and suggest a particularly important role for the lower $H_g(2)$ mode with $\omega \sim 600$ K.

Critical-field data provide another source of information on the superconducting parameters. Measurements of the upper critical field of single crystal K_3C_{60} [54] yield $H_{c2} = 17.5$ T, which implies a coherence length of 45 Å. Analysis of $H_{c2}(T)$ yields a scattering time $\tau \sim (1–2) \times 10^{-14}$ sec, in accordance with transport measurements [27]. Using a penetration depth of roughly 5000 Å, as suggested by muon spin relaxation data [55] and Ginzburg–Landau theory [27, 56], we obtain $\kappa \sim 100$, indicating strong type-II behavior.

A crucial experiment in the development of the BCS theory of superconductivity was the observation of an isotope effect for T_c. For several metals, the isotope parameter $\alpha = -d \ln T_c/d \ln M$ was found to be approximately 0.5, indicating that electron-lattice effects are intimately connected with superconductivity. Values of α less than $\frac{1}{2}$ were found for transition metals and reconciled with the different frequency cutoffs for the phonon and Coulomb interactions, which yield

$$\alpha = \frac{1}{2}\left[1 - \left(\frac{\lambda^*}{\lambda^* - \mu^*}\right)^2\right]. \tag{8.5}$$

There are very few examples of $\alpha > \frac{1}{2}$ [57], and those appear to be explainable by anharmonic phonon effects [58].

At this time, five measured values of α for M_3C_{60} have been reported: $\alpha = 1.4 \pm 0.5$ for 33% ^{13}C replacement [59], $\alpha = 1.7 \pm 0.5$ for 60% ^{13}C replacement [60], $\alpha = 0.37 \pm 0.05$ for $75 \pm 5\%$ ^{13}C replacement [61], $\alpha = 0.30 \pm 0.06$ for 99% ^{13}C replacement [62], and a very recent measurement of $\alpha = 0.2 \pm 0.2$ [63]. One obvious conclusion, if these data are correct, is that lattice effects appear to be related to the superconductivity. The values in the range of 0.3 to 0.4 are consistent with standard models of superconductivity in which α is depressed from $\frac{1}{2}$ by the Coulomb repulsion as described in Eq. (8.5). In particular, an average phonon frequency of 1000 K yields $T_c = 29$ K and $\alpha = 0.37$ for $\lambda = 0.81$ and $\mu^* = 0.19$. Lower average frequencies require stronger coupling and larger μ^*. For example, $\omega \sim 200$ K implies $\lambda = 2.5$ and $\mu^* = 0.31$.

The general question of whether μ^* can be reduced much from μ has been raised [64]. It is argued that the $\log(E_F/E_D)$ factor in μ^* is small and ineffective in reducing μ. In addition, $N(E_F)$ is large because of the narrow bands, and this causes μ, which is proportional to $N(E_F)$, to be large. However, it is likely that since $\mu \sim \langle N(E_F)V \rangle / \varepsilon$, where V is the Coulomb interaction and ε is the dielectric function, the $N(E_F)$ factor in ε will dominate this function and cancel the $N(E_F)$ factor in the numerator. This is characteristic of systems with large $N(E_F)$. The result would be a μ in the range of 0.15 to 0.3.

The experimental measurements of $\alpha \sim 1.5$ [59, 60] provide a greater challenge to conventional theories of superconductivity. We consider three possible explanations for an isotope effect greater than the BCS maximum of 0.5: a strong energy dependence of the electronic density of states near the Fermi level, anharmonic phonons, and isotopically disordered intraball phonons. Electronic structure calculations place the Fermi level on the upper side of a peak in the density of states of width ~ 0.5 eV, as depicted in Figure 8.1. Assuming an average phonon frequency less than 2000 K, a calculation within BCS theory using a Lorentzian form for the density of states [65] yields a maximum change of only ± 0.05 in the isotope effect exponent due to the energy-dependent density of states. An anharmonic phonon with positive quadratic and sixth-order contributions and a negative quartic part can increase the isotope exponent above 0.5 [58, 66]. The effect is most pronounced at weak coupling. Unfortunately, weak coupling implies high-frequency phonons, which are unlikely to be softened by a negative quartic part. Stronger coupling to lower-frequency phonons would imply stronger anharmonicities, which are also unlikely to be physical. In addition, neither of the two previous explanations is consistent with an isotope effect exponent $\alpha \sim 0.4$ upon 100% ^{13}C substitution.

Isotopic disorder is a candidate source of an enhanced isotope effect at intermediate ^{13}C concentrations. At intermediate isotopic substitution, the mass distribution on the C_{60} molecule is disordered [67]. The normal modes of this system will not have the exact symmetries of the isotopically pure material.

Therefore, the electron–phonon coupling function $\alpha^2 F(\omega)$ will spread out as the isotopic disorder allows additional modes to couple to the electrons. However, this broadening of the frequency distribution of the electron–phonon coupling does not produce the additional depression in T_c observed at intermediate isotopic substitutions [68].

Recent measurements [69] of the temperature-dependent resistivity $\rho(T)$ further constrain the theoretical models. The temperature dependence of the resistivity has been fit to the Ziman resistivity formula [70]

$$\rho(T) = \frac{4\pi m}{ne^2 k_B T} \int_0^{\omega_{max}} \frac{\hbar \omega \alpha_{tr}^2 F(\omega)}{(e^{\hbar\omega/k_B T} - 1)(1 - e^{-\hbar\omega/k_B T})} \, d\omega, \qquad (8.6)$$

where $\alpha_{tr}^2 F(\omega)$ is the transport electron–phonon coupling function. This function can be approximated by the electron–phonon coupling function relevant to superconductivity, namely $\alpha^2 F(\omega)$. The resistivity is analyzed using the $\alpha^2 F(\omega)$ from various theoretical models of superconductivity in the fullerenes. These models will be denoted by VZR [40], JD [42], and SLNB [41].

Taking the residual resistivity and overall magnitude of the electron–phonon coupling as adjustable parameters, the best-fit resistivity curves for VZR and SLNB are too flat at low temperature (see Fig. 8.2), suggesting an additional contribution to the electron–phonon scattering from lower-frequency phonons. JD, which emphasizes lower-frequency intraball phonons, yields reasonable agreement with experiment without an additional mode. A lower-frequency contribution to the resistivity could arise from the translational, librational, or alkali-metal phonon modes, as discussed previously. Figure 8.3 shows the resistivity fits with an additional mode of adjustable strength at a frequency of 150 K. The fit is similar for lower-frequency modes in the range from 20 to 200 K. The quality of the fit is not improved if the additional mode is placed at a high frequency, indicating that the agreement is not an artifact of the additional free parameter.

Analysis of the absolute magnitude of the resistivity requires knowledge of the plasma frequency $\omega_p = \sqrt{4\pi n e^2/m}$. Electron energy loss data [71] and screened band-structure calculations [17] suggest $\omega_p/\sqrt{\Sigma} \approx 0.5$ eV. Using this value and a reasonable range of uncertainty for the experimental quantities, we obtain $\lambda \sim 0.3$ to 0.8 for JD, $\lambda \sim 0.7$ to 1.6 for SNLB, and $\lambda \sim 1.4$ to 3.4 for VZR. JD yields λ slightly too low to account for the superconductivity, whereas VZR yields λ slightly too high. These results suggest that a consistent model of electron–phonon-mediated superconductivity in the doped fullerenes should encompass contributions from a range of intraball phonons, with a small contribution from lower-frequency modes of either interball or alkali-metal character.

Although the evidence to date favors the phonon models of superconductivity based on intraball vibrations, the possibility of alkali-metal phonon-mediated superconductivity cannot be entirely ruled out. We consider the possibilities of "rescuing" an alkali-metal-mediated mechanism. The low alkali-

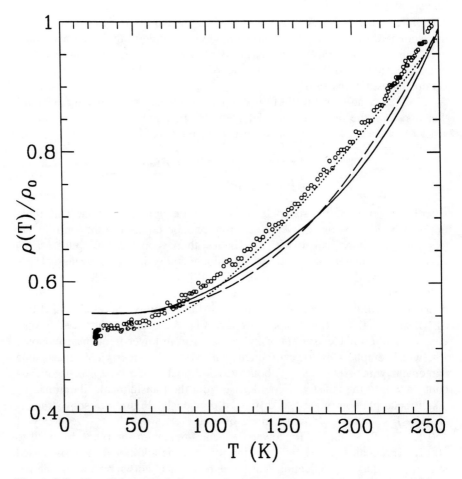

Figure 8.2 Fits of theoretical electron–phonon scattering models to the experimental resistivity ρ (circles). The solid, dashed, and dotted lines are for the models VZR [40], SLNB [41], and JD [42], respectively. The data have been normalized by $\rho_0 = \rho(260 K)$.

metal vibrational frequencies require strong coupling to reproduce the experimental T_c's. The alkali-metal phonons could strongly polarize the adjacent C_{60} molecules, yielding strong coupling [30]. The main objections to such a mechanism are the lack of a change in T_c upon substitution of Rb for K at the same lattice constant and the large carbon isotope effect.

As discussed previously, the alkali-metal atoms probably vibrate in strongly anharmonic potentials, which could decrease the superconducting isotope effect for these modes [58, 66], as seen in PdH(D) [72–74]. In addition, the potential will change form for ions of different radius. These effects could explain the slight increase in T_c upon substitution of Rb for K, although a quantitative

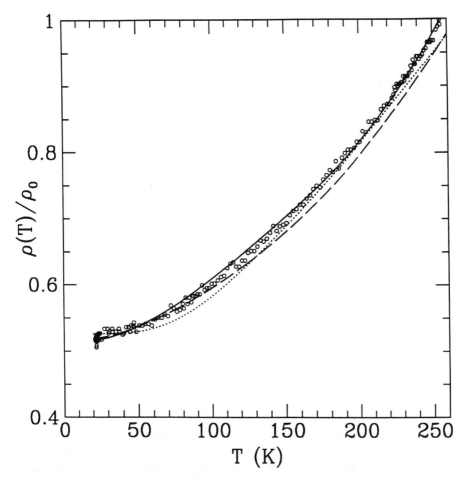

Figure 8.3   Fits of theoretical electron–phonon models with an additional phonon mode at 150 K. The curve assignments are the same as Figure 8.2.

analysis awaits a better knowledge of the alkali-metal atom vibrational potentials. A significant ^{87}Rb to ^{85}Rb or ^{41}K to ^{39}K isotope effect of either sign would lend support to this mechanism.

A mechanism based purely on alkali-metal vibrations cannot account for the substantial carbon isotope effect. One is forced to include a contribution from phonons involving carbon atoms. Consider a model in which interball vibrational motions supplement the alkali-metal atom contributions to the electron–phonon coupling. The M_3C_{60} system is close to a phase transition to the M_4C_{60} bct phase [75, 76]. The bulk motions of the C_{60} molecules corresponding to distortions into this structure could be anharmonic, with a softening of the quadratic potential. This form of potential could enhance the carbon isotope effect [66, 77], yielding $\alpha_{carbon} \approx 0.3$ even though the carbon phonons do not dominate the superconductivity. Such potentials should be pressure dependent, implying an unusual pressure dependence of the carbon isotope effect. Another possible source of an isotope-effect-enhancing anharmonic potential is the very low-frequency librational modes.

Within an alkali-metal-phonon model, the broadening of the intraball Raman-active phonons upon doping would be interpreted as an effect of the orientational disorder of the doped phase. The large experimental value of the gap ratio [52, 53] becomes a natural consequence of strong coupling. The possibility of a large value of the Coulomb repulsion μ^* [30] would not have as profound an effect on T_c as in a weakly coupled superconductor. However, low-frequency electron–phonon coupling to anharmonic alkali-metal phonons should yield a temperature-dependent resistivity that is roughly linear at fairly low temperatures and gains a negative curvature at higher temperatures, at variance with the experimental results. The most likely phonon mechanism for superconductivity remains weak-to-moderate coupling to high-frequency intraball phonons.

The previous discussion has assumed that the electron–phonon interaction is the cause of superconductivity in the doped fullerenes. We now consider the possibility that correlation effects between electrons on the same C_{60} molecule could produce an effective pairing interaction. The standard starting point for a discussion of strong electron correlation is the Hubbard Hamiltonian [78]

$$H = -t \sum_{(NN)} (c_{i\sigma}^{\dagger} c_{j\sigma} + \text{h.c.}) + \frac{U}{2} \sum_{i,\sigma} n_{i\sigma} n_{i-\sigma}, \qquad (8.7)$$

where the first term describes the kinetic energy of the electrons hopping between sites i and j and the second term describes the Coulomb repulsion between electrons of different spins on the same site. The model assumes that the Coulomb interaction is strongly screened, since only an on-site Coulomb term is incorporated.

We consider a Hubbard model where the site index refers to individual carbon atoms on a C_{60} molecule. The pair binding energy is defined as the difference in energy between two C_{60} molecules with three electrons each and one molecule with two and another molecule with four. At first glance, the repulsive nature of the Coulomb interaction appears to guarantee that the electrons will prefer to spread out uniformly between the balls, resulting in a negative pair binding energy. However, strongly correlated electron systems often exhibit subtle behavior. Exact numerical solutions for small model systems [79] and second-order perturbative results for a 60-site system [80, 81] yield a positive binding energy for intermediate values of the ratio U/t. Examination of the spin-charge correlation function on a C_{60} molecule [82] indicates that this pair binding cannot be interpreted in terms of spin-charge separation [81, 83]. Instead, the pair binding results from the dominance of the pair-scattering contributions to the energy [82].

The model allows for a substantial carbon isotope effect through isotopic variation in the hopping matrix element t. The hopping rate between carbon atoms will be a function of the bond length, which will change upon isotopic substitution due to the influence of anharmonic terms on the zero-point motion. The pairing energy is a function of the hopping matrix element, and the transition temperature is a sensitive function of the pairing energy.

$$T_c \sim W e^{-W/E_{pair}}, \tag{8.8}$$

where W is the electronic bandwidth. The change in the C–C bond length upon substitution of ^{13}C for ^{12}C yields a change in T_c of roughly 0.5 K [84], in accord with experiment. The tiny isotopic change in bond length is a factor of ten or twenty smaller than the dynamic changes in bond length due to intraball phonons, suggesting an important role for fluctuations in this model of superconductivity.

The primary difficulty with this model is the assumption that the Coulomb interaction is screened completely over the distance of a carbon–carbon bond. The inclusion of a nearest-neighbor Coulomb repulsion strongly suppresses the pair binding, resulting in no pairing for a nearest-neighbor repulsion on the order of $U/20$ [79]. Perturbative results for a screened Coulomb interaction indicate that the regime of negative pair binding energy for a more realistic system occurs at values of U/t beyond those that are likely to be physically relevant [85]. In fact, a Hubbard-like model is more likely to be relevant to interball hopping and on-ball Coulomb interactions.

8.4 Conclusions

The dynamic nature of this field guarantees significant future changes in current models of the electronic structure and superconducting properties of the fullerenes. Although certain models are consistent with present experimental results, additional data are necessary before we can claim a microscopic understanding of these fascinating systems.

Acknowledgment

We thank D. M. Deaven, P. E. Lammert and D. S. Rokhsar for useful discussions. This research was supported by National Science Foundation Grant No. DMR-9120269 and by the Office of Energy Research, Office of Basic Energy Sciences, Materials Sciences Division of the U.S. Department of Energy under Contract No. DE-AC03-76SF00098. VHC was supported by a National Science Foundation Fellowship and a National Defense Science and Engineering Graduate Fellowship.

References

1. S. Saito, A. Oshiyama, and Y. Miyamoto, *Proceedings of Computational Physics for Condensed Matter Phenomena—Methodology and Applications*, Osaka, Japan, October 21–23, 1991, Ed. M. Imada, S. Miyashita, et al. Springer-Verlag.

2. S. Saito, Fall 1990 Materials Research Society Proceedings, *Clusters and Cluster Assembled Materials,* Boston, MA, ed. R. S. Averback, D. L. Nelson, and J. Bernholc.

3. N. Troullier and J. L. Martins, *Phys. Rev. B* **46,** 1754 (1992).

4. S. Saito and A. Oshiyama, *Phys. Rev. Lett.* **66,** 2637 (1991).

5. S. Saito and A. Oshiyama, *Phys. Rev. B* **44,** 11536 (1991).

6. J. L. Martins and N. Troullier, *Phys. Rev. B* **46,** 1766 (1992).

7. S. C. Erwin and M. R. Pederson, *Phys. Rev. Lett.* **67,** 1610 (1991).

8. V. de Coulon, J. L. Martins, and F. Reuse, *Phys. Rev. B* **45,** 13671 (1992).

9. P. A. Heiney, J. E. Fischer, A. R. McGhie, W. J. Romanov, A. M. Denenstein, J. P. McCauley, Jr., and A. B. Smith III, *Phys. Rev. Lett.* **66,** 2911 (1991).

10. J. H. Weaver, J. L. Martins, T. Komeda, Y. Chen, T. R. Ohno, G. H. Kroll, N. Troullier, R. E. Haufler, and R. E. Smalley, *Phys. Rev. Lett.* **66,** 1741 (1991).

11. M. B. Jost, N. Troullier, D. M. Poirier, J. L. Martins, J. H. Weaver, L. P. F. Chibante, and R. E. Smalley, *Phys. Rev. B* **44,** 1966 (1991).

12. E. Burstein, S. C. Erwin, M. Y. Jiang, and R. P. Messmer, *Physica Scripta* **T41,** in press.

13. S. Saito and A. Oshiyama, *Physica C* **185,** 421 (1992).

14. J. L. Martins and N. Troullier, *Phys. Rev.* **46,** 1766 (1992).

15. M.-Z. Huang, Y.-N. Xu, and W. Y. Ching, *Phys. Rev. B* **46,** 6572 (1992).

16. M.-Z. Huang, Y.-N. Xu, and W. Y. Ching, *J. Chem. Phys.* **96,** 1648 (1992).

17. S. C. Erwin and W. E. Pickett, *Science* **254,** 842 (1991).

18. N. Hamada, S. Saito, Y. Miyamoto, and A. Oshiyama, *Japanese J. Appl. Phys. II Lett.* **30,** L2036 (1991).

19. P. W. Stephens, L. Mihaly, P. L. Lee, R. L. Whetten, S.-M. Huang, R. Kaner, F. Deiderich, and K. Holczer, *Nature* **351,** 632 (1991).

20. M. P. Gelfand and J. P. Lu, *Phys. Rev. Lett.* **68,** 1050 (1992).

21. M. P. Gelfand and J. P. Lu, *Phys. Rev. B* **46,** 4367 (1992).

22. O. Zhou, J. E. Fischer, N. Coustel, S. Kycia, Q. Zhu, A. R. McGhie, W. J. Romanow, J. P. McCauley, Jr., A. B. Smith III, and D. E. Cox, *Nature* **351,** 462 (1991).

23. D. A. Neumann, J. R. D. Copley, W. A. Kamitakahara, J. J. Rush, R. L. Cappelletti, N. Coustel, J. P. McCauley, Jr., J. E. Fischer, A. B. Smith III, K. M. Creegan, and D. M. Cox, *Phys. Rev. Lett,* **67,** 3808 (1991).

24. O. Zhou, G. B. M. Vaughan, Q. Zhu, J. E. Fischer, P. A. Heiney, N. Coustel, J. P. McCauley, Jr., and A. B. Smith III, *Science* **255,** 833 (1992).

25. K. Prassides, J. Tomkinson, C. Christides, M. J. Rosseinsky, D. W. Murphy, and R. C. Haddon, *Nature* **354,** 462 (1991).

26. K. Sinha, J. Menéndez, B. L. Ramakrishna, and Z. Iqbal, preprint.

27. V. H. Crespi, J. G. Hou, X.-D. Xiang, M. L. Cohen, and A. Zettl, *Phys. Rev. B* **46,** 12064 (1992).

28. W. E. Pickett, *Bull. Am. Phys. Soc.* **37,** 455 (1992).

29. J. E. Rowe, R. A. Malic, and E. E. Chaban, *Bull. Am. Phys. Soc.* **37,** 192 (1992).

30. F. C. Zhang, M. Ogata, and T. M. Rice, *Phys. Rev. Lett.* **67,** 3452 (1991).

31. M. S. Dresselhaus, G. Dresselhaus, K. Sugihara, I. L. Spain, and H. A. Goldberg, *Graphite Fibers and Filaments,* Springer Ser. Mat. Sci., Vol. 5, Springer, Berlin, Heidelberg, 1988.

32. J. L. Martins, N. Troullier, and M. Schabel, preprint.

33. P. J. Benning, J. L. Martins, J. H. Weaver, L. P. F. Chibante, and R. E. Smalley, *Science* **252,** 1417 (1991).

34. D. S. Bethune, G. Meijer, W. C. Tang, J. H. Rosen, W. G. Golden, H. Seki, C. A. Brown, and M. S. deVries, *Chem. Phys. Lett.* **179,** 181 (1991).

35. D. E. Weeks and W. G. Harter, *J. Chem. Phys.* **90,** 4744 (1989).

36. P. B. Allen, *Phys. Rev. B* **6,** 2577 (1972).

37. M. G. Mitch, S. J. Chase, and J. S. Lannin, *Phys. Rev. Lett.* **68,** 883 (1992).

38. J. G. Bednorz and K. A. Müller, *Z. Phys. B* **64,** 189 (1986).

39. A. F. Hebard, M. J. Rosseinsky, R. C. Haddon, D. W. Murphy, S. H. Glarum, T. T. M. Palstra, A. P. Ramirez, and A. R. Kortan, *Nature* **350,** 600 (1991).

40. C. M. Varma, J. Zaanen, and K. Raghavachari, *Science* **254,** 989 (1991).

41. M. Schluter, M. Lannoo, M. Needels, and G. A. Baraff, *Phys. Rev. Lett.* **68,** 526 (1992).

42. R. A. Jishi and M. S. Dresselhaus, *Phys. Rev. B* **45,** 2597 (1992).

43. I. I. Mazin, S. N. Rashkeev, V. P. Antropov, O. Jepsen, A. I. Liechtenstein, and O. K. Anderson, *Phys. Rev. B* **45,** 5114 (1992).

44. J. Bardeen, L. N. Cooper, and J. R. Schrieffer, *Phys. Rev.* **106,** 162 (1957); **108,** 1175 (1957).

45. G. M. Eliashberg, *JETP* **11,** 696 (1960).

46. W. L. McMillan, *Phys. Rev.* **167,** 331 (1968).

47. G. Sparn, J. D. Thompson, R. L. Whetten, S.-M. Huang, R. B. Kaner, F. Diederich, G. Grüner, and K. Holczer, *Phys. Rev. Lett.* **68,** 1228 (1992).

48. R. M. Fleming, A. P. Ramirez, M. J. Rosseinsky, D. W. Murphy, R. C. Haddon, S. M. Zahurak, and A. V. Makhija, *Nature* **352,** 787 (1991).

49. A. Oshiyama and S. Saito, *Solid State Comm.* **82,** 41 (1992).

50. V. Z. Kresin, *Bull. Am. Phys. Soc.* **32,** 796 (1987).

51. L. C. Bourne, A. Zettl, T. W. Barbee III, and M. L. Cohen, *Phys. Rev. B* **36,** 3990 (1987).

52. Z. Zhang, C.-C. Chen, and C. Lieber, *Science* **254,** 1619 (1991).

53. L. D. Rotter, Z. Schlesinger, J. P. McCauley, Jr., N. Coustel, J. E. Fischer, and A. Smith III, *Nature* **355,** 532 (1992).

54. J. G. Hou, V. H. Crespi, X.-D. Xiang, W. A. Vareka, G. Briceño, A. Zettl, and M. L. Cohen, *Solid State Comm.,* submitted.

55. Y. J. Uemura et al., *Nature* **352,** 605 (1991).

56. N. R. Werthamer, *Superconductivity,* ed. R. D. Parks, Marcel Dekker, Inc., New York, 1969, p. 338.

57. M. K. Crawford, W. E. Farneth, E. M. McCarron III, R. L. Harlow, and A. H. Moudden, *Science* **250,** 1390 (1990).

58. V. H. Crespi and M. L. Cohen, *Phys. Rev. B* **44,** 4712 (1991).

59. T. W. Ebbeson, J. S. Tsai, K. Tanigaki, J. Tabuchi, Y. Shimakawa, Y. Kubo, I. Hirosawa, and J. Mizuki, *Nature* **355,** 620 (1992).

60. A. A. Zakhidov, K. Imaeda, D. M. Petty, K. Yakushi, H. Inokuchi, K. Kikuchi, I. Ikemoto, S. Suzuki, and Y. Achiba, *Phys. Lett.* **164,** 355 (1992).

61. A. P. Ramirez, A. R. Kortan, M. J. Rosseinsky, S. J. Duclos, A. M. Mujsee, R. C. Haddon, D. W. Murphy, A. V. Makhija, S. M. Zahurak, and K. B. Lyons, *Phys. Rev. Lett.* **68,** 1058 (1992).

62. C.-C. Chen and C. M. Lieber, *J. Am. Phys. Soc.* **144,** 3141 (1992).

63. D. S. Bethune, W. Y. Lee, M. S. deVries, J. R. Salem, W. C. Tang, H. J. Rosen, and C. A. Brown, preprint.

64. P. W. Anderson, preprint.

65. E. Schachinger, M. G. Greeson, and J. P. Carbotte, *Phys. Rev. B* **42,** 406 (1990).

66. V. H. Crespi and M. L. Cohen, *Phys. Rev. B,* submitted.

67. J. M. Hawkins, A. Meyer, S. Loren, and R. Nunlist, *J. Am. Chem. Soc.* **113,** 9394 (1991).

68. D. M. Deaven, P. E. Lammert, and D. S. Rokhsar, preprint.

69. X.-D. Xiang, J. G. Hou, G. Briceño, W. A. Vareka, R. Mostovoy, A. Zettl, V. H. Crespi, and M. L. Cohen, *Science* **256,** 1190 (1992).

70. G. Grimvall, *The Electron–Phonon Interaction in Metals,* North-Holland Publishing Company, 1991, pp. 210–23.

71. E. Sohmen, J. Fink, and W. Krätschmer, *Europhys. Lett.* **17,** 51 (1992).

72. B. Stritzker and W. Buckel, *Z. für Phys.* **257,** 1 (1972).

73. V. B. Ginodman and L. N. Zherikhina, *Sov. J. Low Temp. Phys.* **6,** 278 (1980).

74. V. H. Crespi and M. L. Cohen, *Solid State Comm.* **83,** 427 (1992).

75. P. W. Stephens, L. Mihaly, J. B. Wiley, S.-M. Huang, R. B. Kaner, and F. Diederich, *Phys. Rev. B* **45,** 543 (1992).

76. R. M. Fleming, M. J. Rosseinsky, A. P. Ramirez, D. W. Murphy, J. C. Tully, R. C. Haddon, T. Siegrist, R. Tycko, S. H. Glarum, P. Marsh, G. Dabbagh, S. M. Zahurak, A. V. Makhija, and C. Hampton, *Nature* **352,** 701 (1991).

77. V. H. Crespi and M. L. Cohen, *Phys. Rev. B* **43,** 12921 (1991).

78. J. Hubbard, *Proc. Roy. Soc. Lond. Ser. A* **276,** 238 (1963).

79. S. R. White, S. Chakravarty, M. P. Gelfand, and S. A. Kivelson, *Phys. Rev. B* **45,** 5062 (1992).

80. S. Chakravarty, M. P. Gelfand, and S. A. Kivelson, *Science* **254,** 970 (1991).

81. S. Chakravarty and S. A. Kivelson, *Europhys. Lett.* **16,** 751 (1991).

82. P. E. Lammert and D. S. Rokhsar, preprint.

83. A. J. Heeger, S. Kivelson, J. R. Schrieffer, and W.-P. Su, *Rev. Mod. Phys.* **60,** 781 (1988).

84. S. Chakravarty, S. A. Kivelson, M. I. Salkola, and S. Tewari, preprint; *Bull. Am. Phys. Soc.* **37,** 524 (1992).

85. W. E. Goff and P. Phillips, *Bull. Am. Phys. Soc.* **37,** 355 (1992).

9

Electronic Structure of the Alkali-Intercalated Fullerides, Endohedral Fullerenes, and Metal-Adsorbed Fullerenes

Steven C. Erwin

Man muß versuchen, bis zum Äußersten ins Innere zu gelangen. Der Feind des Menschen ist die Oberfläche.
(Samuel Beckett)

You got to love it from the inside out—otherwise it will kill you.
(Danny Laporte—World Champion 250 cm³ Motor Cross)

9.1 Introduction

Our understanding of the electronic structure of the C_{60} fullerenes and their condensed-phase derivatives has advanced largely through a recent convergence of theoretical and experimental efforts. Much early experimental work has concentrated on synthesis, structural characterization, and physical properties. Interest in electronic properties grew rapidly only after methods for synthesizing gram quantities of C_{60} were developed in 1990 [1], and then dramatically after the discovery in 1991 of superconductivity at 18 K in the potassium-doped fulleride $K_x C_{60}$ [2]. Conversely, early theoretical work concentrated on the nature of the electronic states and their distribution in molecular C_{60}, and was largely semiempirical. Only recently has the availability of solvent-free samples led to reliable x-ray structural characterizations of the solid phases, a prerequisite for first-principles theoretical studies of electronic structure. At the same time, a wide variety of experimental techniques (photoemission, optical absorption, electron energy-loss spectroscopy, nuclear magnetic resonance) has been used to probe the electronic states of the fullerenes and fullerides. Detailed comparison of theoretical and experimental results has led to a consistent description of the electronic structure of these materials, although important differences still remain to be addressed.

In this chapter the focus will be theoretical, and limited to three broad topics: the alkali-intercalation fulleride crystals, endohedral fullerene complexes (atoms encapsulated in C_{60} molecules), and adsorption of C_{60} molecules on simple metal surfaces. The first topic has been the subject of a great many experimental investigations, and will be treated here with first-principles electronic-structure methods. The latter two are just beginning to receive significant experimental attention, and the discussion here will be in terms of simple models that provide an understanding of the most important aspects of those systems.

9.2   Electronic Structure of K_xC_{60} ($x = 0, 3, 4, 6$) Fulleride Crystals

In this section, first-principles band-structure calculations for the zero-temperature stoichiometric phases of A_xC_{60} ($x = 0, 3, 4, 6$) are described. The calculations were all performed with A = potassium. Although results for the Rb_xC_{60} series have been shown (both theoretically and experimentally) to be qualitatively very similar to K_xC_{60}, the electronic behavior of Li- and Na-doped fullerides is quite different [3], and may reflect the proximity of the A_xC_{60} materials to a metal–insulator transition. In this regard, no claim is made or implied here concerning the merits of a band description relative to other methods. Indeed, in light of experimental evidence of important electron–electron correlations and of intrinsic orientational disorder in these materials, one of the goals of this section is to understand the extent to which a Bloch description of electron states in A_xC_{60} is valid.

9.2.1   Computational Method

All the calculations described in this section use the local-density approximation (LDA) to density-functional theory. We solve the Kohn–Sham equations using a local-orbital method, in which the Bloch basis functions are expanded on a fixed set of Gaussian-orbital basis functions. The Primitive Gaussian functions are contracted to form linear-combination-of-atomic-orbitals (LCAO) basis functions corresponding to all the occupied, and several unoccupied, atomic states; this set of basis functions is augmented by primitive Gaussian functions until the basis set is sufficiently complete to give converged eigenvalue spectra. The choice of local-orbital functions allows compact basis sets to be used (~ 15 basis functions per atom in this case) to perform accurate all-electron calculations without the need for pseudopotentials.

Minimization of the LDA total energy then leads to a standard eigenvalue problem, in which core, valence, and conduction states are self-consistently relaxed on equal footing. The form of the potential and electronic charge density is completely general and without shape approximations. Electron correlation effects are included through the Ceperly–Alder exchange-correlation functional.

For the carbon basis sets, four s-type and three p-type LCAO functions were formed from thirteen primitive Gaussian functions with exponents in the range 0.2481 to 9470; for the potassium basis sets, five s-type and four p-type functions were formed from fourteen Gaussian exponents in the range 0.1000 to 128908. This local-orbital method, as well as the basis sets, have been extensively tested, and give eigenvalue spectra and charge distributions essentially indistinguishable from state-of-the-art linearized augmented-plane-wave methods. Further details of the computational method, and numerical comparisons with other methods, are given in Ref. [4].

For each of the fullerides considered here, the charge density and potential were iterated to self-consistency using the Γ point for Brillouin-zone integration; because of the large unit cells, this is sufficient for calculating band structure and charge densities. No assumptions about rigid-band behavior were made: Each calculation was performed independently, using identical methods and basis sets. Band structures were calculated by direct diagonalization at $\sim$50 k points along high-symmetry directions. Densities of states were calculated by the linear tetrahedron method, using $\sim$2500 k points in the irreducible 1/24 of the zone; the energies at these points were calculated by direct diagonalization at $\sim$75 inequivalent k points, followed by Fourier interpolation to the denser meshes.

9.2.2   Solid C_{60}

Although the undoped phase of solid C_{60} might seem the simplest starting point for a discussion of the doped ($x = 3$, 4, 6) phases, its structural description is quite complex. At 249 K, solid C_{60} undergoes a first-order phase transition, from a room-temperature orientationally disordered face-centered-cubic (fcc) structure, to a low-temperature orientationally ordered simple-cubic (sc) structure with four C_{60} molecules per unit cell [5]. The effect of this transition on electronic properties is not well understood. Measurements of the temperature dependence of the photoconductivity of C_{60} films show a pronounced peak in the dark current near 250 K [6], suggesting that the low-temperature phase may have a slightly smaller band gap than the room-temperature phase, but this interpretation is complicated by the observation that even at 14 K, considerable orientational disorder ($\sim$30%) persists in the nominally ordered phase [7].

Disordered structures are beyond the reach of current first-principles electronic-structure methods. We take a pragmatic approach by assuming an ordered structure that is computationally manageable. For solid C_{60}, a good starting point is an ordered fcc structure with one fullerene per unit cell. The fullerene molecule is assumed to be unchanged from the gas phase, with a cage radius of 3.57 Å and two different bond lengths, 1.40 and 1.46 Å. We choose an orientation given by aligning the twofold axes of the C_{60} with the Cartesian axes of the lattice, so that the resulting point group is T_h (24 symmetry operations), and we use a lattice constant (14.2 Å) corresponding to the room-temperature fcc Bravais lattice.

The resulting energy band structure for solid C_{60} is shown in Figure 9.1. The ground state is insulating, with a direct gap at X of 1.23 eV. The valence band (VB) holds ten electrons, and arises from C_{60} molecular h_u states broadened into a band state 0.63 eV wide. The conduction band (CB) is formed from molecular states of t_{1u} symmetry, broadened into a band 0.55 eV wide. These results are all within 0.10 eV of a recent LCAO calculation [8] and a

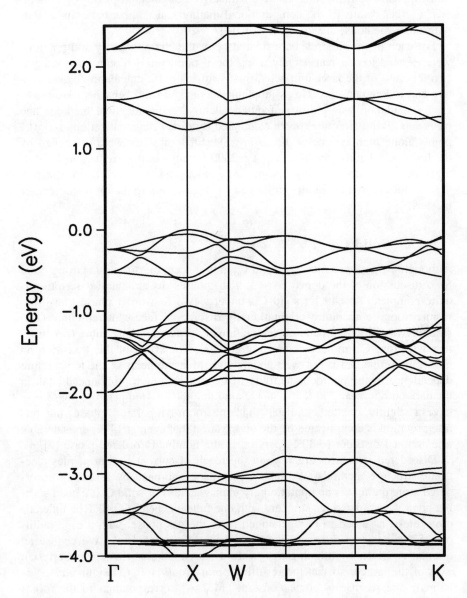

Figure 9.1   Theoretical self-consistent band structure for solid C_{60}.

recent plane-wave pseudopotential calculation [9]. They are quite different, however, from an early Gaussian-based pseudopotential calculation [10], which found a gap 0.27 eV larger and a VB width 0.21 eV smaller (symptomatic of an underconverged basis set), as well as qualitatively different features in the band structure. Total energy studies [9] have shown that the cohesive energy and bulk modulus of solid C_{60} are characteristic of a van der Waals bonding mechanism, and are in good agreement with compressibility experiments.

9.2.3 K_6C_{60}

We turn next to the saturation-doped phase, K_6C_{60}, the band structure for which was first discussed in Ref. [11]. Six alkali metals per fullerene cannot be accommodated in an fcc lattice, and consequently this phase exists only in a body-centered-cubic (bcc) Bravais lattice [12]. Each C_{60} may be viewed as surrounded by a cage of 24 K atoms, within a conventional cubic cell of lattice constant 11.39 Å. Room-temperature x-ray diffraction studies indicate orientationally ordered C_{60} molecules with bond lengths that are unchanged from the undoped and gas phases. The calculated energy band structure, shown in Figure 9.2, predicts an insulating ground state, in agreement with photoemission and optical absorption (see Section 9.2.5), with an indirect gap of 0.48 eV, VB maximum at Γ, and CB minimum at N. The VB is of t_{1u} symmetry at Γ, and is 0.53 eV wide. Although the BZ for the bcc lattice is different from the fcc lattice of C_{60}, the close correspondence of the bands is clear: In particular, the threefold lowest-lying CB of C_{60} has been filled by six electrons (from K $4s$ states) to become the highest VB of K_6C_{60}. Further evidence in support of this interpretation comes from analysis of the orbital character of the VB states in K_6C_{60}, which are almost completely derived from C basis functions, with only a small (less than 4%) admixture of K character. These findings strongly suggest that, in contrast to the undoped phase, bonding in K_6C_{60} is almost entirely ionic, with each C_{60} molecule binding six excess electrons. The -6 oxidation state of C_{60} is energetically very unstable in the gas phase (see Section 9.3.1); the additional energy to ionize six $4s$ electrons from K atoms results in a total energy cost of 62.1 eV/unit cell for complete charge transfer (assuming a dilute lattice). Energetic stability arises only from the Coulomb interaction (Madelung energy) between the charged constituents: For point charges, this energy gain amounts to 71.4 eV/unit cell, more than enough to make the charge transfer energetically favorable.

Although the charge transfer from K is essentially complete, the distribution of this charge on the C_{60} is not uniform, for reasons that are partly electrostatic and partly quantum mechanical. One way to characterize this nonuniformity is to calculate the total amount of charge Q_n associated with each atom n. For K_6C_{60}, there are three inequivalent C atoms (C_1, C_2, C_3) and just one type of K atom. Equipartitioning of the K $4s$ electron would lead to uniform charges $C_1^{eq} = C_2^{eq} = C_3^{eq} = 6.1$ and $K^{eq} = 18$. Mulliken population analysis, which uses the LDA eigenvectors to compute the self-consistent charge distribution,

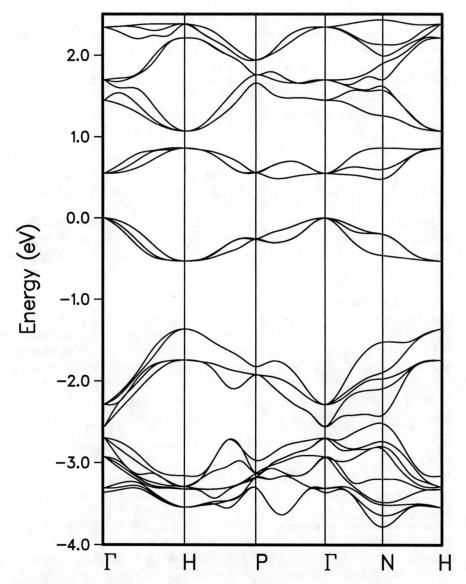

Figure 9.2 Theoretical self-consistent band structure for K_6C_{60}.

shows a slight deviation from equipartitioning: a surplus of 0.02 electrons on C_1 and a deficit of 0.02 and 0.03 electrons on C_2 and C_3, respectively. (Although these numbers seem small relative to the total charge per atom, they should be viewed as deviations from the nominal value of 0.10 excess electrons.) This trend is also reflected in core-level shifts of ~0.1 eV. The origin of this nonuniformity is simply the electrostatic attraction between the excess charge on the C_{60} and the nearby positive K^+ ions: Since the C_1–K distance is the shortest, it is electrostatically most favorable (in principle) to localize 0.50 excess electrons on each of the twelve C_1 atoms. Localization incurs a quantum penalty by raising the kinetic energy, and so the self-consistent charges settle for a compromise between equipartitioning and complete localization (see Table 9.1). We have previously proposed [11] that the suppression of orientational disorder observed in K_6C_{60} arises from the electrostatic coupling between excess charge on the C_{60} and the K^+ ions and, in particular, that deviation from equipartitioning implies a barrier to free rotation. This nonuniformity, which may be manifested experimentally in C core-level spectroscopy, is a key element of this interpretation: Without it there is no evidence for electrostatic coupling between the C_{60} and K^+ ions, and hence no barrier (or only a very small barrier) to rotation.

Finally, we can get a detailed picture of where the excess charge resides by plotting, in Figure 9.3(a), the valence electron density in the plane of one of the C_{60} hexagons. The hexagon contains the three inequivalent C atoms (C_1 at the bottom, C_2 in the middle, and C_3 at the top of the figure). The alternation of bond lengths within the hexagon leads to a slight excess of electron density in the shorter bonds (C_1–C_1 and C_2–C_3) relative to the longer bonds (C_1–C_2 and C_3–C_3). In order to understand how the presence of K^+ ions affects the distribution of excess charge within this hexagon, we subtract off the self-consistent valence density from a (hypothetical) bcc C_{60} crystal, and show the residual density in Figure 9.3(b). If the six extra electrons were distributed uniformly on each C atom, this plot would show a threefold symmetry, and each C would show identical contours. This is clearly not the case: Relatively large amounts of charge have actually been removed from the bonding regions of C atoms nearest to the K^+ ions. Apparently, one effect of K incorporation in forming K_6C_{60} is to weaken some of the bonds in the C_{60} molecule. This effect has recently been confirmed by other calculations [13].

Finally, Figure 9.4 shows the charge density from states in the filled t_{1u} band. The radially directed C p_z-like states are plainly evident. Only very small

Table 9.1 Charge on each of the four inequivalent atoms in the K_6C_{60} unit cell. Q_i^{eq} denotes the (ideal) charge on each atom that would result from full ionization of K and equipartitioning of the excess charge on the 60 C atoms. Q_i^{sc} is the self-consistent charge on each atom, calculated from the LDA eigenfunctions. Q_i^M is the (ideal) charge distribution that minimizes the Madelung energy, subject to constraints discussed in the text

Atom	Equivalent atoms/cell	Q_i^{eq}	Q_i^{sc}	Q_i^m
C_1	12	6.10	6.12	6.50
C_2	24	6.10	6.08	6.00
C_3	24	6.10	6.07	6.00
K	6	18.00	18.17	18.00

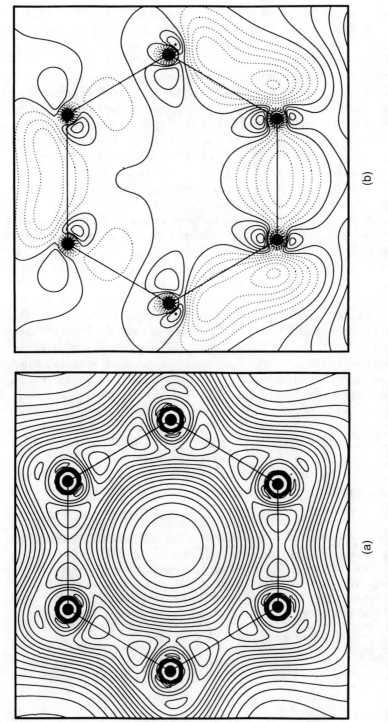

(a)

(b)

Figure 9.3 (a) Self-consistent valence-electron density for K_6C_{60}, in the plane of a hexagon containing the three inequivalent carbon atoms C_1, C_2, C_3. Contours are separated by 0.020 a.u., which is also the value of the lowest contour. (b) Same as above, with the valence density from a hypothetical bcc (undoped) C_{60} crystal subtracted off. Solid contours denote charge surplus, and dotted contours denote charge deficit. The contour spacing is 0.005 a.u.

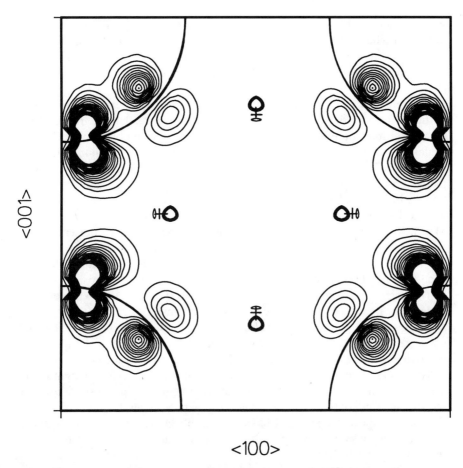

<001>

<100>

Figure 9.4 Self-consistent density from the filled t_{1u} band of K_6C_{60}, in a plane containing the centers of four C_{60} molecules (at the corners of the plot area) and four K^+ ions (marked by crosses). The C_{60} cages are indicated by quarter circles.

mixing of K p functions is evident, in agreement with the finding of total charge transfer. Also, the electrostatic attraction between charge on the C_{60} and the K^+ ions is manifested here as a polarization of the charge toward those ions.

9.2.4 K_3C_{60}

Since the first discovery of superconductivity in potassium-doped C_{60} [2], much attention has focused on the series of A_3C_{60} phases, which is now known to be superconducting for A = K, Rb, and Cs/Rb alloys [14], but insulating for A = Li and Na [15]. Although a wide variety of superconducting mechanisms has been proposed to explain the relatively high transition temperatures observed, there is a growing body of theoretical and experimental evidence to support standard BCS electron–phonon coupling. Here, the discussion will be limited to what can be learned from electronic-structure calculations regarding both the normal and superconducting states. A detailed treatment of normal-state transport properties within a Fermi-liquid picture has been given in Ref. [16].

Rietveld analysis of powder x-ray data for K_3C_{60} indicates an fcc lattice (lattice constant 14.24 Å) with K atoms filling the available tetrahedral and octahedral interstitial sites [17]. The best fit was obtained by placing C_{60} molecules randomly in one of two orientations related by 90° rotation, each populated by 50%. In a sense, this degree of orientational order is midway between the (room-temperature) freely-rotating fullerenes of solid C_{60} and the orientationally ordered fullerenes of K_6C_{60}. As with C_{60}, we choose a single orientation and explore the consequences of that choice. Pseudopotential calculations of Martins and Troullier [18] are in excellent agreement with the present results. The first account of this work originally appeared in Ref. [19].

The band structure for uniorientational K_3C_{60} is shown in Figure 9.5. The bands are nearly identical to those of solid C_{60}, with the three extra electrons half-filling the t_{1u} band. The width of this band is 0.61 eV, approximately 10% wider than in solid C_{60}. Figure 9.6 shows the density of states (DOS) for the t_{1u} band; the calculated Fermi-level DOS for both spins is $N(\varepsilon_F) = 13.2$ states/ eV cell. Two of the three bands cross the Fermi level, so that the Fermi surface has two sheets, as shown in Figure 9.7. One sheet forms a roughly spherical hole pocket, centered at Γ. Small outward protrusions, along the eight equivalent $\langle 111 \rangle$ directions, make contact with similar inward protrusions from the second sheet, a consequence of a band degeneracy along the $\langle 111 \rangle$ directions. The second sheet is open, and has considerably more structure. Thin double tubes extend outward along Cartesian axes to make contact with the square faces of the fcc Brillouin zone. In a repeated-zone representation (Fig. 9.8), these tubes connect to next-nearest-neighbor zones, creating two symmetry-equivalent sheets that are completely interlinked, but that never touch.

In the presence of a magnetic field, electrons in the normal state follow equations of motion dictated by the shape of orbits on the Fermi surface. A number of possible electron and hole orbits can be identified here, some of which are sketched in Figure 9.9. These orbits can be quantified by calculating the frequencies of de Haas–van Alphen oscillations, given by $f = (\hbar c/e)A$, where A is the cross-sectional area enclosed by an extremal orbit, and by the corresponding cyclotron effective masses $m^*/m = (\hbar^2/2\pi m)\, \partial A/\partial E$. The numerical results, tabulated in Table 9.2, show very small frequencies (a consequence of the lattice real-space cell) and a wide range of effective masses, from 0.6 (for orbits around the thin connecting tubes) to 8.4 (for orbits normal to $\langle 110 \rangle$ on the ''body'' of the second sheet).

Gradients of the electronic energy bands enter into both the normal theory of metals and the BCS theory of superconductivity. In order to calculate band gradients, we have previously used a Fourier-interpolation method of Pickett et al. [20], fitting the first-principles energies at 61 k points to a symmetrized

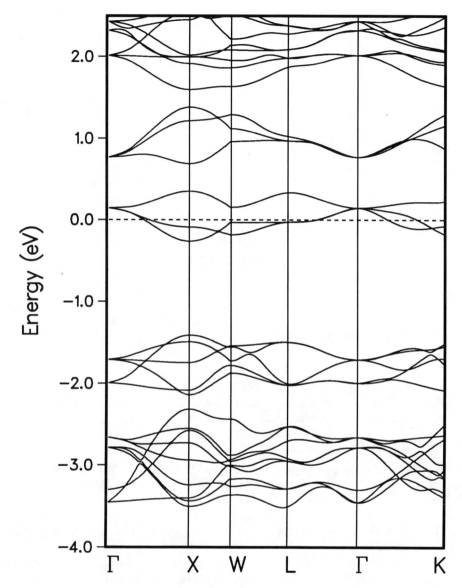

Figure 9.5 Theoretical self-consistent band structure for K_3C_{60}.

combination of a large number of plane waves. The extra variational freedom in this fit is then used to minimize a "roughness" factor, allowing accurate interpolation to much denser k-point meshes, and accurate evaluation of the Bloch electron velocities, $\mathbf{v}_{n\mathbf{k}} = \mathbf{\nabla}_k \varepsilon_{n\mathbf{k}}$ (where $\varepsilon_{n\mathbf{k}}$ is the energy at $\mathbf{k}$ of the nth band). The density-of-states-weighted magnitude of this quantity, averaged over all sheets of the Fermi surface, then defines the Fermi velocity, v_F, an important

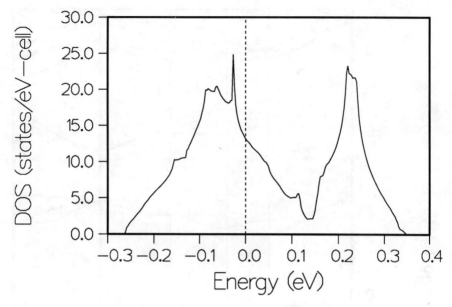

Figure 9.6 Theoretical density of states for K_3C_{60}, assuming a uniorientational structure. The dotted line marks the Fermi level.

quantity in the theory of metals. For K_3C_{60}, this has the value $v_F = 1.8$ m $\times$
10^7 cm s^{-1}, considerably smaller than the free-electron value of 5.7 m $\times 10^7$
cm s^{-1}. Another important quantity is the Drude plasma frequency, $\Omega_p^2 =$
$(4\pi/3)e^2 v_F^2 N(\varepsilon_F)$, for which we find $\hbar\Omega_p = 1.2$ eV. In BCS theory, this quantity determines the clean-limit London penetration depth, $\Lambda = c/\Omega_p = 1600$
Å. The corresponding value for a dirty superconductor is larger by a factor $(1 + \xi/l)^{1/2}$, where l is the mean free path and ξ is the intrinsic coherence length;
the resulting value is roughly 3000–3500 Å; this is in quite reasonable agree-
ment with the reported values of 2400 Å (lower critical-field measurements
[21]), 4800 Å (muon-spin relaxation [22]), and 6000 Å (^{13}C nuclear magnetic
resonance [23]).

Finally, as we have previously reported [19], one can use the measured
coherence length, $\xi = 26$ Å [21], along with our calculated Fermi velocity to
calculate the superconducting gap parameter $\Delta = \hbar v_F/\pi\xi = 14$ meV and BCS
ratio $2\Delta/k_B T_c = 17$. Although this is significantly larger than the measured
BCS ratios of 5.3 (tunneling measurements [24]) and 3–5 (infrared reflectivity
[25]), they are all consistent with strong coupling. The quantitative discrepancy
may be partly due the assumption of an orientationally ordered K_3C_{60}: Gelfand
and Lu [26] have shown that random disorder rearranges the spectral density,
eliminating the sharp structure in the DOS of Figure 9.6, but leaves the bandwidth unchanged and the electron states itinerant. However, since v_F is essentially a measure of the bandwidth, it is unlikely that a factor of $\sim$3 discrepancy

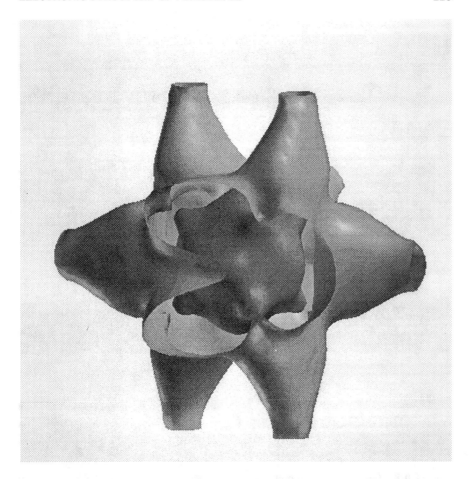

Figure 9.7 The two sheets of the K_3C_{60} Fermi surface, each enclosing unoccupied states. Part of the second sheet is cut away to reveal the first sheet inside.

could be accounted for in this way. Another difficulty may lie with the interpretation of the measured coherence length: Palstra et al. [27] have argued that the measured ξ may be sensitive to granularity effects, and that the intrinsic coherence length is actually 150 Å. None of these issues has been definitively settled as of this writing.

9.2.5   Photoemission Results

Several comprehensive reviews of photoemission studies on A_xC_{60} have already appeared; for excellent detailed discussions, see especially Refs. [28] and [29]. Here we provide only a short summary of key results and a comparison to calculated results. Other experimental probes of electronic structure, notably

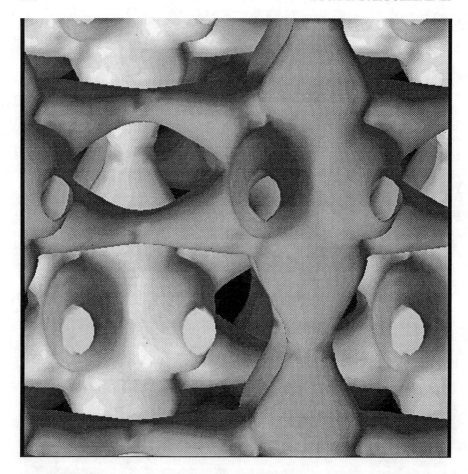

Figure 9.8 Repeated-zone representation for the second sheet of the K_3C_{60} Fermi surface. The two symmetry-equivalent pieces are multiply connected along Cartesian axes; although interpenetrating, they never touch.

optical absorption [30], electron energy-loss spectroscopy [31], and NMR [32], are generally in agreement with the findings of photoemission, and so will not be discussed here.

The utility of photoelectron spectroscopy rests upon two approximations: (1) the independent-particle model, and (2) constant matrix elements for transitions between initial and final states. To the extent that these assumptions hold, the distribution of photons (for photoemission) or electrons (for inverse photoemission) reflects the electronic density of states.

For reference, Figure 9.10 shows the densities of states calculated here for the phases $x = 0, 3, 4, 6$; for convenience, all are referenced to the centroid of the t_{1u} band (which is the actual Fermi level only for K_3C_{60}). No other energy shifts have been made in these curves. There is a remarkable similarity across

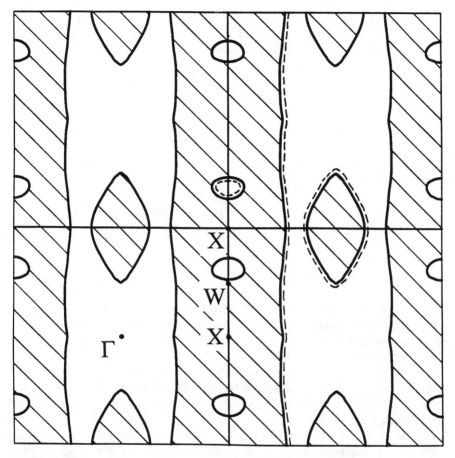

Figure 9.9 Repeated-zone representation, in the $k_z = 0$ plane, of the second Fermi-surface sheet of K_3C_{60}; occupied regions are cross-hatched. Several extremal electron and hole orbits are indicated by dotted curves.

the entire series. Apparently the effect of K incorporation on the distribution of electronic states is small (but not negligible), serving to a first approximation only as an electron source. It is worth noting, however, that the gap between the top of the h_u band and the bottom of the t_{1u} band decreases with t_{1u} filling (this is the band gap for solid C_{60}, but simply a gap between two filled bands for K_6C_{60}), from 1.23 eV for C_{60} to 0.85 eV for K_6C_{60}. Other gaps, between both occupied and unoccupied bands, also tend to decrease as more K is incorporated. Most of this effect is due to broadening of the bands with increasing x, rather than with shifting of their centroids. The effect is largest for bands further from ε_F: Indeed, the width of the t_{1u} is relatively constant around 0.5 eV, while the width of the band centered at 1 eV actually *decreases*, from 0.63 eV for C_{60} to 0.38 eV for K_6C_{60}.

Table 9.2 De Haas–van Alphen frequencies f and cyclotron effective masses m^*/m for the external orbits in K_3C_{60}. The orbits for $\langle 100 \rangle$ fields on the second Fermi-surface (FS) sheet are sketched in Figure 9.9

FS sheet	Orbit center	Particle	Field direction	f (MG)	m^*/m
1	Γ	hole	$\langle 100 \rangle$	64	2.2
1	Γ	hole	$\langle 110 \rangle$	70	3.1
2	$\sim(W + X)/2$	hole	$\langle 100 \rangle$	6	0.6
2	Γ	electron	$\langle 100 \rangle$	46	1.9
2	Γ	hole	$\langle 110 \rangle$	178	8.4

Results from both photoemission and inverse photoemission measurements for K_xC_{60} are summarized in Figure 9.11, reproduced from Benning et al. [28] For the undoped phase, the predicted band gap of 1.2 eV is quite far from the measured threshold energy of 2.6 eV. The difference can be viewed in several ways. The first is that the ground-state LDA band structure cannot be reliably used to predict the values of excited-state properties such as the band gap. The second (complementary) view emphasizes the importance of electron–electron correlation [29]. Since photoemission measures the energy required to create a *separated* electron–hole pair, the difference between the photoemission threshold and the HOMO–LUMO excitation threshold is a direct measure of this separation energy (the correlation energy). Indeed, recent cluster-based LDA calculations show that this energy is ~1 eV [33]. In the solid, transition energies calculated on the basis of a ground-state energy spectrum are not expected to reproduce this effect.

Upon doping with K, a well-defined Fermi edge develops, growing with increasing K exposure until full occupation is reached at $x = 6$, at which point the now-filled band moves away from the Fermi level. This behavior is in good qualitative agreement with the predictions of band theory, although the occupied width (~1 eV) of the t_{1u} band near half-filling is substantially greater than the total calculated width of 0.6 eV (which is relatively unchanged upon doping). Other changes are evident in the unoccupied-state spectra [29]: In particular, the separation between the LUMO + 1 and LUMO + 2 bands (referred to the undoped phase) increases with incorporation of K, from 1.0 eV for C_{60} to 1.3 eV for K_3C_{60} to 1.6 eV for K_6C_{60}. This effect is not reproduced by the band calculations.

Nevertheless, the major aspects of the calculated electronic structure are in good agreement with the photoemission data. The picture of progressive filling of the t_{1u} band, to produce insulating C_{60} and K_6C_{60} states and a metallic K_3C_{60} state, is well confirmed experimentally and theoretically. A rigid-band model, although valid as a first approximation, does not describe the detailed changes in either the calculated or measured densities of states, some of which are large. For a more detailed comparison of electronic-structure calculations with photoemission data over the whole VB spectrum, see Ref. [9].

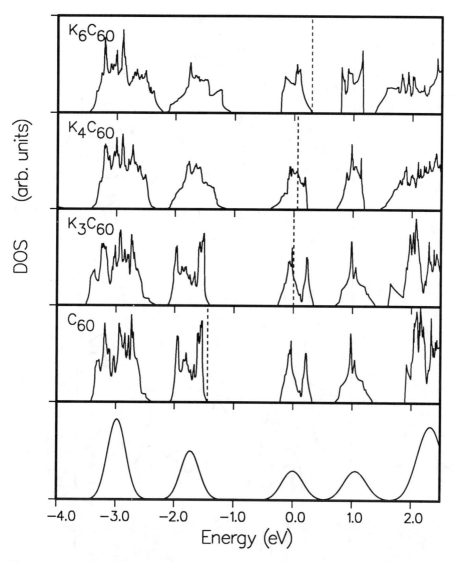

Figure 9.10   Theoretical densities of states for K_xC_{60}, $x = 0$, 3, 4, 6. The bottom panel shows the corresponding (Gaussian-broadened) molecular C_{60} eigenvalues. Within each panel, the valence-band maximum (C_{60}, K_6C_{60}) or Fermi level (K_3C_{60}, K_4C_{60}) is marked by a dotted line.

9.2.6 K_4C_{60}: Failure of the Band Picture?

In contrast to the other pure phases, very little is known about K_4C_{60}. X-ray powder patterns have been indexed with a body-centered tetragonal (bct) structure [34], which is the only noncubic symmetry in the intercalant series $x =$

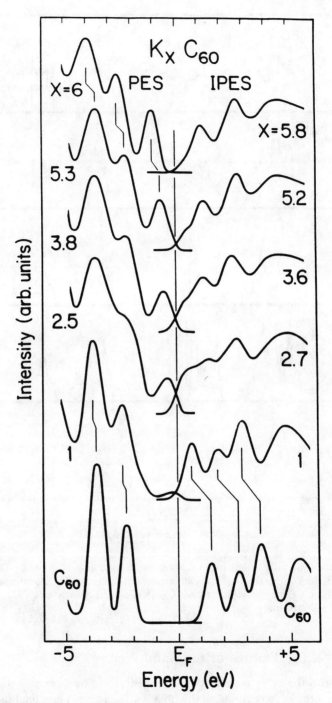

Figure 9.11 Photoemission and inverse photoemission spectra for K_xC_{60}, reproduced from Benning et al., Ref. [28].

0, 3, 4, 6. The bct lattice parameters for K_4C_{60} are $a = 11.886$ Å and $c = 10.774$ Å. Since C_{60} lacks a fourfold axis, the symmetry of an ordered structure with $c/a \neq 1$ can be at most orthorhombic: The symmetry can only be tetragonal if (1) there is an accidental degeneracy of the a and b axes, or (2) the C_{60} molecules are orientationally disordered (or rapidly rotating), at least about the c axis. Since disorder (but not spinning) is observed in the $x = 3$ phase, a disordered scenario may be the most likely scenario for K_4C_{60}, possibly with two randomly populated orientations, as for K_3C_{60}.

No evidence of superconductivity in K_4C_{60} has been found to date. Normal-state conductivity can be probed by NMR: According to the Korringa relaxation mechanism, metallic behavior is characterized by a constant value of $T_1 T$, where T_1 is the ^{13}C relaxation time and T is the temperature. NMR measurements of $T_1 T$ in K_3C_{60} show only a very weak temperature dependence [35], in agreement with photoemission, optical absorption, and band theory. For K_4C_{60}, $T_1 T$ is strongly temperature dependent, indicating an insulating ground state. This finding provides the focus of our discussion.

In Figure 9.12 the LDA band structure for K_4C_{60} is shown, based on a bct lattice with one C_{60} per unit cell, and for which the twofold axis of the C_{60} is aligned with the side of the tetragonal cell. Since the bct Brillouin zone is somewhat unfamiliar, it is convenient to think of the bct conventional unit cell as most closely resembling that of K_6C_{60}, with a slight ($\sim 10\%$) compression along the c axis. This elongates the BZ by a similar factor and introduces more faces, but the basic shape is essentially the same. Indeed, the labeling of high-symmetry k points is almost unchanged, with Γ, H, P, X, M of the bct zone corresponding to Γ, H, P, N, H of the bcc zone, respectively.

It is clear from Figure 9.12 that band theory predicts the uniorientational bct K_4C_{60} structure to have a metallic ground state. There has been some speculation, using rigid-band arguments based on the theoretical DOS for the t_{1u} band of K_3C_{60}, that the Fermi level for K_4C_{60} would fall near the DOS minimum (see Fig. 9.6) that separates the lower two bands from the upper one, possibly leading to a band insulator. This is clearly not the case. Indeed, because of the low symmetry of the bct K_4C_{60} structure (there are only eight point-group operations, arising from three Cartesian mirror planes), all remaining band degeneracies are lifted; this is clear on comparison of the K_4C_{60} bands to those for K_6C_{60}, particularly at the Γ and H points. The band splittings are substantial (roughly 0.1 eV at the zone center and as much as 50% of the bandwidth on the zone faces), leading to a DOS with fewer sharp features than for the cubic symmetries (see Fig. 9.10) and a Fermi-level DOS $N(\varepsilon_F) = 18.3$ states/eV cell, nearly 40% larger than for K_3C_{60}.

This result is strongly at odds with the NMR finding of an insulating ground state. While it is tempting to blame the artificial assumptions made here regarding orientational order, similar assumptions for the $x = 0$ and 3 phases lead to band descriptions in rather good agreement with experiment, and there is no fundamental reason to expect otherwise for the $x = 4$ phase. Likewise, explanations for the insulating behavior of K_4C_{60} based on C_{60} Jahn–Teller

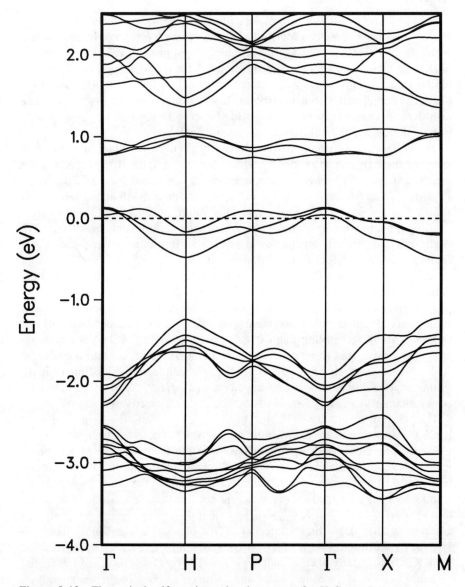

Figure 9.12   Theoretical self-consistent band structure for K_4C_{60}.

distortion or a Mott-insulator transition would seem to apply equally to K_3C_{60}, where no such explanation is required.

One possible resolution would be the formation of a charge-density wave (CDW), the simplest version of which would lead to a commensurate modulation of the lattice and a doubling of the unit cell. It is plausible that such a distortion would be energetically driven: By forming a normal insulating ground state, the gap energy would be gained. As motivation for this scenario, note that the bct structure with $c/a = 0.90$ implies a stronger electronic coupling along the c axis than within the ab plane. This lends the K_4C_{60} structure a qualitative resemblance to an array of one-dimensional metallic chains, which are generally subject (by the Peierls theorem) to a periodic distortion along the chain axis and consequent opening of a gap.

More quantitative evidence can be found in the K_4C_{60} Fermi surface. It is well known that Fermi-surface nesting can lead to electronic instabilities and formation of a CDW or spin-density-wave ground state: The classic example is Cr, which undergoes a Fermi-surface-driven transition to an antiferromagnetic ground state. For K_4C_{60}, three bands cross the Fermi level, and hence the Fermi surface, shown in Figure 9.13, has three sheets. Two form closed hole pockets centered at Γ, and show no nesting. The third forms a quasi-two-di-

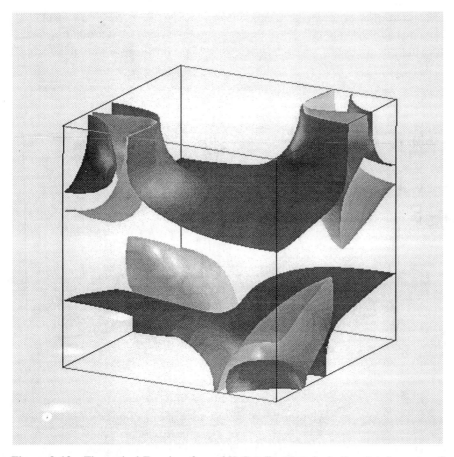

Figure 9.13 Theoretical Fermi surface of K_4C_{60}. Four tetrahedrally related corners of the cube correspond to Brillouin-zone centers in a repeated-zone scheme.

mensional sheet that extends over most of the *ab* plane, and is characterized by large flat portions in regions away from Γ. These flat sheets form approximately parallel surfaces, strongly reminiscent of the flat parallel Fermi-surface sheets of a one-dimensional metallic chain. To quantify the nesting, and identify possible nesting vectors $\mathbf{Q}$, we calculate

$$\xi_{nm}(\mathbf{Q}) = N^{-1} \sum_{\mathbf{k}} \delta(\varepsilon_{n,\mathbf{k}} - \varepsilon_F)\, \delta(\varepsilon_{m,\mathbf{k}+\mathbf{Q}} - \varepsilon_F), \qquad (9.1)$$

where n and m are band indices. Figure 9.14 shows a plot of $\xi_{33}(\mathbf{Q})$ in the $(Q_x, 0, Q_z)$ plane, and reveals a single moderately strong nesting vector centered at $\mathbf{Q} = (0, 0, \pi/c)$ (some spread about $Q_x = 0$ is also evident, and arises from the slightly curved parts of this Fermi-surface sheet). This is precisely the nesting vector required to give a period-doubling CDW along the c axis. Of course, Fermi-surface nesting is neither a necessary nor sufficient condition for CDW formation, but it does lend theoretical support to such a scenario.

The question of whether such a CDW state in fact leads to an insulating ground state is simple to answer in principle, but notoriously difficult in practice. In principle, one looks for lattice periodicities that lower the total energy,

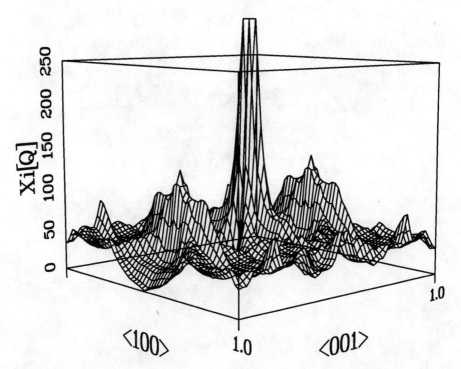

Figure 9.14 Surface plot of $\xi_{33}(\mathbf{Q})$ in the $(Q_x, 0, Q_z)$ plane. The plotting plane covers a rectangular area bounded by $Q_x = [-3/2\ \pi/a, +3/2\ \pi/a]$ and $Q_z = [-3/2\ \pi/c, +3/2\ \pi/c]$.

and then calculates the band structure at the minimum-energy geometry. In practice, the smallest-period CDW (a doubled unit cell of K_4C_{60}) is quite expensive to treat even with LDA methods; furthermore, there is some theoretical evidence suggesting that local approximations to the correlation energy are too crude to describe the energetics driving CDW formation, so that even in principle the LDA approach might fail [36]. Nevertheless, one can *assume* a lattice distortion and calculate the electronic structure to examine the effect on the bands (which should be well described by LDA). We have done this for a cell doubling along the c axis, assuming a lattice modulation of 5%. Although the ground-state band structure is not insulating per se, the value of $N(\varepsilon_F)$ is reduced by ~50%, and the calculated conductivity is reduced by roughly an order of magnitude. Since the K_xC_{60} materials are known to be close to a Mott-insulator transition [29, 35], such a CDW might well drive K_4C_{60} below the Mott conductivity limit, and could thus be manifested as an insulating ground state.

9.3   Endohedral Fullerenes: $A@C_{60}$

In this section, a simple model is presented for describing the energetics of endohedral fullerene molecules, $A@C_{60}$, focusing in particular on charge transfer between the endohedral atom, A, and the C_{60} cage. The present model cannot describe the detailed electronic structure of the endohedral complex: It simply provides a qualitative understanding of the relevant energetics. Although quantitative predictions are made for the stable charge states of the endohedral fullerene complexes, this is not the primary goal here. The most important findings are qualitative: (1) for endohedral neutral atoms the on-center geometry is an energy minimum, while for endohedral charged ions the energy minimum is off center; (2) higher charge states can be stabilized, for certain atoms, by moving (further) off center; (3) for charge-transfer endohedrals, the condensed phase (not yet synthesized experimentally) is predicted to show unusual orientational-ordering properties not seen in solid C_{60}.

9.3.1   A Simple Energetics Model
On-Center Case.

Consider first the simplest case, in which a single neutral atom A is constrained to sit at the center of an isolated neutral C_{60} molecule. In the lowest-level approximation, negligible overlap is assumed between the occupied-state wave functions of the A atom and C_{60} molecule. The resulting energy cost, $E_t(Q)$, for the Q-electron charge-transfer reaction

$$A + C_{60} \rightarrow A^{Q+}@C_{60}^{Q-} \tag{9.2}$$

is then completely decoupled into three contributions: (1) $E_{IP}(Q)$, the sum of the first Q ionization potentials of A, (2) $E_{EA}(Q)$, the sum of the first Q electron affinities of C_{60}, and (3) $E_{int}(Q)$, the classical interaction energy of the two oppositely charged objects. (Only charge transfer from A to C_{60} is considered here.) For the moment, Q will be treated as a continuous variable, and only in the range $0 \leq Q \leq 3$. Given analytic expressions for these three energy terms, the most stable charge state, Q_0, is found by minimizing $E_t(Q)$, subject the constraint of integral Q_0. The resulting predictions can then be compared to existing Hartree–Fock and LDA calculations and to available experimental evidence; discrepancies will point the way to improvements to the simple model.

For ionization potentials of the endohedral atom, standard experimental values are used for the first three electron removal energies I_1, I_2, I_3. The relations $E_{IP}(1) = I_1$, $E_{IP}(2) = I_1 + I_2$, $E_{IP}(3) = I_1 + I_2 + I_3$ determine the three coefficients of a cubic polynomial in Q:

$$E_{IP}(Q) = aQ + bQ^2 + cQ^3 \tag{9.3}$$

In this way, $E_{IP}(Q)$ reproduces the experimental removal energies for integral Q, and smoothly interpolates between those values for nonintegral Q.

The first electron affinity of C_{60} has been measured by several groups [37], all of which find a value 2.75 eV. The second electron affinity has been estimated, on the basis of experimental evidence for a stable 2– charge state, to be positive with a magnitude of several tenths of an eV. Nothing is known experimentally about the third electron affinity. For the present work, first-principles LDA results of Pederson et al. [33, 38] are used, in which the total energy was calculated explicitly as a function of charge state for $0 \leq Q \leq 6$; they find that $E_{EA}(Q)$ is accurately described by a quadratic in Q:

$$E_{EA}(Q) = \alpha + \beta(Q - Q_m)^2 \tag{9.4}$$

where $\alpha = -3.34$, $\beta = 1.51$, $Q_m = 1.49$. This equation predicts first and second electron affinities within 0.1–0.2 eV of the experimental values. The quadratic form of Eq. (9.4) is easily understood as the energy required to charge a (classical) spherical capacitor, with the added feature that $Q = Q_m$ is the lowest-energy charge state.

The resemblance of the C_{60} "charging energy" (an intrinsically quantum-mechanical property) to the charging energy of a capacitor suggests that the same model will work for the interaction energy $E_{int}(Q)$, the more so since the assumption of decoupling makes $E_{int}(Q)$ an intrinsically classical property. Hence, we take

$$E_{int}(Q) = -Q^2/R, \tag{9.5}$$

where R is the C_{60} cage radius. This is the simplest possible form for the interaction energy. A more realistic treatment would divide the charge Q into 60 equal parts located on the atomic sites of the C_{60}; the resulting interaction energy is within 0.01 eV of Eq. (9.5). For this reason, more sophisticated versions of $E_{int}(Q)$ are unnecessary.

It is now straightforward to compute $E_t(Q) = E_{IP}(Q) + E_{EA}(Q) + E_{int}(Q)$ for most of the atoms in the periodic table. In Table 9.3, the following quantities

Table 9.3 Predicted stable charge states for on-center endohedral complexes $A@C_{60}$, from minimization of $E_t(Q)$, as discussed in Section 9.3.1. Q_0 is the value of Q that minimizes E_t; Q_0^i is the (non-negative) integral value of Q that minimizes E_t; $\Delta E_t^{\pm 1}$ the energy of the Q_0^i state relative to the $Q_0^i \pm 1$ state, in eV; Q_{HF}^i is the (integral) charge transfer calculated by the Hartree–Fock method (Ref. [43]). A dash in the column for Q_0 indicates that no minimum exists in E_t; a dash in the columns for $\Delta E_t^{\pm 1}$ indicates the experimental data are unavailable

A	Z	Q_0	Q_0^i	ΔE_t^{-1}	ΔE_t^{+1}	Q_{HF}^i	A	Z	Q_0	Q_0^i	ΔE_t^{-1}	ΔE_t^{+1}	Q_{HF}^i
Li	3	0.5	1	1.6	63.6		Tc	43	0.6	0	—	0.3	
Be	4	1.5	0	—	2.3		Ru	44	0.4	0	—	0.3	
B	5	0.4	0	—	1.3		Rh	45	0.4	0	—	0.4	
C	6	—	0	—	4.2		Pd	46	0.2	0	—	1.3	
N	7	−0.5	0	—	7.5		Ag	47	0.4	0	—	0.5	
O	8	0.1	0	—	6.6	0	Cd	48	—	0	—	2.0	
F	9	—	0	—	10.4	0	In	49	0.6	1	1.2	6.8	
Ne	10	−0.8	0	—	14.5		Sn	50	1.0	0	—	0.3	
Na	11	0.5	1	1.9	35.2		Sb	51	−0.3	0	—	1.6	
Mg	12	1.5	0	—	0.6		Te	52	0.1	0	—	2.0	
Al	13	0.6	1	1.0	6.2		I	53	—	0	—	3.4	
Si	14	—	0	—	1.1		Xe	54	—	0	—	5.1	
P	15	−0.7	0	—	3.5		Cs	55	—	1	3.1	13.0	1
S	16	0.1	0	—	3.3		Ba	56	—	2	2.1	—	2
Cl	17	—	0	—	5.9		La	57	2.0	2	1.0	2.0	2
Ar	18	−1.2	0	—	8.7		Ce	58	1.9	2	1.2	3.1	
K	19	0.6	1	2.7	19.5	1	Pr	59	1.9	2	1.5	4.5	
Ca	20	1.6	2	0.2	33.8	1	Nd	60	—	2	1.4	—	
Sc	21	1.4	1	0.5	0.7		Pm	61	—	2	1.2	—	
Ti	22	1.2	1	0.2	1.5		Sm	62	—	2	1.0	—	
V	23	1.0	1	0.3	2.6		Eu	63	—	2	0.8	—	2
Cr	24	0.7	1	0.3	4.4		Gd	64	—	1	0.9	0.0	
Mn	25	0.9	0	—	0.4	0	Tb	65	—	2	0.6	—	
Fe	26	—	0	—	0.8		Dy	66	—	2	0.4	—	
Co	27	—	0	—	0.8		Ho	67	—	2	0.3	—	
Ni	28	0.3	0	—	0.6		Er	68	—	2	0.2	—	
Cu	29	0.4	0	—	0.7		Tm	69	1.6	2	0.0	6.6	
Zn	30	—	0	—	2.4		Yb	70	1.5	1	0.8	0.1	
Ga	31	0.6	1	1.0	8.4		Lu	71	—	1	1.6	1.8	
Ge	32	—	0	—	0.9		Hf	72	0.5	1	0.0	2.8	
As	33	−0.4	0	—	2.8		Ta	73	—	0	—	0.9	
Se	34	0.1	0	—	2.7		W	74	—	0	—	1.0	
Br	35	—	0	—	4.8		Re	75	—	0	—	0.9	
Kr	36	—	0	—	7.0		Os	76	—	0	—	1.7	
Rb	37	0.6	1	2.9	15.2		Ir	77	—	0	—	2.1	
Sr	38	1.6	2	1.1	26.5		Pt	78	—	0	—	2.0	
Y	39	1.5	1	0.6	0.2		Au	79	—	0	—	2.2	
Zr	40	1.1	1	0.2	1.0		Hg	80	—	0	—	3.4	
Nb	41	0.7	1	0.1	2.2		Tl	81	0.6	1	0.9	8.3	
Mo	42	0.5	0	—	0.1	Pb	82	0.9	0	—	0.4		
							Bi	83	0.4	0	—	0.3	

are tabulated for each atom: Q_0 is defined as the value of Q that minimizes E_t: Q_0^i, the *integral* value of Q that minimizes E_t; $\Delta E_t^{-1} = E_t(Q_0^i - 1) - E_t(Q_0^i)$, the stabilization energy relative to the $Q_0^i - 1$ state; and $\Delta E_t^{+1} = E_t(Q_0^i + 1) - E_t(Q_0^i)$, the additional energy required to transfer one electron more than Q_0^i. The majority of atoms are most stable for $Q = 0$ configurations. Seventeen atoms are most stable in $Q = 1$ states, including all the alkali metals, a number of early transition metals, and several Group III atoms. Only fourteen atoms prefer $Q = 2$ states, including Ca, Sr, Ba, La, and most of the lanthanide rare earths. In this on-center model, no atoms lead to stable $Q = 3$ charge states. Of particular interest, however, is the magnitude of the energy required to transfer one electron beyond Q_0^i. For several of the $Q = 2$ stable states, these values are quite small, for example, 2.0 eV (La), 3.1 eV (Ce), and 4.5 eV (Pr). For the $Q = 1$ stable states, they are as low as 0.1 eV (Yb), 0.2 eV (Y), and 0.7 eV (Sc). In the next section, we consider the possibility that the off-center geometry leads to an energy gain sufficient to transfer another electron beyond Q_0^i.

Off-Center Case.

For atoms sitting off center, several new effects may be considered, such as polarization of the excess charge on the C_{60}, the increasing importance of A-C_{60} orbital overlap, changes in the ionization potentials of the atom due to the resulting electric field (and its gradient), etc. By including the first two of these effects, a general feature of the C_{60} endohedral complex is reproduced, namely that the on-center geometry is an energy minimum for neutral atoms and an energy maximum for ions. This follows directly from the simple considerations to follow, and has been demonstrated explicitly in Hartree–Fock calculations. In the interest of keeping the model simple while retaining the essential physics, only these first two effects will be treated.

We start by generalizing the model of the previous section to include the effect of C_{60} charge polarization. The charge-transfer process leads to electrons occupying the C_{60} LUMO states; since these are relatively delocalized, we may crudely approximate the charge $-Q$ as confined to the surface of a perfectly conducting sphere. (The remaining electrons, up to and including the HOMO level, are considered frozen in this approximation.) For a point charge Q held at a distance $d < R$ from the sphere center, the resulting surface charge asymmetry for a perfectly conducting sphere is given by [39]

$$\sigma(\theta) = \frac{Q}{4\pi R^2} \left(\frac{R}{d}\right) \frac{1 - (R/d)^2}{[1 + (R/d)^2 - 2(R/d)\cos\theta]^{3/2}} \tag{9.6}$$

where θ is defined with respect to the axis formed by the sphere center and the endohedral atom. Note that this expression correctly reduces to $\sigma = -Q/4\pi R^2$

in the limit $d \to 0$. The interaction energy between the off-center charge at d and the induced C_{60} surface density now generalizes to

$$E_{int}(Q, d) = -\frac{Q^2}{R} \left(\frac{1}{1 - (d/R)^2} \right) \qquad (9.7)$$

$E_{int}(Q, d)$ is maximized for $d = 0$, reducing in that limit to Eq. (9.5); for nonzero d, this term represents an energy gain corresponding to the attraction between the point charge Q and the increasingly polarized surface density.

The asymmetry of the surface density also enters into the electron affinity of the C_{60}. Consider the asymmetry as arising from the electric field of the off-center point charge. In general, the energy of a perfectly conducting sphere in an electric field is given to lowest order by $E = -p\varepsilon/2$, where p is the induced dipole moment and ε is the field strength. In our case, the dipole moment is given by the surface integral

$$p = \int dS \, (R \cos \theta)\sigma(\theta) = -Qd \qquad (9.8)$$

(This also shows that, for a perfectly conducting sphere, the endohedral complex has vanishing monopole *and* dipole moments.) The electric field varies over the sphere surface, so it is convenient to introduce instead the polarizability γ, defined by $p = \gamma\varepsilon$. The C_{60} electron affinity may now be generalized to [33]

$$E_{EA}(Q, d) = \alpha + \beta(Q - Q_m)^2 - p^2/2\gamma \qquad (9.9)$$

where, for a conducting sphere, $p = -Qd$ and $\gamma = R$. This formula also represents an energy gain for nonzero d, again reducing to Eq. (9.4) for $d = 0$.

We turn now to the second new effect, namely the increased orbital overlap between off-center atoms and C_{60} HOMO and LUMO wave functions. This will be treated in a very approximate fashion, using atom–atom pair potentials of the Lennard–Jones form:

$$E_{overlap}(d) = \sum_{i=1}^{60} U_{LJ}(\mathbf{r}_i - \mathbf{r}_A(d)) \qquad (9.10)$$

$$U_{LJ}(r) = 4\varepsilon \left[\left(\frac{\sigma}{r} \right)^{12} - \left(\frac{\sigma}{r} \right)^6 \right] \qquad (9.11)$$

The parameters ε and σ may be derived from atomic polarizabilities [40], and are taken here to be independent of Q.

The total energy function, $E_t(Q)$, of the previous section is now generalized to

$$E_t(Q, d) \equiv E_{int}(Q, d) + E_{EA}(Q, d) + E_{overlap}(d) \qquad (9.12)$$

Figure 9.15 shows the Q-dependent energy gain for the endohedral complex

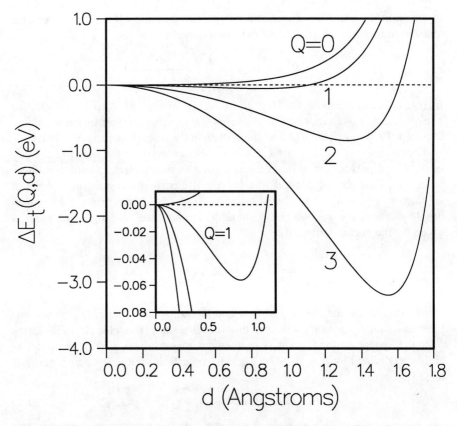

Figure 9.15 Energy gain for off-center endohedral atoms in a C_{60} molecule, as a function of assumed charge state. Lennard–Jones parameters for Ne–C interaction have been used for these curves; other parameters give slightly different results.

with an off-center atom, $\Delta E_t(Q, d) \equiv E_t(Q, 0)$; for this figure, parameters for Ne–C interaction [41] have been used to define $U_{LJ}(r)$. Several features are noteworthy. For $Q = 0$, the minimum energy is for $d = 0$, since only the Lennard–Jones term contributes. For $Q > 0$, the on-center position is an local maximum, as mentioned. For $Q = 1$, the minimum is at $d = \sim 0.8$ Å, and the energy gain is only 0.06 eV. For $Q = 3$, the minimum moves out to ~ 1.6 Å, and the energy gain is over 3 eV; this is enough to stabilize La, and possibly Ce, in $Q = 3$ charge states.

General Features of the Condensed State.

Although modeling the charged fullerene as a perfectly conducting sphere may provide a qualitative, and even semiquantitative, description of endohedral energetics, a more realistic picture is needed to describe fullerene–fullerene interaction in the condensed state. To this end, imagine replacing the perfectly conducting sphere by an "imperfectly" conducting sphere. The perfect screening of Eqs. (9.6) and (9.8) is then modified, with the result that the endohedral complex has a net dipole moment. In the simplest treatment, the moment $p = Qd$ of Eq. (9.8) is replaced by $p' = fQd$, where $0 < f < 1$, so that $f = 1$ corresponds to perfect screening and $f = 0$ to no screening. The dipole moment of the complex is then $(1 - f)Qd$. This also modifies the expressions for $E_{int}(Q, d)$ and $E_{EA}(Q, d)$, possibly in a complex, nonlinear fashion, but that effect is not considered here.

In solid C_{60}, fullerene–fullerene interactions are adequately described by van der Waals and high-multipole Coulomb forces, both of which are important for descriptions of order–disorder phenomena. For an endohedral condensed state that involves atoms stable in $Q > 0$ charge states, dipole–dipole Coulomb interactions may predominate. Assuming this to be the case, one can investigate the orientational ordering of a lattice of dipoles. Considering each pair of dipoles separately, the interaction energy is minimized when the two dipoles are antiparallel within planes and (more or less) parallel between planes. Considering only nearest-neighbor interactions, this tendency will lead to a unique ground-state configuration only if the lattice is bipartite (i.e., can be divided into two equivalent sublattices). This raises an interesting question: Since C_{60} crystallizes into an fcc lattice, one expects the same for endohedral complexes with $Q = 0$ states. But the fcc lattice is not bipartite, and so for $Q > 0$ endohedrals, the dipole–dipole interaction leads to a situation reminiscent of a frustrated antiferromagnet. Whether this frustration will lead to a phase transformation is open to question. At the very least, it will play an important role in order–disorder effects.

9.3.2 LDA and Hartree–Fock Calculations

There have been relatively few quantum-chemical calculations for endohedral complexes. Early LDA calculations [42] for on-center La@C$_{60}$ used Mulliken population analysis to find a formal charge transfer of $Q_{Mull} = 2.85$, in contradiction to our finding of $Q_0^i = 2$. It is well known, however, that Mulliken methods can be unreliable when there is substantial orbital overlap. This is demonstrated by a later Hartree–Fock (HF) calculation [43] for on-center La@C$_{60}$, which gave a very different value of $Q_{Mull} = 0.79$; a more detailed analysis of the occupied HF molecular orbital states gave the result $Q = 2$, in agreement with our model. Recent highly converged LDA results [44] for La@C$_{60}$ also predict $Q = 2$.

Total energies of various electronic configurations have been calculated by Chang et al. [43] using HF methods, for a number of on-center endohedral atoms: O, F, K, Ca, Mn, Cs, Ba, La, Eu, U. Their predicted stable charge states are tabulated in Table 9.3. All but one agree with the predictions of our simple model; the single exception of Ca is due to the very small energy difference (0.2 eV in our model) between the $Q = 1$ and 2 states.

Several groups have examined the energetics of off-center endohedrals. Cioslowski and Fleischmann [45] have used HF methods to calculate the off-center stability of Na^+; the minimum-energy position is at $d = 0.66$ Å, and the corresponding stabilization energy is 0.04 eV. Similar results were found by Schmidt et al., who have used all-electron LDA methods to find an equilibrium point at $d = 0.7$ Å and a stabilization energy of 0.09 eV. Both of these results are in good qualitative agreement with our results for Ne^+ (cf. Section 9.3.1 and Fig. 9.15) of $d = 0.8$ Å and an energy gain of 0.06 eV.

9.3.3 Experimental Results

Experimental evidence for the existence of fullerene endohedrals dates back to the initial discovery and synthesis of C_{60}. The first such complex discovered was $La@C_{82}$, whose stability was argued to arise from two electrons being transferred to the C_{82} cage, producing a closed electronic shell [46]. Support for this scenario is found in the observed natural abundance of (empty) C_{84} molecules, whose stability is also attributed, at least in part, to a closed shell.

This view was recently contradicted, however, by an electron paramagnetic resonance (EPR) study of $La@C_{82}$, which measured the La hyperfine coupling constant consistent with a $Q = 3$ charge state [47]. Although our model does not treat C_{82} explicitly, one can argue that since the quantities $1/\beta$ [in Eq. (9.4)] and R [in Eq. (9.5)] both scale with the cage radius, the opposing energetics of $E_{IP}(Q)$ and $E_{EA}(Q)$ will partially cancel, and so the energetics of a slightly larger cage will be largely the same as for C_{60}. The experimental evidence for a $Q = 3$ state is then supported by the discussion of Section 9.3.1, where we argued that $La@C_{60}$ has a stable $Q = 3$ charge state, in which the La sits off center by 1–2 Å.

In another recent experiment, $Fe@C_{60}$ has been synthesized by contact-arc vaporization and characterized by Mössbauer spectroscopy; the result is a Mössbauer shift consistent with neutral Fe [48]. This finding can also be understood on the basis of Table 9.3, which shows Fe to be stable in the $Q = 0$ state; moreover, stabilization in the $Q = 1$ state requires gaining an additional 0.8 eV, an order of magnitude larger than off-center charge polarization effects provide.

Finally, we mention a possible contradiction between experiment and the present model. Recent synthesis and EPR characterization of $Sc@C_{82}$ (which, by the argument presented, was treated as equivalent to $Sc@C_{60}$) show some evidence for a $Q = 3$ ground state [49]. This is in contrast to our model prediction of $Q_0^i = 2$ for an off-center geometry (the required extra 0.7 eV is easily supplied by polarization effects). An additional 8.3 eV is required to stabilize the $Q = 3$ state, substantially more than is available from off-center effects. The EPR results do show, however, that the electron spin may be more closely associated with the Sc atom than is the case for $La@C_{82}$, so that a simple picture of integral charge transfer may be inappropriate.

9.4   Fullerenes Adsorbed on Metal Surfaces

Fullerene monolayers have been successfully deposited on a variety of substrates, including gallium arsenide [50], silicon [51], and a number of metals
(Ag, Au, Mg, Cr, Bi) [52]. For semiconductor substrates, Ohno et al. [50]
have found that there is very little charge transfer between substrate and C_{60},
and the bonding is primarily van der Waals: For example, for thin fullerene
layers on GaAs(110), 0.02 electrons are transferred for n-type substrate doping,
while there is no measurable transfer for p-type substrates. For C_{60} monolayers
on metal substrates, the interaction is stronger: The C_{60} LUMO-derived level
is partially occupied, resulting in significant metal-to-fullerene charge transfer
and pinning of the LUMO level near the Fermi level. However, the close resemblance of the photoemission spectra to that of pure C_{60} indicates that LUMO–
metal hybridization effects, as well as polarization of the transferred charge,
may be sufficiently small to justify a perturbative treatment.

In the spirit of Section 9.3.1, a model is described here for the binding
energy and charge transfer between an ideal simple metal substrate and a single
adsorbed C_{60} fullerene molecule. The model neglects a number of interactions
that, although possibly important for sensitive probes (e.g., surface-enhanced
Raman scattering), can be justifiably ignored in a first treatment. As with the
model for endohedral fullerenes, the focus is on the most important contributions to the C_{60}–metal binding energy, and for the most part neglects the detailed electronic structure of both adsorbate and substrate. The primary result
is a numerical relationship between metal work function and charge transfer,
showing that, in contrast to the alkali-intercalated fulleride solids of Section
9.2, metal-adsorbed fullerene molecules cannot accept more than two electrons, regardless of the metal work function. An earlier account of this work
has appeared elsewhere [53].

9.4.1   A Simple Energetics Model

We want to find the stable charge state(s) and associated binding energy of a
C_{60} molecule adsorbed on a metal surface, where the metal is defined solely
by the value of its work function, W. In this way, any covalent bonding to the
substrate is excluded from consideration; this is a more complicated effect than
the present model can describe. The binding energy, with respect to infinite
separation from the metal, will be described by a family of curves, $E(Q, z)$,
where Q is the total charge transferred from the metal to the C_{60}, and z is the
distance from the center of the C_{60} to the (idealized) abrupt metal surface.
Although it is easiest to think of Q as being an integer, this will not be a
requirement, and a range of allowed values of Q will emerge naturally from
the solution.

The Unpolarized Molecule.

The solution will be built up in stages, improving the level of approximation at each stage. We start by completely ignoring any interaction between the C_{60} and the metal. For the total energy of the isolated (i.e., gas-phase) C_{60} molecule with excess charge Q, the LDA result of Eq. (9.4) is used:

$$E^{(1)}(Q, z) = E_{EA}(Q) = \alpha + \beta(Q - Q_m)^2 \qquad (9.13)$$

where the superscript (1) indicates the first level of approximation, and the values of α, β, and Q_m are given in Section 9.3.1.

In the second level of approximation, we account for the fact that a charged object near a metal surface interacts with its image charge (at this level of treatment, the metal is considered to be perfectly conducting, semi-infinite, and without internal structure). Two simplifying assumptions are made here: (1) Polarization of the excess charge Q is ignored, and (2) the excess charge is treated as though it were smeared out into a spherical shell, which is equivalent to a point charge at the sphere center. This allows a simple Coulomb term to be included in the binding-energy expression, so that

$$E^{(2)}(Q, z) = E_{EA}(Q) + \tfrac{1}{2} Q^2/2z \qquad (9.14)$$

At typical equilibrium separations (values of z), the binding energies given by $E^{(2)}(Q, z)$ are within 5% of the full solution to be arrived at; however, it is important to note that Eq. (9.14) does not actually predict an equilibrium separation at all, since it has no repulsive term.

To obtain an equilibrium separation, $z_0(Q)$, a repulsive interaction must be included. This is done by assuming a Q-independent van der Waals interaction, $E_{vdW}(z)$, between the C_{60} and the metal, that is, a stiff repulsive core and long-range, weakly attractive tail, with a minimum value occurring around 3–4 Å. We again use a Lennard–Jones potential $U_{LJ}(r)$, the form for which is given in Eq. (9.11), and a sum pairwise over atoms to obtain

$$E_{vdW}(z) = \sum_{j} \sum_{i=1}^{60} U_{LJ}(|\mathbf{r}_i - \mathbf{r}_j|) \qquad (9.15)$$

where i runs over the atoms in the C_{60} molecule, and j runs over all atoms in the metal (assumed to be simple cubic for this purpose). This gives have a fairly realistic (third-level) approximation to the binding energy:

$$E^{(3)}(Q, z) = E_{EA}(Q) + \tfrac{1}{2} Q^2/2z + E_{vdW}(z) \qquad (9.16)$$

Although the van der Waals term is crucial for obtaining a minimum in the binding-energy curves, it actually contributes very little to the magnitude of the binding energy itself (of order 0.5 eV at equilibrium, compared to the first two terms, which contribute of order 5 eV per electron transferred).

Polarization Corrections.

Further corrections to the binding energy of Eq. (9.16) can now be considered. The first two terms on the right-hand side of Eq. (9.16) are exact only in the absence of polarization. For a molecule near a metal surface, such effects may be important. To correct for polarization, a term is appended to Eq. (9.16) that depends on the distribution of excess charge:

$$E^{(4)}(Q, z) = E_{EA}(Q) + \tfrac{1}{2} Q^2/2z + E_{vdW}(z) + E_{pol}(\{q_i\}, z) \quad (9.17)$$

in which q_i denotes the excess charge on atom i. The polarization term contains both quantum corrections to $E_{EA}(Q)$ as well as electrostatic corrections to $Q^2/2z$. Our basic strategy is to compute the quantum correction using Hückel theory, and the electrostatic correction using classical electrostatics.

For the electrostatic correction, it is simpler to calculate the entire electrostatic contribution as whole, rather than separating it into the $Q^2/2z$ term plus polarization corrections. To this end, define

$$\bar{E}_{elec}(\{q_i\}, z) = \frac{1}{2} \sum_i \frac{q_i^2}{R_C} + \frac{1}{2} \sum_{i \neq j} \frac{q_i q_j}{|\mathbf{r}_i - \mathbf{r}_j|} + \frac{1}{2} \sum_{i,k} \frac{q_i q_k}{|\mathbf{r}_i - \mathbf{r}_k|} \quad (9.18)$$

where R_C is the effective radius of a single C atom (arbitrarily chosen to be the expectation value of a hydrogenic C $2p$ orbital), i and j run over the 60 C atoms, and k runs over the 60 C atoms in the image molecule. The q_i are constrained to be positive, and must sum to Q. The question of how the q_i are themselves determined will be deferred until later.

There is a difficulty with this definition of $\bar{E}_{elec}(\{q_i\}, z)$. Although the first two terms correctly include classical electrostatic polarization effects, they also include quantum contributions already accounted for by $E_{EA}(Q)$. This is easy to see in the unpolarized case, for which the first two terms of Eq. (9.18) contribute a term proportional to Q^2; this is the same Q^2 term that appears in Eq. (9.14). It is also easy to see how to correct Eq. (9.18), by subtracting off this nonpolarized contribution:

$$E_{elec}(\{q_i\}, z) = \frac{1}{2} \sum_i \frac{q_i^2 - q^2}{R_C} + \frac{1}{2} \sum_{i \neq j} \frac{q_i q_j - q^2}{|\mathbf{r}_i - \mathbf{r}_j|} + \frac{1}{2} \sum_{i,k} \frac{q_i q_k}{|\mathbf{r}_i - \mathbf{r}_k|} \quad (9.19)$$

where $q \equiv Q/60$. The third term does not duplicate any earlier contribution (it replaces the less accurate $Q^2/2z$), and so it does not require patching up.

We turn now to the quantum correction. In the simple Hückel description of C_{60}, a 60 × 60 Hamiltonian is constructed, representing the π-electron network. There are three possible values for the Hamiltonian matrix elements: The value α is assigned to all diagonal matrix elements, β to all off-diagonal elements between nearest-neighbor atoms, and 0 everywhere else. We choose α and β so that the Hückel HOMO and LUMO levels are equal to their LDA counterparts. Within this framework, the total energy is then a simple eigen-value sum:

$$E_{\text{Hückel}} = \sum_j n_j \varepsilon_j \qquad (9.20)$$

where n_j is the occupancy of the state with eigenvalue ε_j.

To include the effect of charge polarization on the eigenvalues (and total energy), we assume that excess charge q_i associated with atom i will change the α for that atom by an amount proportional to q_i; that is, we replace each α by $\alpha_i \equiv \alpha - kq_i$. The off-diagonal values are left unchanged. This results in new, $\{q_i\}$-dependent eigenvalues and occupation numbers, and thus a $\{q_i\}$-dependent total energy:

$$E_{\text{Hückel}}(\{q_i\}, z) = \sum_j n_j(\{q_i\}) \varepsilon_j(\{q_i\}) \qquad (9.21)$$

Note that there is an implicit dependence of $E_{\text{Hückel}}$ on z, since the $\{q_i\}$ depend on z. The constant k, which defines α_i, is chosen so that the Hückel output atomic charges (the sum of the squares of the occupied-eigenvector components) are equal to the input charges q_i. In this way we impose a self-consistency requirement on the charges, mimicking the approach of density-functional theory.

As defined, $E_{\text{Hückel}}(\{q_i\}, z)$ clearly double counts some of the nonpolarized quantum-mechanical energy already accounted for in $E_{\text{EA}}(Q)$. To remove it, and retain only the polarization contribution, we define

$$\Delta E_{\text{Hückel}}(\{q_i\}, z) \equiv E_{\text{Hückel}}(\{q_i\}, z) - E_{\text{Hückel}}(\{q_i = q\}, z) \qquad (9.22)$$

Together, the electrostatic and quantum polarization corrections are quite small, of order 0.1 eV at equilibrium, justifying the perturbative treatment. Now we have all the pieces necessary to account for the binding energy of a C_{60} molecule chemisorbed by van der Waals forces onto a simple metal surface, including a reasonable accounting of the full effects of charge transfer (Coulomb attraction, electrostatic polarization, and quantum-mechanical polarization):

$$E(Q, z) = E_{\text{EA}}(Q) + E_{\text{elec}}(\{q_i\}, z) + \Delta E_{\text{Hückel}}(\{q_i\}, z) + E_{\text{vdW}}(z) \qquad (9.23)$$

Finally, we turn to the $\{q_i\}$. With the total energy defined explicitly in terms of the $\{q_i\}$, a reasonable procedure is to choose the $\{q_i\}$ so as to minimize $E(Q, z)$ for each possible value of Q and z. This again gives the model a flavor reminiscent of density-functional calculations; in the present case, one has a "discrete-charge functional" to be minimized with respect to all possible charge distributions.

Binding Energies and Electron Transfer.

Binding-energy curves $E(Q, z)$ are shown in Figure 9.16, for Q in the range $0 \leq Q \leq 3$. Although the binding-energy expression in Eq. (9.23) is a con-

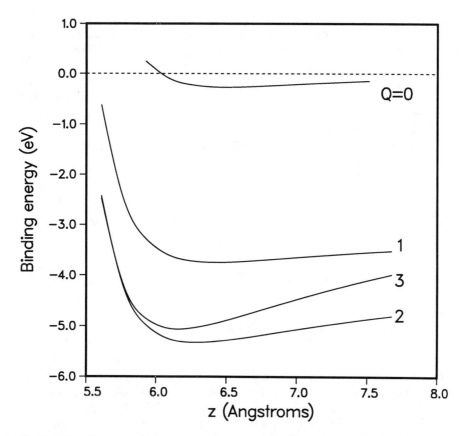

Figure 9.16 Theoretical binding-energy curves for a C_{60} molecule adsorbed on a simple metal surface, for charge states of 0, 1, 2, and 3. The adsorption distance, z, is measured from the metal surface to the center of the C_{60} cage.

tinuous function of Q, we restrict ourselves here to considering integer values. As expected on the basis of the Coulomb attraction of the fullerene to its image, equilibrium adsorption distances decrease with the magnitude of charge transferred: $z_{eq}(Q) = 6.50, 6.43, 6.27, 6.16$ Å for $Q = 0, 1, 2, 3$, respectively. The corresponding binding energies at equilibrium are $E(Q, z_{eq}(Q)) = -0.27$, $-3.75, -5.33, -5.06$ eV for $Q = 0, 1, 2, 3$, respectively. This behavior is easily understood as the result of competition between the Q^2 dependence of the Coulomb binding energy and the z^{-12} hard-core repulsion from the Lennard–Jones term.

Within this model, in order for the transfer of Q electrons to be energetically favorable, the following condition must be satisfied:

$$E(Q - 1, z_{eq}(Q - 1)) - E(Q, z_{eq}(Q)) > W^*_{loc} \qquad (9.24)$$

where W^*_{loc} is the local work function near the adsorbed molecule. For work functions less than roughly 3.5 eV, a transfer of one electron is expected; for W^*_{loc} of order 1.5 eV or less, two electrons may be transferred. Transfer of more than two electrons appears extremely unlikely, regardless of the value of the work function. This prediction stands in strong contrast to the alkali-inter-calated fulleride solids, which show as many as six electrons transferred per fullerene; the difference is due primarily to the less efficient electrostatic screening afforded by the surface geometry. However, we again caution that the present description is tentative, and does not account for work-function modification by the overlayer, fullerene–metal hybridization effects, or interaction between neighboring adsorbate molecules within the overlayer.

9.4.2   Experimental Results

At the time of writing, there are no experimental results for fullerene films on alkali- or other simple-metal surfaces. For adsorption on noble-metal surfaces, a variety of interesting phenomena has been observed, such as scanning tunneling microscopy images that show variations of the π-electron density on every other carbon atom.

Although noble-metal substrates are expected to show sizable covalent interactions with absorbed fullerenes, some qualitative features of the adsorption energetics may still be adequately described. Surface-enhanced Raman spectroscopy (SERS) has recently been used as a sensitive probe to characterize the charge transfer from Ag, Cu, and Au substrates [52]. Shifts in the $A_g(2)$ modes of -28.3, -23.2, and -15.4 cm^{-1} for Ag, Cu, and Au, respectively, are consistent with a reduction of charge transfer with increasing work function ($W = 4.3$, 4.7, and 5.1 eV for Ag, Cu, and Au, respectively). This is also in accord with the charge-transfer condition of Eq. (9.24) and the results of Figure 9.16, provided covalent effects are negligible. Indeed, the SERS shifts for Ag and Cu are similar to those found for alkali-intercalated thin films A_xC_{60} with $2 < x < 3$, suggesting that Ag and Cu substrates show charge transfer as large as 2–3 electrons. This finding is at odds with the predictions of our model, and may be an indication that covalent interaction is important for the noble metals.

Acknowledgments

It is a pleasure to acknowledge collaborations with M. R. Pederson, W. E. Pickett, R. P. Messmer, E. Burstein, and M. Y. Jiang. Partial support is acknowledged from the Laboratory for Research on the Structure of Matter at the University of Pennsylvania and from the Office of Naval Research. Most of the calculations were performed on the IBM 3090 at the Cornell National Supercomputing Facility.

References

1. W. Krätschmer, L. D. Lamb, K. Fostiropoulos, and D. R. Huffman, *Nature* **347**, 354–58 (1990).

2. A. F. Hebard, M. J. Rosseinsky, R. C. Haddon, D. W. Murphy, S. H. Glarum, T. T. M. Palstra, A. P. Ramirez, and A. R. Kortan, *Nature* **350**, 600 (1991).

3. C. Gu, F. Stepniak, D. M. Poirier, M. B. Jost, P. J. Benning, Y. Chen, T. R. Ohno, J. L. Martins, J. H. Weaver, J. Fure, and R. E. Smalley, *Phys. Rev. B* **45**, 6348–51 (1992).

4. S. C. Erwin, M. R. Pederson, and W. E. Pickett, *Phys. Rev. B* **41**, 10437 (1990).

5. P. A. Heiney, J. E. Fischer, A. R. McGhie, W. J. Romanow, A. M. Denenstein, J. P. McCauley, and A. B. Smith, *Phys. Rev. Lett.* **66**, 2911–14 (1991).

6. J. Mort, M. Machonkin, R. Ziolo, M. Machonkin, D. R. Huffman, and M. I. Ferguson, *Appl. Phys. Lett.*, to appear; J. Mort, R. Ziolo, M. Machonkin, D. R. Huffman, and M. I. Ferguson, *Chem. Phys. Lett.*, to appear.

7. P. A. Heiney, *J. Phys. Chem. Solids*, to appear.

8. Y. N. Xu, M.-Z. Huang, and W. Y. Ching, *Phys. Rev. B*, to appear.

9. N. Troullier and J. L. Martins, *Phys. Rev. B*, to appear.

10. S. Saito and A. Oshiyama, *Phys. Rev. Lett.* **66**, 2637–40 (1991).

11. S. C. Erwin and M. R. Pederson, *Phys. Rev. Lett.* **67**, 1610–13 (1991).

12. O. Zhou, J. E. Fischer, N. Coustel, S. Kycia, Q. Zhu, A. R. McGhie, W. J. Romanow, J. P. McCauley, Jr., A. B. Smith III, and D. E. Cox, *Nature* **351**, 462–64 (1991).

13. W. Andreoni, F. Gygi, and M. Parrinello, *Phys. Rev. Lett.* **68**, 823–26 (1992).

14. M. J. Rosseinsky, A. P. Ramirez, S. H. Glarum, D. W. Murphy, R. C. Haddon, A. F. Hebard, T. T. M. Palstra, A. R. Kortan, S. M. Zahurak, and A. V. Makhija, *Phys. Rev. Lett.* **66**, 2830–32 (1991); K. Tanigaki, T. W. Ebbesen, S. Saito, J. Mizuki, J. S. Tsai, Y. Kubo, and S. Kuroshima, *Nature* **352**, 222–23 (1991).

15. M. J. Rosseinsky, D. W. Murphy, R. M. Fleming, R. Tycko, A. P. Ramirez, T. Siegrist, G. Dabbagh, and S. E. Barret, *Nature* **356**, 416–68 (1992).

16. S. C. Erwin and W. E. Pickett, *Phys. Rev. Lett.*, submitted.

17. P. W. Stephens, L. Mihaly, P. L. Lee, R. L. Whetten, S.-M. Huang, R. Kaner, F. Deiderich, and K. Holczer, *Nature* **351**, 632–34 (1991).

18. J. L. Martins and N. Troullier, *Phys. Rev. B* (to appear).

19. S. C. Erwin and W. E. Pickett, *Science* **254**, 892–95 (1991).

20. W. E. Pickett, H. Krakauer, and P. B. Allen, *Phys Rev. B* **38**, 2721–26 (1988).

21. K. Holczer et al., *Phys. Rev. Lett.* **67**, 271 (1991).

22. Y. J. Uemura et al., *Nature* **352**, 605 (1991).

23. R. G. Tycko, G. Dabbagh, M. J. Rosseinsky, D. W. Murphy, A. P. Ramirez, and R. M. Fleming, *Phys. Rev. Lett.* **68**, 1912–15 (1992).

24. Z. Zhang, C.-C. Chen, and C. M. Lieber, *Science* **254**, 1619–21 (1991).

25. L. D. Rotter, Z. Schlesinger, J. P. J. McCauley, N. Coustel, J. E. Fischer, and A. B. I. Smith, *Nature* **355**, 532–34 (1992).

26. M. P. Gelfand and J. P. Lu, *Phys. Rev. Lett.* **68**, 1050–53 (1992).

27. T. T. M. Palstra, R. C. Haddon, A. F. Hebard, and J. Zaanen, *Phys. Rev. Lett.* **68**, 1054–57 (1992).

28. P. J. Benning, D. M. Poirer, T. R. Ohno, Y. Chen, M. B. Jost, F. Stepniak, G. H. Kroll, J. H. Weaver, J. Fure, and R. E. Smalley, *Phys. Rev. B* **45**, 6899–913 (1992).

29. J. H. Weaver, *Acc. Chem. Res.*, to appear.

30. T. Pichler, M. Matus, J. Kürti, and H. Kuzmany, *Solid State Commun.* **81**, 859–62 (1992).

31. E. Sohmen, J. Fink, and W. Krätschmer, *Europhys. Lett.* **17**, 51–55 (1992).

32. W. H. Wong, M. E. Hanson, W. G. Clark, G. Grüner, J. D. Thompson, R. L. Whetten, S.-M. Huang, R. B. Kaner, F. Diederich, P. Petit, J.-J. André, and K. Holczer, *Europhys. Lett.* **18**, 79–84 (1992).

33. M. R. Pederson and A. A. Quong, *Phys. Rev. B,* submitted.

34. R. M. Fleming, M. J. Rosseinsky, A. P. Ramirez, D. W. Murphy, J. C. Tully, R. C. Haddon, T. Siegrist, R. Tycko, S. H. Glarum, P. Marsh, G. Dabbagh, S. M. Zahurak, A. V. Makhija, and C. Hampton, *Nature* **352**, 701–3 (1991).

35. D. W. Murphy, M. J. Rosseinsky, R. M. Fleming, R. Tycko, A. P. Ramirez, R. C. Haddon, T. Siegrist, G. Dabbagh, J. C. Tully, and R. E. Walstedt, preprint.

36. A. W. Overhauser, in *Charge Density Phenomena in Potassium,* J. T. Devreese and F. Brosens, Eds., Plenum, New York, 1983, pp. 41–65.

37. S. H. Yang, C. L. Pettiette, J. Conceicao, O. Cheshnovsky, and R. E. Smalley, *Chem. Phys. Lett.* **139**, 233 (1987).

38. M. R. Pederson, S. C. Erwin, W. E. Pickett, K. A. Jackson, and L. L. Boyer, in *Proc. International Symp. on the Physics and Chemistry of Finite Systems,* Richmond, VA, 1991.

39. J. D. Jackson, *Electrodynamics,* Wiley, New York, 1975, p. 56.

40. A. G. Bezus, M. Kocirik, and A. A. Lopatkin, *Zeolites* **4**, 346 (1984); W. A. Steele, in *The Interaction of Gases with Solid Surfaces,* Vol. 3 of *The International Encyclopedia of Physical Chemistry and Chemical Physics,* D. D. Eley and F. C. Tompkins, Eds., Pergamon, Oxford, 1974.

41. A. L. R. Bug, A. Wilson, and G. A. Voth, preprint.

42. A. Rosèn and B. Wästberg, *J. Am. Chem. Soc.* **110**, 8701–3 (1988).

43. A. H. H. Chang, W. C. Ermler, and R. M. Pitzer, *J. Chem. Phys.* **94**, 5004–10 (1991).

44. A. A. Quong, M. R. Pederson, and S. C. Erwin, preprint.

45. J. Cioslowski and E. D. Fleischmann, *J. Chem. Phys.* **94**, 3730–34 (1991).

46. Y. Chai, T. Guo, J. Changming, R. E. Haufler, L. P. F. Chibante, J. Fure, L. Wang, J. M. Alford, and R. E. Smalley, *J. Phys. Chem.* **95**, 7564–68 (1991).

47. R. D. Johnson, M. A. de Vries, J. Salem, D. Bethune, and C. S. Yannoni, *Nature* **355**, 239–40 (1992).

48. T. Pradeep, G. U. Kulkarni, K. R. Kannan, T. N. Guru Row, and C. N. R. Rao, *J. Am. Chem. Soc.* **114**, 2272–73 (1992).

49. C. S. Yannoni, M. Hoinkis, M. S. de Vries, D. S. Bethune, J. R. Salem, M. S. Crowder, and R. D. Johnson, *Science,* to appear.

50. T. R. Ohno, Y. Chen, S. E. Harvey, G. H. Kroll, J. H. Weaver, R. E. Haufler, and R. E. Smalley, *Phys. Rev. B,* to appear.

51. Y. Z. Li, M. Chander, J. C. Patrin, and J. H. Weaver, *Phys. Rev. B,* to appear.

52. S. J. Chase, W. S. Bacsa, M. G. Mitch, and L. J. Pilione, preprint.

E. Burstein, S. C. Erwin, M. Y. Jiang, and R. P. Messmer, *Physica Scripta,* to appear.

Exo- and Endohedral Fullerene Complexes in the Gas Phase

Helmut Schwarz, Thomas Weiske, Diethard K. Böhme, and Jan Hrušák

The decisive role of mass spectrometry as an analytical method in the characterization of fullerenes and its derivatives is well documented [1]. Two recent review articles are also dedicated exclusively to the mass spectrometry of fullerenes [2]. Since the articles by Ross et al. [2] also adequately treat the physico-chemical aspects of the chemistry of fullerene ions, a discussion of these aspects need not be included here. The reality is that the first [3] and second [4] ionization energies, as well as the electron affinities [5], are known for several fullerenes, that it has been possible to produce C_{60}^{2-} as a "naked" ion in the gas phase [6], that the proton affinities of C_{60} and C_{70} have been established [7], that ion–molecule reactions of fullerene ions have been studied [7, 8], and that results of molecular-beam experiments have been reported that explore fundamental aspects of unimolecular [9], collision- [9c, 10], surface- [11], and laser-induced [12a,b] dissociation processes of fullerene ions and mobilities of $C_x^{+ \cdot}$ ions [13] as well as the neutralization of $C_{60}^{+ \cdot}$ [14]. These investigations, as well as aspects of the exciting chemistry of fullerenes [15], will be recalled to the extent that they are needed for the clarification of what perhaps is the most interesting aspect of fullerene chemistry, the possibility of placing something inside the hollow space of fullerenes. Can fullerene molecules serve as molecular containers and so provide a potential for *endohedral* chemistry—chemistry inside a cage, as predicted by a variety of theoretical investigations [16]? As will become evident later, the discussion of the properties of endohedral fullerene complexes, for which the symbol $M@C_x$ has been proposed [17], also requires an analysis of *exohedral* fullerenes that have been modified on the "outer skin" of the carbon cluster. A discussion of these latter complexes will be given first.

10.1   Exohedrally Modified Fullerene Ions

All of the "derivatization" of fullerenes carried out in solution so far take place, without exception, on the outer surface [15], irrespective of whether it is a matter of producing organometallic derivatives [18], of the methylation or silylation of polyanions [19], the hydrogenation [20], fluorination [21], chlorination [22], or bromination [22] of fullerenes, the acid-catalyzed fullerenation of aromatics [23], the addition of amines [24], radicals [25], or carbenes [26], or the epoxidation of fullerenes [27]. The gas-phase chemistry of fullerenes, such as C_{60} reacting with electrophiles E^+ ($E^+ = CH_3^+$, $C_2H_5^+$, $C_4H_9^+$), also takes place on the outer skin—how could it be otherwise! However, the observation in single-collision experiments that the adduct of C_{60} with t-$C_4H_9^+$ for example, loses mainly C_4H_8, is interesting. It could indicate that new C–C bonds *do not* result in the formation of the adduct and that the $C_{60}(t$-$C_4H_9^+)$ adduct is more appropriately viewed as a proton-bound dimer of the type $C_{60}\cdots H^+\cdots C_4H_8$. The observation that the proton remains with the C_{60} cluster in collisional excitation is not surprising since the proton affinity (PA) of C_{60} is higher than that of C_4H_8 [7]. $C_{60}H^+$ is itself extraordinarily resistant against charge transfer and also fragmentation: Proton transfer to bases B proceeds only if PA(B) > 205 kcal mol^{-1}, and collision-induced dissociation to generate $C_{60}^{+\cdot}$ occurs under multiple collisions corresponding to center-of-mass collision energies of >23 eV [7]. The ease with which $C_{60}H^+$ is formed, and its considerable reluctance to break up, bestow substantial weight to Kroto's suggestion [28] that this species could be responsible for the diffuse interstellar bands [29].

The metal atoms in the $C_{60}K_3^+$ ions produced by the laser-induced vaporization of doped graphite are assumed by the authors to be situated on the outside surface of the clusters [30]. Freiser et al. [31] have recently demonstrated that "naked" transition-metal ions M^+ (M = Fe, Co, Ni, Cu, Rh, La, VO) can be "glued" to the exterior of C_{60}. M^+ and ML^+ (L = ligand) were allowed to react with C_{60} in gas-phase experiments near *thermal* energies, and the resulting MC_{60}^+ ions were characterized mass spectrometrically. As is expected for *exohedral* complexes (see also Section 10.2), collisional excitation of mass-selected MC_{60}^+ ions leads exclusively to the removal of the metal atom; the C_{60} cluster itself remains intact [Eq. (10.1)].

$$exo\text{-}MC_{60}^+ \longrightarrow \begin{array}{l} M + C_{60}^{+\cdot} \\ M^+ + C_{60} \\ [MC_{60-m}]^+ + C_m \end{array} \qquad (10.1)$$

Freiser's findings require that the metal atom be coordinated on the outside surface of the cluster. If this were a case of *endohedral* bonding, the breakdown of C_{60} should occur unless the transition *endo* → *exo* is assumed to be facile. Such an assumption is not justified given the radii of M^+ and the calculated activation energy of >6 eV for the Li^+ transition $Li@C_{60}^+ \rightarrow exo\text{-}LiC_{60}^+$ [16e]. The next section will show that the properties of clusters of the type $M@C_x^+$ are fundamentally different from the systems investigated by Freiser.

10.2 Endohedral Fullerene Complexes, $M@C_x^+$

The examples described in the previous section show that the production of endohedral complexes cannot be achieved with "conventional" chemical approaches. Fundamentally different methods must be developed, and in their realization two entirely different principles can be envisaged: (1) Atoms M that are accidentally present in the "inner side" as the shell of the cluster closes are "welded into" the fullerene molecule as it forms in the laser vaporization of graphite. (2) Atoms M penetrate the shell of the fullerene in a high-energy bimolecular collision. Once the atoms are physically trapped, they are barred from leaving their cage by an extremely high barrier. Both concepts have actually been realized, and the most important aspects of the properties of endohedral complexes and also the mechanism of the penetration will now be discussed.

That laser vaporization of graphite produces not only fullerenes [32], but also *metals* trapped in fullerene cages, was realized almost simultaneously with the "discovery" of C_{60} by the Smalley group [33], which observed that the mass spectra of graphite doped with $LaCl_3$ and intensely bombarded with an ArF laser contain signals for LaC_n^+ as well as for $C_n^{+\cdot}$. While Smalley et al. [33] took their results to indicate the formation of a spherical $C_x^{+\cdot}$ cluster containing a metal atom in its hollow interior, the Exxon group [30, 34] originally took the stand that the clusters were empty and that the metal was "attached" to the outer surface. This controversy has been resolved in the meantime, since it could be shown with two independent experiments that not only LaC_{60}^+ [33], but also other metal-containing fullerenes of the type MC_x^+ (M = La, Ni, K, Rb, Cs, Y, U; $x \geq 60$) produced by laser vaporization of doped graphite [17a, 35], actually do exist as endohedral compounds. This follows directly from the fact that reactions of MC_x^+ with O_2, NH_3, N_2O, or H_2O do not occur in the gas phase [Eq. (10.2)], while externally bonded metal-cluster ions, for example, the YC_{60}^+ ions produced by ion–molecule reactions, are oxidized [Eq. (10.3)]. Apparently the "skin of carbon" protects the inherently reactive metal from a chemical interaction with the world outside. Related results were recently reported by McElvany [36] for the generation and characterization of yttrium–fullerene complexes $Y@C_x^+$.

$$Y@C_x^+ + N_2O \rightarrow \text{no reaction} \qquad (10.2)$$
$$exo\text{-}YC_{60}^+ + N_2O \rightarrow YO^+ + C_{60} + N_2 \qquad (10.3)$$

Before dealing with the second argument in favor of an endohedral structure $M@C_x^+$, it is necessary to mention the singular characteristics of fullerene cations when they are activated in the gas phase. *Even*-numbered clusters $C_x^{+\cdot}$ with $x \geq 32$ have been observed to fragment with the loss of energy-rich [37] C_2 molecules, viz. *even*-numbered C_m fragments ($m = 2, 4, 6, \ldots$), irrespective of the manner of activation. Whether C_m is lost intact, or as multiples C_2 units, is currently unresolved, although MNDO calculations are in favour of a concerted loss of C_m [37d]. However, it is important to note that the thermodynamically favored loss of C_3 units, which dominates in the breakup of smaller clusters $C_x^{+\cdot}$ ($x \leq 30$), does *not* appear in the spectra of fullerene cations and dications. The explanation for this dichotomy apparently lies in the special cage structure of the fullerenes. Loss of C_m ($m = 2, 4, 6, \ldots$), according to the "shrink-wrap" mechanism [38] (Scheme 10.1 for the loss of C_2), produces a "repaired" closed cage only when *even-numbered* C_m units break off. Since closed-fullerene structures presumably are accessible only down to $C_{32}^{+\cdot}$ (however, see [13]), it is not surprising that the reaction channel [Eq. (10.4)] dominates in the regime $C_x^{+\cdot}$ ($x \geq 32$), whereas C_3 is split off as a neutral molecule for $x \leq 30$ [Eq. (10.5)].

$$C_x^{+\cdot} \xrightarrow{x \geq 32} C_{x-m}^{+\cdot} + C_m, \quad m = 2, 4, \ldots \tag{10.4}$$

$$C_x^{+\cdot} \xrightarrow{x \leq 30} C_{x-3}^{+\cdot} + C_3 \tag{10.5}$$

There is agreement between estimates [12a] as well as several experimental findings [9d,f] concerning the energetic requirements for the C_2 loss from $C_{60}^{+\cdot}$. Five *half* C–C bonds are broken in step $2 \rightarrow 3$ of Scheme 10.1; assuming that C_2 is produced and the cage closes again, one can estimate that the binding energy of C_2 to the cluster is adequately given by the relationship $5 \times \frac{1}{2} \times D_{CC} - D_o(C_2)$. It follows that ~4 eV are required for every loss of C_2. This result agrees with the experimental data of Lifshitz [9f] (4 ± 0.5 eV) for the binding energy of C_2 in $C_{60}^{+\cdot}$, viz. C_{60}^{2+} [39]. That considerably more than 4 eV must be pumped into the molecule, in the breakup of C_{60}^+; for example, to bring about dissociation in the time window determined by the experiment, is not surprising, given the size of the molecule. The fact that *even-numbered* ful-

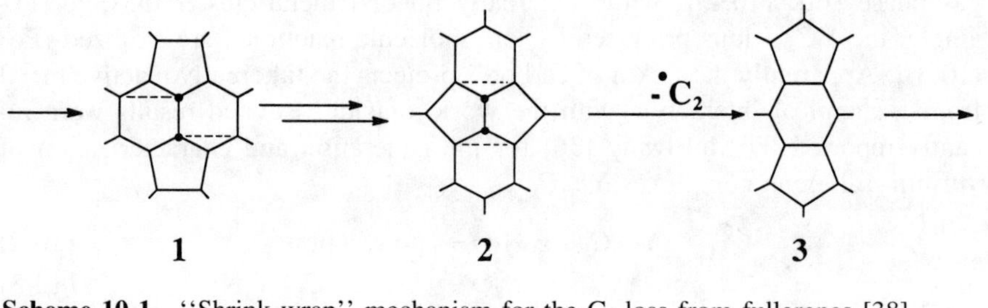

Scheme 10.1   "Shrink-wrap" mechanism for the C_2 loss from fullerenes [38].

lerenes $C_x^{+\cdot}$ ($x \geq 32$) break up with production of *even-numbered* fragments only after considerable activation (>18 eV) is decisive. This behavior characterizing fullerenes also distinguishes Smalley's metal fullerenes MC_x^+, which, in contrast to Freiser's exohedral complexes [Eq. (10.1)], break apart to split off C_m only after considerable energy has been added [Eq. (10.6)].

$$M@C_x^+ \longrightarrow \begin{array}{l} [M@C_{x-m}]^+ + C_m, \quad m = 2, 4, 6, \ldots \\ M + C_x^{+\cdot} \\ M^+ + C_x \end{array} \qquad (10.6)$$

Further proof that the metals are wrapped in a coat of carbon in the laser vaporization of doped graphite follows from the details of the "shrink-wrap" mechanism. The dismantling in the breakup of *empty* $C_x^{+\cdot}$ clusters ends at $C_{32}^{+\cdot}$, which is incapable of losing C_2 upon further activation; the $C_{32}^{+\cdot}$ simply breaks apart instead. If, on the other hand, the $C_x^{+\cdot}$ cluster contains a metal in its empty space, the endpoint of the breakup through "*vaporization*" of C_m ($m = 2, 4, 6, \ldots$) can be expected to lie *above* $C_{32}^{+\cdot}$. In fact, the shrinkage really does end already at C_{44}^+ when a calcium atom is inside the cage and at $C_{48}^{+\cdot}$ with the somewhat larger Cs atom. Excitation of $K@C_{44}^+$, viz. $Cs@C_{48}^+$, can no longer lead to closed endohedral clusters smaller in size in the evaporation of C_m because of the ionic radius of the metal; the complex bursts apart and loses the metal instead.

Recently it has become possible to produce macroscopic amounts of endohedral metal fullerenes [17a, 40a,c]. The results for $La@C_{82}$ are particularly noteworthy. XPS analysis indicates a formal charge of $3+$ for the central La atom (two electrons are given off to the C_{82} cage in the formation of what is presumably a stable C_{82}^{2-} dianion, and the third electron is delocalized in the inner space). The compound itself can be extracted in boiling; moist toluene and the carbon cluster protects the metal from attack by, for example, H_2O or O_2. Similar results are also reported for yttrium complexes for which a doubly doped $Y_2@C_{82}$ was also prepared. Based on an EXAFS study, however, this conclusion has been questioned, and it was suggested that the compound rather corresponds to an *exohedral* dimer $C_{82}Y-X-Y-C_{82}$, where -X- is a bridging carbon or oxygen species [40f]. ESR spectroscopic data of $M@C_{82}$ (M = Sc, Y and La) were however interpreted in terms of endohedral complexes [40g]. Obviously, more detailed work is required to resolve these conflicting findings and interpretations. The generations of $La_2@C_{80}$ [40b] and of $Sc_3@C_{82}$ [40d, h] were also described recently, and laser vaporization studies [40e] indicate the formation of $U@C_{28}$. Numerous rare-earth species in carbon cages ($M@C_n$: M = Ce, Nd, Sm, Eu, Gd, Tb, Dy, Ho, or Er) were generated by using the carbon-arc evaporation technique and studied mass spectrometrically [40i].

If the contact-arc vaporization of graphite is carried out in a helium atmosphere spiked with $Fe(CO)_5$, the mass spectra of the resulting toluene-extracted soot contains signals that are interpreted as an indication of the generation of $Fe@C_{60}$, and this assignment is aided by [57]Fe Mössbauer spectroscopy [41a]. However, these findings have not been reproduced by others [41b].

For many reasons one does not assign a high probability to achieving successfully the second variation for producing an endohedral complex—the penetration of an intact fullerene molecule with an atom (or molecule). On the one
hand the penetration of the skin of the cluster will require a high activation
energy and, as shown by the examples discussed in Section 10.2, will not be
realizable with conventional methods at thermal energies. An alternative is provided by a molecular-beam experiment in which highly accelerated fullerenes
are shot through a stationary gas atmosphere N in which not only elastic, but
also inelastic collisions can take place with the result that (1) the activation
energy required for penetration is delivered as center-of-mass energy (E_{cm}), and
(2) the neutral species N is "captured" by the projectile $C_x^{+\cdot}$. However, first-
principle considerations and the overwhelming profusion of collision experiments of this kind in which N was not captured do not exactly generate the
courage to try such an experiment, and, as a matter of fact, there is no documented example of a bimolecular reaction in which an observable encounter
complex "PN" would have been achieved in a high-energy collision. Collision-induced dissociation into a diversity of products F_i occurs instead, without
N being incorporated.

$$\vec{P} + N \rightarrow [PN]^* \begin{array}{l} \longrightarrow \vec{F}_i + \vec{N} \\ \\ \longrightarrow [\overrightarrow{F_iN}] + \vec{F}_j \end{array} \qquad (10.7)$$

Fullerene and its ions (the latter are employed in experiments of this kind
for practical reasons since ions are easier to manipulate) do not seem to be an
exception in that exclusive loss of C_m ($m = 2, 4, 6, \ldots$) was observed in all
the collision-induced processes [9e,f,i, 10] carried out earlier. Fragments containing the collision partner N were never reported [Eq. (10.8)]. An example
from our own work for the reaction of $C_{60}^{+\cdot}$ with D_2 is shown in Figure 10.1.

$$C_x^{+\cdot} + N \rightarrow [C_xN^+]^* \begin{array}{l} \longrightarrow C_{x-m}^{+\cdot} + C_m + N \\ \\ \not\longrightarrow [C_{x-m}]^{+\cdot} + C_m \end{array} \qquad (10.8)$$

That a penetration experiment does not have to be fundamentally hopeless,
in spite of the poor prognosis, will be demonstrated with a series of findings,
all of which are connected with the extraordinary properties of these clusters.
(1) In contrast to "normal" molecules, *no* collision-induced dissociation occurs
when $C_x^{+\cdot}$ ions ($x = 60, 70, 84$) are shot at Si or C surfaces with a speed
exceeding 20,000 km h^{-1}. The ions are reflected as if they were "rubber balls"
[11]. An extraordinarily high resilience and the ability to accommodate large
amounts of energy as vibrational excitation seem to be characteristic features
of fullerenes. (2) The potential to accommodate large amounts of energy (up
to 30 eV) without dissociating was also demonstrated in other experiments [7,
9d, 12]. Furthermore, taking into account that one is able to vary E_{cm} over a
wide range with a choice of the kinetic energy of $C_x^{+\cdot}$ and the mass of N, the
opportunity exists of finding an "energy window" for optimal penetration.

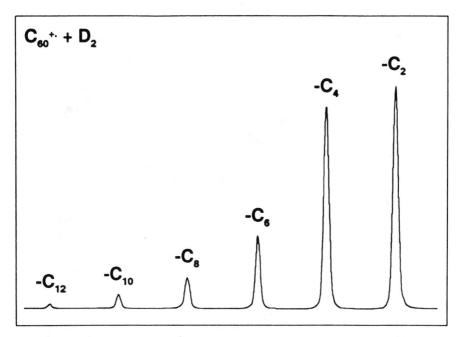

Figure 10.1   CA mass spectrum of $C_{60}^+/D_2(E_{lab} = 8$ keV); linked scan with $B/E =$ constant.

Indeed, we were able recently to demonstrate in our laboratory [42] that noble gases (He and Ne) can be injected into singly and multiply charged C_x^{n+} fullerene molecules (x = 60, 70; n = 1–3), and so to provide evidence for the formation of endohedral carbon-cluster noble-gas compounds by an unprecedented high-energy bimolecular reaction. Related and other experiments were later reported by several groups [4c, 43]. Here, we shall discuss the major aspects without describing in detail the techniques and machines used as this technical information is available from the original papers. For the experiments conducted at TU Berlin, a brief description of the central aspects should suffice. The collision experiments were performed with a four-sector $B(1)E(1)B(2)E(2)$ mass spectrometer (B stands for magnetic and E for electric sector) [44]. This is a modified VG Instruments ZAB mass spectrometer (Fig. 10.2), which has been built by AMD Intectra, W-2833 Harpstedt, Germany, by combining the BE part of a ZAB-HF-3F machine (MS I) with an AMD 604 double-focusing mass spectrometer (MS II) by a system of einzel lenses. C_x^{n+} ions (x = 60, 70; n = 1–3) were generated by electron-impact ionization at 100 eV of the vapor of C_{60} or C_{70} (the temperature of the solid-probe tip was approximately 550°C) at a low pressure of 10^{-6} mbar and a source-block temperature of 270°C. The ions were accelerated up to 8 keV maximum kinetic energy, mass selected with $B(1)E(1)$, and then allowed to collide with neutral gases N at room temperature in a collision chamber located in the field-free

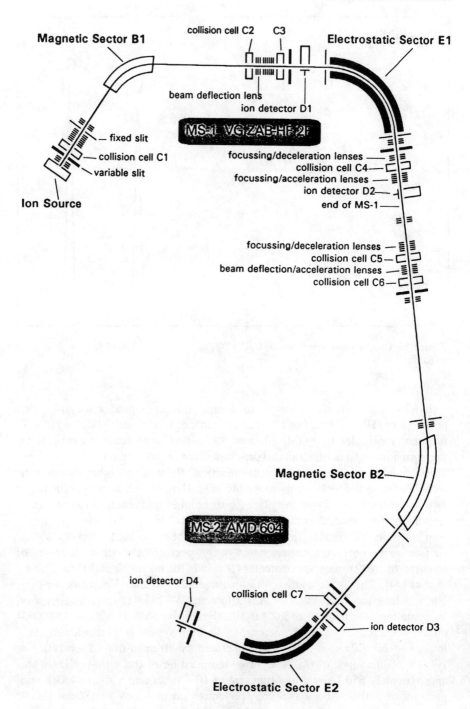

Figure 10.2 Schematic diagram of the *BEBE* tandem mass spectrometer at the TU
Berlin [44].

region between $E(1)$ and $B(2)$. In typical experiments the primary ion beam was reduced by 60–70%. The resulting two-dimensional ion-current surface was explored with a $B(2)/E(2) = $ constant linked scan (sampling all fragment ions arising from the same precursor ion) [45].

In distinct contrast to Figure 10.1 (collision of 8-keV $C_{60}^{+\cdot}$ with D_2), a fundamentally different, unexpected, and unprecedented result is obtained when helium is used as a collision gas (Fig. 10.3). In addition to the signals corresponding to losses of C_m, new signals appear at *higher* masses at all signals; the mass difference Δm is equal to 4 when ^{4}He is used [Fig. 10.3(a)] and 3 when ^{3}He is employed [Fig. 10.3(b)].

The differences that appear with D_2 and He cannot be accounted for by energetic effects since E_{cm} is identical for these two gases; rather, other factors (presumably size and shape of N) are responsible. Totally analogous results are obtained with $C_{70}^{+\cdot}$. While helium is incorporated (appearance of $[C_{70-m}He]^{+\cdot}$; Fig. 10.4), the other collision partners (D_2, O_2, Ar, and SF_6) only lead to collision-induced elimination of C_m. The observation that the $C_{60}He^{+\cdot}$ fragment arises from $C_{70}^{+\cdot}$ (Figure 10.3; loss of formally C_{10}) is particularly noteworthy, as is the fact that the abundance of this signal does not conform to the monotonous decrease of signal intensities reported in Figure 10.3. Obviously, $C_{60}He^{+\cdot}$ possesses a unique stability, and if Smalley's argument [38a, 46] that loss of "C_m" units is accompanied by a "self-repair" of the cluster cage (shrinkwrap mechanism) applies, the $C_{60}He^{+\cdot}$ ion generated in the bimolecular reaction of $C_{70}^{+\cdot}$ with He corresponds to the endohedral fullerene–noble-gas complex of $C_{60}^{+\cdot}$.

Experiments aimed at incorporating noble gases other than helium were also successful with neon. As demonstrated in Figure 10.5, the *even*-numbered fragment ions $C_{60-m}^{+\cdot}$ ($m = 0, 2, 4, 6, \ldots$) generated in the reaction with Ne are accompanied by "satellite" signals at higher masses corresponding to the incorporation of Ne, except for $m = 2$ and 4, for which the satellite signal cannot be assigned with confidence. The cross section for the reaction with Ne is smaller by a factor of $\sim$10 than in the case of He, and the origin of this effect will be discussed further.

Gross and co-workers have reported [4c, 43b] that high-energy ($E_{cm} = 421$ eV) collisions of C_{60}^{+} with argon causes "incorporation" of Ar and decomposition to form a series of *odd*-carbon species $C_xAr^{+\cdot}$ ($x = 49, 51, 53$, and 55). As the kinetic energy loss of the projectile in the formation of $C_xAr^{+\cdot}$ is huge [4c] and perhaps too big to be accepted by a conventional electrostatic analyzer as used in our machine, it is no surprise that these species were not observed by using the Berlin *BEBE* tandem mass spectrometer. When krypton and xenon were used, no products indicative of the incorporation of the heavier noble gases were observed [4c].

The successful incorporation of He in fullerene radical cations, $C_x^{+\cdot}$, together with the propensity of fullerenes to form multiply charged fullerene cations C_x^{n+} ($n \geq 2$) [4a,b, 9e,f, 10b,c] prompted us to perform high-energy beam experiments of C_{60}^{n+} ($n = 2, 3$) with He. The injection of helium into C_{60}^{2+}

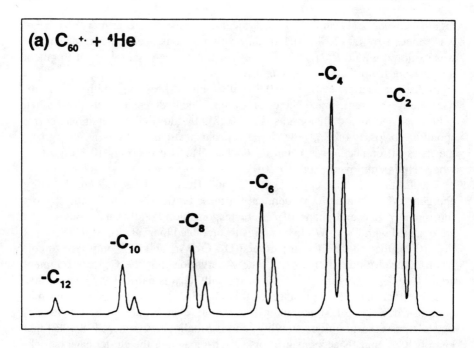

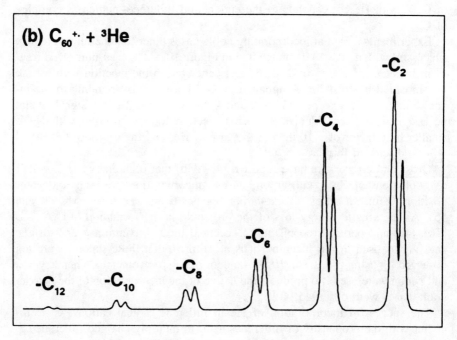

Figure 10.3 CA mass spectra for $C_{60}^{+\cdot}$ with (a) ^{4}He and (b) ^{3}He E_{lab} = 8 keV; linked scan with $B(2)/E(2)$ = const.

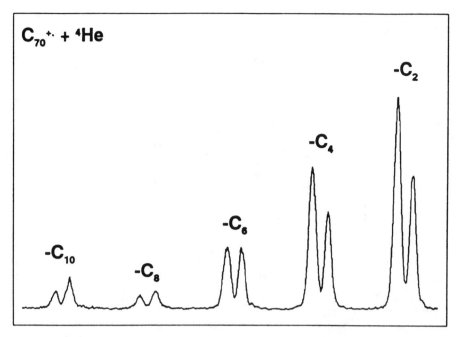

Figure 10.4   CA mass spectrum of $C_{70}^{+\cdot}/{^4}He$ (E_{lab} = 8 keV); linked scan with $B(2)/E(2)$ = constant.

was investigated at laboratory energies of 6, 8, 10, and 16 keV, and the results obtained at 6 keV are shown in Figure 10.6, where they are compared with D_2 (top). While for D_2 there was no evidence for retention of the collision gas at the sensitivity of the experiments, helium retention was observed at all four kinetic energies, decreasing in efficiency with increasing energy. The accompanying fragmentation that occurs by the loss of *even*-numbered carbon units is in agreement with results obtained in collision-induced dissociation experiments conducted elsewhere in which O_2 [9e, 10a,b] or Xe [10c] were used as the collision gases without observing incorporation of these gases. In our experiments up to 16 carbon atoms were lost from C_{60}^{2+} at 6 keV and 20 carbon atoms at 16 keV laboratory energies.

The retention of helium by C_{60}^{3+} was monitored at a kinetic energy of 9 keV (primary ions were accelerated through a potential of 3 keV). The experiments were more difficult due to the lower signal intensities, but Figure 10.7 clearly demonstrates the fragmentation with a loss of 16 carbon atoms in *even*-numbered carbon units, which has previously been reported without evidence for the inclusion of the collision gas [10a,b]. In the present experiment, it is obvious that He is retained in the cluster fragment ions [Fig. 10.7(b)], while, at the sensitivity of the experiment, D_2 is not [Fig. 10.7(a)].

Before addressing the central question of the *location* of the noble gases in the fullerene ion, that is, *endo-* versus *exohedral* carbon clusters, and the mech-

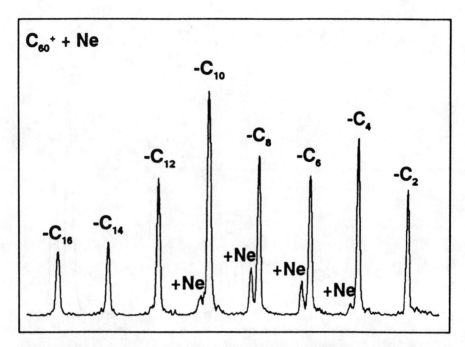

Figure 10.5 CA mass spectra of C_{60}^{+}/Ne (E_{lab} = 3 keV); linked scan with $B(2)/E(2)$ = constant.

anism of penetration of the carbon shell, we will briefly discuss the results of high-energy beam experiments aimed at generating $C_{60}He^{+\cdot}$ *without* fragmentation.

As already mentioned, although first-principles considerations may suggest that the probability of detecting the *intact* product $C_{60}N^{n+}$ of a high-energy, bimolecular reaction of C_{60}^{n+} with a collision gas N is quite small [47], the chances of actually generating $C_{60}N^{n+}$ are not that remote provided low-mass stationary target gases N and appropriate laboratory energies of the projectile ions C_{60}^{n+} are used to ensure relatively small center-of-mass energies, E_{cm}. However, there are several principal and practical obstacles to overcome in order to observe the formation of an *intact* adduct complex $C_xN^{+\cdot}$ (x = 60, 70; N = He, Ne), which are a result of the energetics of the high-energy bimolecular reaction. For example, if He is taken up by $C_{60}^{+\cdot}$ in a perfectly inelastic collision, the $C_{60}He^{+\cdot}$ product has a reduced kinetic energy by the amount of E_{cm} (e.g., 44 eV in an 8-keV experiment). At the same time its mass is higher than the original projectile $C_{60}^{+\cdot}$, making it difficult to observe the intact $C_{60}He^{+\cdot}$ ion in certain linked scan modes [4c, 42e, 43a,b,d]. In addition, interference signals occur due to inelastic scattering of $C_{60}^{+\cdot}$ with He, and the cross section of the noble-gas attachment is strongly dependent on the laboratory energy of the projectile [43d]. However, these difficulties were eventually overcome, and

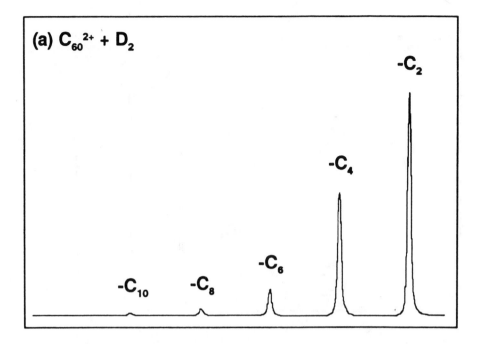

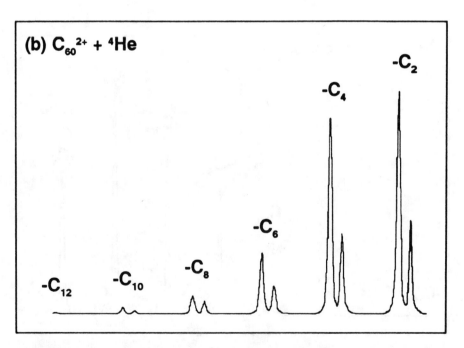

Figure 10.6 CA mass spectra for C_{60}^{2+} with (a) D_2 and (b) ^{4}He $E_{lab} = 6$ keV; linked scan with $B(2)/E(2) = $ constant.

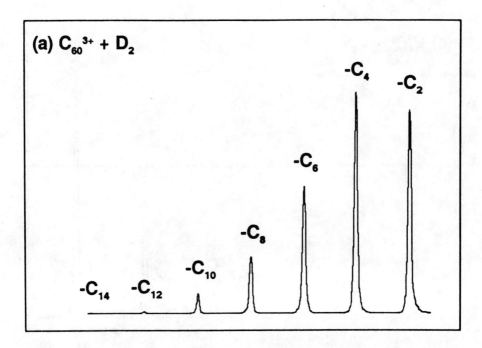

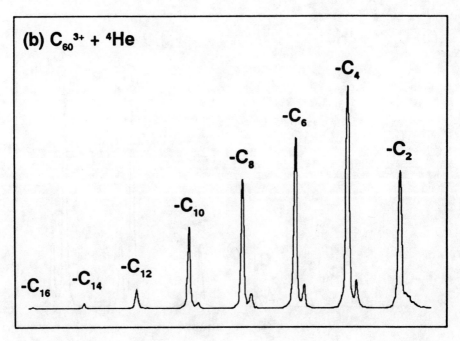

Figure 10.7 CA mass spectra for C_{60}^{3+} with (a) D_2 and (b) ^4He $E_{lab} = 9$ keV; linked scan with $B(2)/E(2) =$ constant.

Ross and Callahan [43a] were the first to demonstrate unambiguously the *direct formation* of $C_{60}He^{+\cdot}$ in keV collisions of $C_{60}^{+\cdot}$ with helium. This result was later confirmed by Caldwell et al. [4c, 43b].

Our measurement also demonstrates that retention of He by $C_{60}^{+\cdot}$ is achievable at laboratory energies of 4, 5, 6, and 8 keV, is optimal at kinetic energies in the range from 5 to 6 keV, and falls at higher and lower energies. When the linked scan with $B/E =$ constant is chosen optimally with full recognition of possible complications due to the presence of an energy-loss peak, the $C_{60}He^{+\cdot}$ is easily visible and separated from the energy-loss peak for $C_{60}^{+\cdot}$, as shown in Figure 10.8(a) for a laboratory energy of 6 keV. It should be mentioned that under *multiple-collision conditions,* the group of Gross [4c, 43b] demonstrated the incorporation of *two* helium atoms in $C_{70}^{+\cdot}$. In a spatially and temporally separate double-collision experiment, the Berlin group was recently able to inject a ^{3}He and a ^{4}He atom in $C_{60}^{+\cdot}$ to generate $^{3}He^{4}He@C_x^{+\cdot}$ ($x \leq 60$) [42f], and the spectrum is given in Figure 10.9.

As seen in Figure 10.8(b), no significant, if any, addition of D_2 and $C_{60}^{+\cdot}$ is apparent at 6 keV, nor was it observed at 3, 4, and 8 keV. The latter observation is at variance with the findings of Gross et al. [4c, 43b]. They report for the high-energy reaction of $C_{60}^{+\cdot}$ with D_2 the formation of weak, though clearly recognizable, signals at m/z 724 and m/z 722, which they assign to the formation of $C_{60}D_2^{+\cdot}$ and $C_{60}D^+$. In line with our findings (Fig. 10.1), the apparent noninclusion [4c, 43b] of deuterium (D or D_2) in *all* fragment ions suggests that the deuterium is *not* trapped *inside* the cage in the Gross experiment but rather bound externally and split off in the course of fragmentation. It should be recalled that collisional activation of protonated *exohedral* fullerenes, $C_{60}H^+$ and $C_{70}H^+$, also does not give rise to any fragment ions containing a proton; rather, a hydrogen atom is lost to generate $C_x^{+\cdot}$ ($x = 60, 70$) and, upon higher excitation, $C_{x-m}^{+\cdot}$ ($m = 2, 4, 6, \ldots$) fragments are formed [7].

We have also conducted collisional experiments of C_{60}^{2+} with He and D_2 at a laboratory energy of 6 keV, as shown in Figure 10.10. Again, there is clear evidence for the formation of an intact $C_{60}He^{2+}$ adduct ion [Fig. 10.10(a)], while similar experiments with D_2 did not show adduct formation [Fig. 10.10(b). On sensitivity grounds, these experiments could not be performed with C_{60}^{3+}.

As already discussed in the context of $M@C_x^{+\cdot}$ ($M =$ metal atom) clusters, the endpoint of C_m losses is metal dependent; this observation lent strong support for the endohedral structure hypothesis for complexes generated by laser vaporization of doped graphite. For the noble-gas fullerene complexes $M@C_{60}^{+\cdot}$ generated in bimolecular penetration processes, Caldwell et al. [4c] demonstrated the existence of a similar effect. For He-containing product ions, the lowest carbon number is $C_{44}He^{+\cdot}$; this endpoint is, perhaps fortuituously, the same as for the $K@C_x^+$ ($C_{44}K^+$) and may be due to the comparable van der Waals radii of He (1.4 Å) and K^+ (1.4 Å). Neon has a van der Waals radius of 1.5 Å, which is intermediate between the radii of K^+ (1.4 Å) and Cs^+ (1.7 Å). The lowest, unambiguously identified product in the neon series is $C_{46}Ne^{+\cdot}$; this is intermediate between $C_{44}K^+$ and $C_{48}Cs^+$, the lowest-carbon-number ful-

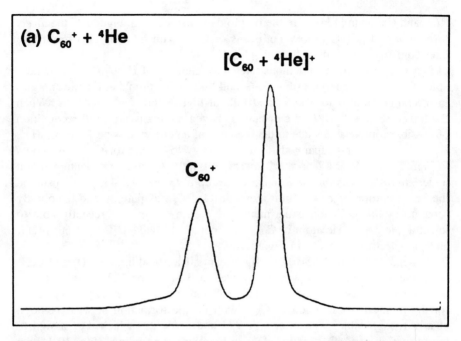

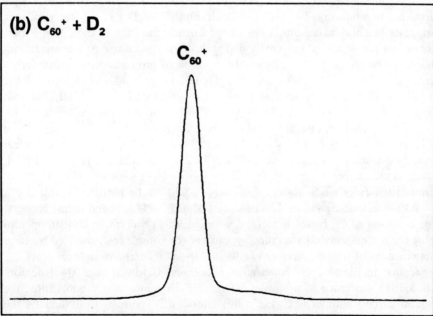

Figure 10.8 (a) $B(2)/E(2)$ = constant linked scan in the vicinity of the center of m/z 724 with an initial kinetic energy of 6 keV for $C_{60}^{+\cdot}$. The large peak on the right represents the $C_{60}He^{+\cdot}$ adduct ion, and the peak on the left arises from cutting the energy-loss tail of $C_{60}^{+\cdot}$ in the $B(2)$, $E(2)$ two-dimensional surface. (b) Similar collision experiments with D_2 as a target gas. There is no real evidence for a $C_{60}D_2^{+\cdot}$ adduct ion on the right of the energy-loss peak.

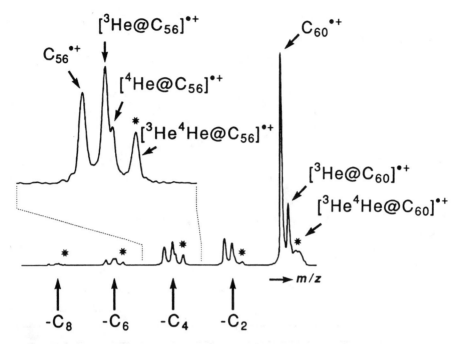

Figure 10.9 Sequential inclusion of ^{3}He and ^{4}He via high-energy bimolecular reactions of C_{60}^+ according to: $C_{60}^+ \rightarrow {}^3He@C_{60}^+ \rightarrow {}^3He^4He@C_{60}^+$. A star indicates the inclusion of both ^{3}He *and* ^{4}He [42f].

lerenes containing these metals. Argon has a van der Waals radius of 1.8 Å, and the lowest Ar-containing product ion corresponds to $C_{49}Ar^+$, slightly larger than the $C_{48}Cs^+$ system.

The precise location of the helium atom in the C_xHe^{n+} clusters ($n \geq 1$), of course, cannot be determined with the experiments described, but there are indirect indications in addition to the findings discussed in the previous part that we are almost certainly dealing with intracavity bonding rather than with cluster ions in which helium is attached to the outer sphere of the fullerene cage. (1) If the helium atom formed an exohedral complex, one would normally expect that higher noble gases, due to their larger polarizabilities, would also form cluster adducts with C_x^+. This is not observed experimentally. Rather, the cross section, σ, for adduct formation follows the qualitative order $\sigma_{He} > \sigma_{Ne} > \sigma_{Ar}$ with no adduct formation for Kr and Xe. (2) The process $C_{60}^+ \rightarrow C_{58}^+ + C_2$ requires at least 4–6 eV [9d,f, 12a, 39], while the upper binding energy of He to $C_6H_6^+$, for example, does not exceed 0.2 eV according to MP2/6-31G**//MP2/3-21G* ab initio MO calculations [42b,e, 48]. Consequently, if the helium atom was not trapped inside the cage but bound externally, the thermal energy of the cluster should be already high enough to detach a helium atom. Upon activation, on thermochemical and kinetic grounds helium rather than C_m should be eliminated from the C_xHe^+ fullerenes. This is

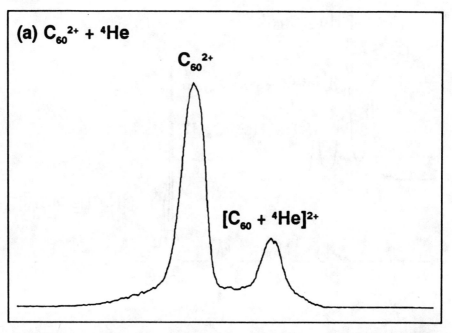

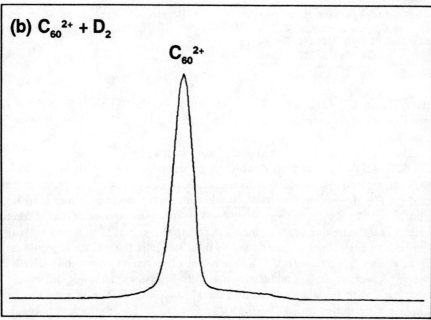

Figure 10.10   (a) $B(2)/E(2) =$ const. linked scan in the vicinity of the $C_{60}He^{2+}$ adduct ion with an initial kinetic energy of 6 keV with ^4He as a target gas. The small peak on the right represents the $C_{60}He^{2+}$ adduct ion, and the large peak on the left arises from cutting the energy-loss tail of C_{60}^{2+} in the $B(2)$, $E(2)$ two-dimensional surface. (b) Similar experiments with D_2 as a target gas. There is no real evidence for a $C_{60}D_2^{2+}$ adduct ion on the right.

not observed experimentally. (3) Mass-selected $C_xHe^{+\cdot}$ ions ($x = 60, 58$) lose not He, but C_2 unimolecularly. This behavior contrasts with Freiser's externally bound $C_{60}M^+$ systems [31], which upon collision-induced dissociation generates $C_{60}^{+\cdot}$ and/or M^+ and *no* $[C_{60-m}M]^+$ fragments. However, the behavior of the noble-gas fullerene complexes is perfectly analogous to Smalley's and McElvany's endohedral neutral-containing $M@C_x^{+\cdot}$ complexes.

However, the staunchest critics may only be convinced that the noble-gas–fullerene complex ion has an endohedral structure when the corresponding *neutral* species can be shown to exist. Experimental evidence that neutral noble-gas–fullerene complexes have a finite lifespan would leave no doubt that the noble-gas atom is physically trapped inside the cavity. This was recently accomplished [49] by subjecting $C_{60}He^{+\cdot}$ to a neutralization–reionization mass spectrometry (NRMS) [50], in which a beam of $C_{60}He^{+\cdot}$ ions could be successfully neutralized to neutral $C_{60}He$ species that were subsequently reionized. The experimental conditions impose a *minimum* lifespan of ~90 μs on the neutral $C_{60}He$ complex, and the results show definitively that both the neutral complex and the precursor ion must possess the endohedral structure, represented as $He@C_{60}$ and $He@C_{60}^{+\cdot}$, respectively [51].

All NR experiments were carried out using the four-sector mass spectrometer [44] schematically shown in Figure 10.2. The experimental evidence summarized in the spectra shown in Figure 10.11 was obtained from the following experiments:

A mixture of C_{60} and C_{70} (containing >90% C_{60}) was evaporated by heating the sample of 550°C, and the vapor was ionized by electron impact (70 eV). The resulting ions are accelerated to 5000 eV, and magnet $B(1)$ is scanned over the m/z range 714–734 under conditions of double focusing [$m/\Delta m = 1000$, with ion detection at $D(2)$]. This yields the partial mass spectrum displaying the molecular ion region of $C_{60}^{+\cdot}$ shown in Figure 10.11(a).

The $C_{60}He^{+\cdot}$ ions are generated by collisions of the high-energy $C_{60}^{+\cdot}$ ions ($E_{kin} = 5000$ eV) with quasistationary ^{4}He target atoms in collision cell $C(3)$ (transmission 40%). The center-of-mass collision energy of the C_{60}/He complex makes this reaction endothermic by ~27 eV. Therefore, the accelerating voltage of the source-generated $C_{60}^{+\cdot}$ ions used for the synthesis of $C_{60}He^{+\cdot}$ is raised to 5027 eV so that $C_{60}He^{+\cdot}$ ions produced by collision with ^{4}He in $C(3)$ will pass through the electric sector $E(1)$, which remains set for the transmission of 5000 eV ions.

Under the conditions of the first experiment but with ^{4}He in collision cell $C(3)$ and a raised accelerating voltage, a scan of $B(1)$ yields the partial spectrum shown in Figure 10.11(b): As in the first experiment, $B(1)$ transmits the various isotopomers of $C_{60}^{+\cdot}$, but because of the kinetic energy constraint imposed by $E(1)$, this sector will primarily transmit the $C_{60}He^{+\cdot}$ particles generated in $C(3)$ to detector $D(2)$. [The diffuse background in Fig. 10.11(b) arises from ions that have lost kinetic energy in inelastic collisions between $C_{60}^{+\cdot}$ and He for which helium attachment has not occurred.]

This is demonstrated by the spectrum shown in Figure 10.11(c): Under the

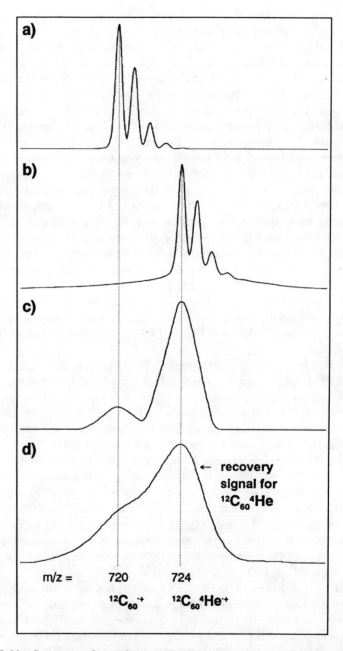

Figure 10.11 Sequence of experiments (details in text): (a) molecular ion region of $^{12}C_{60-x}{}^{13}C_x{}^+$ ($x = 0$–4); (b) "synthesis" of $^{12}C_{60}He^+$; (c) "purification" of m/z 724; (d) NR experiment on $C_{60}{}^+$ and $C_{60}He^+$. (d) represents an accumulation of ~400 spectra; for each spectrum ~6.3 ions were detected. The ratio of the number of ions in (c) to (d) is calculated as 6×10^4:1.

conditions that lead to the spectrum shown in Figure 10.11(b), $B(1)$ is set for optimal transmission of $^{12}C_{60}^{+\cdot}$ ions [m/z 724 in Fig. 10.11(b)]. Then $B(2)$ is scanned over the appropriate mass range and $C_{60}^{4}He^{+\cdot}$ is separated from the mixture of $C_{60}^{4}He^{+\cdot}$ (85%) and $^{12}C_{60}^{+\cdot}$ (15%) ions transmitted by $E(2)$. The voltages applied to sectors $E(1)$ and $E(2)$ were such that optimal transmission occurred for ions of 5000 eV translational energy and the signals were measured at $D(4)$.

The presence of the "doublet" signal Figure 10.11(c) is an enormous advantage for the NR experiment in that the $C_{60}^{+\cdot}$ component acts as a "mass marker." This avoids the problem of imprecise mass assignment that occasionally arises when using the "linked-scan" method. For the neutralization of the $C_{60}^{+\cdot}$ and $C_{60}He^{+\cdot}$ ions, see Figure 10.11(d), an experiment analogous to that described under Figure 10.11(c) is performed as follows: (1) To obtain the highest possible yield of neutrals in the electron transfer processes, a neutralization gas is used whose ionization energy (IE) lies close to the IE of C_{60} (7.6 eV) [13] (condition of near-resonant electron transfer). From a great many neutralization experiments on $C_x^{+\cdot}$ clusters [14], it followed that $(CH_3)_3N$ (IE = 7.8 eV, [52]) is one of the best reducing agents for $C_{60}^{+\cdot}$. Thus collision cell $C(5)$ was charged with $(CH_3)_3N$ at a pressure that reduced the beam transmission to ~10%. All unreacted ions were removed from the beam by maintaining the deflector electrode at 1000 V so that only the beam of neutral species entered collision cell $C(6)$. In this cell a fraction of the neutral species is reionized by collision-induced ionization with $(CH_3)_3N$ as the target gas. (2) Since the energy required for the collision-induced reionization is taken from the translational energy of the m/z 724 ion beam, the voltage applied to sector $E(2)$ was reduced slightly so that the $C_{60}He^{+\cdot}$ signal was detected at optimal transmission.

As shown in Figure 10.11(d), the recovery signal for $C_{60}He$ shows up very clearly. This demonstrates unequivocally the existence of a species with the composition $C_{60}He$ for which the endohedral structure $He@C_{60}$ is the only conceivable physical arrangement. We note in passing that this result makes neutral $He@C_{60}$ the first identified noble-gas–carbon compound [53].

Considering the vertical nature of the electron transfer processes in neutralization–reionization experiments [54], it follows that the $He@C_{60}^{+\cdot}$ ionic complex also has the endohedral structure. In addition, the noble-gas–fullerene complexes $NC_x^{+\cdot}$ (N = He, Ne; x = 52–70) exhibit the same characteristics as $He@C_{60}^{+\cdot}$ in gas-phase experiments, and thus they must also have the endohedral structure.

Let us now return to the question of how much energy is required to penetrate the carbon shell and to bring a noble gas inside the $C_x^{+\cdot}$ cage. We have sought [48] to answer this question by performing model calculations in which the noble gas is forced through the center of a C_6H_6 or $C_6H_6^{+\cdot}$ plane. At the 3-21G*//3-21G* level of theory, the transition structure for the perpendicular passage of helium through C_6H_6 is 12.5 eV higher in energy than completely separated C_6H_6 and He. At higher levels of theory, this barrier drops for the neutral system to 11.2 eV (MP2/3-21G*//MP2/3-21G*) and 10.7 eV (MP2/6-31G**//MP2/3-21G*); for the ionic system $C_6H_6^+$/He, at the highest level of theory used we obtain a barrier of 9.4 eV. As this energy barrier is clearly smaller than the kinetic energy that is available to the collision complex $C_x^{+\cdot}$/He (>30 eV for $x = 60$, 70), the penetration of the cluster shell and the inclusion of a helium atom inside the cage is energetically possible. For the injection of Ne, preliminary MO calculations result in a barrier that is at least 2 eV higher than the one calculated for the helium system, and for the injection of Ar even larger barriers (>15 eV) have to be overcome. Recent measurements [43e] on the collision energy dependence of He and Ne capture by $C_{60}^{+\cdot}$ are in pleasing agreement with our model calculations [48] in that the threshold for the He captures lies at 6 ± 2 eV in the center-of-mass reference frame, whereas that for Ne lies higher at 9 ± 1 eV.

The alternative that penetration of the carbon shell does not involve the hexagons of $C_x^{+\cdot}$ but the smaller five-membered rings is unlikely. For example, recent molecular-mechanics calculations indicate [43d] that the amounts of energy required for He to penetrate into the cavity of $C_{60}^{+\cdot}$ through the center of five- and six-membered rings are 13.1 and 9.35 eV, respectively. Similarly, MNDO calculations [16e] for the migration of Li^+ suggest, again, a clear energetic preference to penetrate the hexagon face as compared with the five-membered ring (7.3 versus 10.2 eV).

In conclusion, endohedral fullerene complexes do indeed exist, and they can be made by two entirely different techniques. While the laser vaporization of doped graphite provides a convenient route to generate metal-containing systems, the ion bombardment technique is the method of choice to include noble gases. Recent, detailed molecular-dynamics simulations by Mowrey, Ross, and Callahan [43d] have provided a picture of the trapping process that is likely to guide further experiments. In addition, collisions of beams of Li^+, Na^+, and Ne^+ ions with *neutral* C_{60} also give rise to the formation of endohedral complexes [55] and *mixed* HeLa@C_{82}^+ has been obtained by reacting a beam of La@C_{82}^+ (generated by laser vaporization of doped graphite) with helium atoms [56]. Most recently, helium encapsulation experiments were used to distinguish fullerene ions which obey the isolated pentagon rule from those which do not [57]. It was demonstrated that those $C_x^{+\cdot}$ clusters which form perfect fullerene networks give rise to observable He@$C_x^{+\cdot}$ clusters in high-energy bimolecular collisions. In contrast, the penetration of fullerene ions which violate the isolated pentagon rule, e.g. $C_{58}^{+\cdot}$ and $C_{68}^{+\cdot}$, is associated with the evaporation of C_m ($m = 2$, 4) [57]. Collision experiments of derivatized fullerene ions, e.g. $C_{60}D^+$ or $C_{60}(H)CH_3^{+\cdot}$, with helium also result in penetration, in the course of which all ligands are evaporated from the cluster surface [58].

Acknowledgment

We thank the Deutsche Forschungsgemeinschaft and the Fonds der Chemischen Industrie for the continuing support of our work. D. K. B. thanks the Alexander-von-Humboldt Foundation for a

Humboldt Senior Award. We are especially grateful to Dr. Wolfgang Krätschmer, Heidelberg, for the generous gift of fullerene samples, Prof. J. K. Terlouw, McMaster University, Hamilton, for decisive contributions to the successful realization of difficult neutralization–reionization experiments with $C_{60}^{+\cdot}$ and $He@C_{60}^{+\cdot}$, and our colleagues Dr. S. L. Anderson, Dr. E. E. B. Campbell, Dr. M. L. Gross, Dr. C. Lifshitz, Dr. M. M. Ross, Dr. R. E. Smalley, and Dr. W. Thiel for sending us manuscripts prior to publication.

References

1. (a) R. N. Thomas, in preparation; (b) The entire March issue 1992 of *Acc. Chem. Res.* deals with the chemistry and physics of fullerenes.

2. (a) S. W. McElvany, M. M. Ross, and J. H. Callahan, *Acc. Chem. Res.* **25,** 162 (1992); (b) S. W. McElvany and M. M. Ross, *J. Am. Chem. Soc. Mass Spectrom.* **3,** 268 (1992).

3. (a) D. L. Lichtenberger, K. W. Nebesny, and C. D. Ray, *Chem. Phys. Lett.* **176,** 203 (1991); (b) J. A. Zimmermann, J. R. Eyler, S. B. H. Bach, and S. W. McElvany, *J. Chem. Phys.* **94,** 3556 (1991); (c) R. K. Yoo, B. Ruscic, and J. Berkowitz, *J. Chem. Phys.* **96,** 911 (1992).

4. (a) S. W. McElvany, M. M. Ross, and J. H. Callahan, *Mat. Res. Soc. Symp. Proc.* **206,** 697 (1991); (b) C. Lifshitz, M. Iraqi, T. Peres, and J. Fischer, *Rapid Commun. Mass Spectrom.* **5,** 238 (1991); (c) K. A. Caldwell, D. E. Giblin, and M. L. Gross, *J. Am. Chem. Soc.* **114,** 3743 (1992); (d) H. Steger, J. de Vries, W. Kamke, and T. Drewello, *Chem. Phys. Lett.,* in press; (e) S. Petrie, G. Javahery, J. Wang, and D. K. Bohme, *J. Phys. Chem.* **96,** 6121 (1992).

5. (a) S. H. Yang, C. L. Peltiette, J. Conceicao, O. Cheshnovsky, and R. E. Smalley, *Chem. Phys. Lett.* **139,** 233 (1987); (b) R. F. Curl and R. E. Smalley *Science* **242,** 1017 (1988).

6. (a) P. A. Limbach, L. Schweikhard, K. A. Cowen, M. T. McDermott, A. G. Marshall, and J. V. Coe, *J. Am. Chem. Soc.* **113,** 6795 (1991); (b) R. L. Hettich, R. N. Compton, and R. H. Ritchie, *Phys. Rev. Lett.* **67,** 1242 (1991).

7. (a) S. W. McElvany and J. H. Callahan, *J. Phys. Chem.* **95,** 6186 (1991); (b) T. Weiske and H. Schwarz, unpublished results.

8. L. S. Sunderlin, J. A. Paulino, J. Chow, B. Kahn, D. Ben-Amotz, and R. R. Squires, *J. Am. Chem. Soc.* **113,** 5489 (1991).

9. (a) P. P. Radi, T. L. Bunn, P. R. Kemper, M. E. Molchan, and M. T. Bowers, *J. Chem. Phys.* **88,** 2809 (1988); (b) P. P. Radi, M. E. Rincon, M. T. Hsu, J. Brodbelt-Lustig, P. Kemper, and M. T. Bowers, *J. Phys. Chem.* **93,** 6187 (1989); (c) P. P. Radi, M. T. Hsu, J. Brodbelt-Lustig, M. Rincon, and M. T. Bowers, *J. Chem. Phys.* **92,** 4817 (1990); (d) P. P. Radi, M. T. Hsu, M. T. Rincon, P. R. Kemper, and M. T. Bowers, *Chem. Phys. Lett.* **174,** 223 (1990); (e) D. R. Luffer and K. H. Schram, *Rapid Commun. Mass Spectrom.* **4,** 552 (1990); (f) C. Lifshitz, M. Iraqi, T. Peres, and J. E. Fischer, *Int. J. Mass Spectrom. Ion Processes* **107,** 565 (1991); (g) D. Schröder and D. Sülzle, *J. Chem. Phys.* **94,** 6933 (1991); (h) M. T. Bowers, P. P. Radi, and M. T. Hsu, *J. Chem. Phys.* **94,** 6934 (1991); (i) T. Drewello, K.-D. Asmus, J. Stach, R. Herzschuh, and C. S. Foote, *J. Phys. Chem.* **95,** 10551 (1991).

10. (a) R. J. Doyle, Jr., and M. M. Ross, *J. Phys. Chem.* **95,** 4954 (1991); (b) A. B. Young, L. M. Cousins, and A. G. Harrison, *Rapid Commun. Mass Spectrom.* **5,** 226 (1991); (c) D. Ben-Amotz, R. G. Cooks, L. Dejarma, J. C. Gunderson, S. H. Hoke II, B. Kahr, G. L. Payne, and J. M. Wood, *Chem. Phys. Lett.* **183,** 149 (1991).

11. (a) R. D. Beck, P. St. John, M. M. Alvarez, F. Diederich, and R. L. Whetten, *J. Phys.*

Chem. **95,** 8402 (1991); (b) R. C. Mowrey, D. W. Brenner, B. I. Dunlap, J. W. Mintmire, and C. T. White, *J. Phys. Chem.* **95,** 7138 (1991).

12. (a) S. C. O'Brien, J. R. Heath, R. F. Curl, and R. E. Smalley, *J. Chem. Phys.* **88,** 220 (1988); (b) S. Maruyama, M. E. Lee, R. E. Haufler, Y. Chai, and R. E. Smalley, *Z. Phys. D.* **19,** 409 (1991); (c) C. Z. Wang, C. H. Xu, C. T. Chan, and K. M. Ho, *J. Phys. Chem.* **96,** 3563 (1992).

13. G. v. Helden, M.-T. Hsu, P. R. Kemper, and M. T. Bowers, *J. Chem. Phys.* **95,** 3835 (1991).

14. T. Wong, J. K. Terlouw, T. Weiske, and H. Schwarz, *Int. J. Mass Spectrom. Ion Processes* **113,** R23 (1992).

15. Review: H. Schwarz, *Angew. Chem. Int. Ed. Engl.* **31,** 293 (1992).

16. (a) A. Rosen and B. Wästberg, *J. Am. Chem. Soc.* **110,** 8701 (1988); (b) J. Cioslowski, *J. Am. Chem. Soc.* **113,** 4139 (1991); (c) J. Cioslowski and E. D. Fleischmann, *J. Chem. Phys.* **94,** 3730 (1991); (d) A. H. H. Chang, W. C. Ermler, and R. M. Pitzer, *J. Chem. Phys.* **94,** 5004 (1991); (e) D. Bakewies and W. Thiel, *J. Am. Chem. Soc.* **113,** 3704 (1991); (f) B. Wästberg and A. Rosen, *Phys. Scripta* **44,** 276 (1991); (g) J. Cioslowski, *Spectroscopic and Computational Studies of Inclusion Compounds.*, Kluwer Publishers, Dordrecht, 1992, in press; (f) P. P. Schmidt, B. I. Dunlap, and C. I. White, *J. Phys. Chem.* **95,** 10537 (1991); (g) M. Kolb and W. Thiel, *J. Comput. Chem.*, submitted.

17. Y. Chai, T. Guo, C. Jin, R. E. Haufler, L. P. F. Chibante, J. Fure, L. Wang, J. M. Alford, and R. E. Smalley, *J. Phys. Chem.* **95,** 7564 (1991).

18. (a) J. M. Hawkins, T. A. Lewis, S. D. Loren, A. Meyer, J. R. Heath, Y. Sabato, and R. J. Saykally, *J. Org. Chem.* **55,** 6250 (1990); (b) J. M. Hawkins, A. Meyer, T. A. Lewis, S. Loren, and F. J. Hollander, *Science* **252,** 312 (1991); (c) P. J. Fagan, J. C. Calabrese, and B. Malone, *Science* **252,** 1160 (1991); (d) A. J. Balch, V. J. Catalano, and J. W. Lee, *Inorg. Chem.* **30,** 3980 (1991); (e) A. L. Balch, V. J. Catalano, J. W. Lee, M. M. Olmstead, and S. R. Parkin, *J. Am. Chem. Soc.* **113,** 8953 (1991); (f) R. S. Koefod, M. F. Hudgens, and J. R. Shapley, *J. Am. Chem. Soc.* **113,** 8957 (1991); (g) P. J. Fagan, J. C. Calabrese, and M. Malone, *J. Am. Chem. Soc.* **113,** 9408 (1991).

19. J. W. Bausch, G. K. Sarya Prakash, G. A. Olah, D. S. Tse, D. C. Lorents, Y. K. Bae, and R. Malhatra, *J. Am. Chem. Soc.* **113,** 3205 (1991).

20. (a) R. E. Haufler, J. Conceicao, L. P. F. Chibante, Y. Chai, N. E. Byrne, S. Flanagan, M. M. Haley, S. C. O'Brien, C. Pan, Z. Xiao, W. E. Billups, M. A. Ciufolini, R. N. Hauge, J. L. Margrave, L. J. Wilson, R. F. Curl, and R. E. Smalley, *J. Phys. Chem.* **94,** 8634 (1990); (b) For the gas-phase hydrogenation (and alkylation) of ionized C_{60}, see D. Schröder, D. K. Böhme, T. Weiske, and H. Schwarz, *Int. J. Mass Spectrom. Ion Processes* **116,** R13 (1992); (c) S. Petrie, G. Javahery, J. Wang, D. K. Bohme, *J. Am. Chem. Soc.* **114,** 6268 (1992).

21. (a) H. Selig, C. Lifshitz, T. Peres, J. E. Fischer, A. R. McGie, W. J. Romanow, J. P. McCandey, Jr., and A. B. Smith III, *J. Am. Chem. Soc.* **113,** 5475 (1991); (b) J. H. Holloway, E. G. Hope, R. Taylor, G. J. Langley, A. G. Avent, T. J. Dennis, J. P. Hare, H. W. Kroto, and D. R. M. Walton, *J. Chem. Soc. Chem. Commun.* 966 (1991).

22. G. A. Olah, I. Bucsi, C. Lambert, R. Aniszfeld, N. J. Trivedi, D. K. Sensharma, and G. K. Surya Prakash, *J. Am. Chem. Soc.* **113,** 9385 (1991).

23. G. A. Olah, I. Bucsi, C. Lambert, R. Aniszfeld, N. J. Trivedi, D. K. Sensharma, and G. K. Surya Prakash, *J. Am. Chem. Soc.* **113,** 9387 (1991).

24. A. Hirsch, Q. Li, and F. Wudl, *Angew. Chem. Int. Ed. Engl.* **30,** 1309 (1991).

25. P. J. Krusic, E. Wasserman, K. F. Preston, J. R. Morton, and P. N. Keizer, *Science* **254**, 1183 (1991).

26. T. Suzuki, Q. Li, K. C. Khemani, F. Wudl, Ö. Almarson, *Science* **254**, 1186 (1991).

27. (a) K. M. Creegan, J. L. Robbins, W. K. Robbins, J. M. Millar, R. D. Sherwood, P. J. Tindall, D. M. Cox, J. P. McCaulry, Jr., D. R. Jones, T. T. Gallagher, and A. M. Smith III, *J. Am. Chem. Soc.* **114**, 1103 (1992); (b) Y. Elemas, S. Silverman, C. Shen, M. Kao, C. S. Foote, M. M. Alvarez, and R. L. Whetten, *Angew. Chem. Int. Ed. Engl.* **31**, 000.

28. H. W. Kroto, A. W. Allafs, and S. P. Balm, *Chem. Rev.* **91**, 1213 (1991).

29. H. W. Kroto, *Science* **242**, 1139 (1988).

30. D. M. Cox, K. C. Reichmann, A. Kaldor, *J. Chem. Phys.* **88**, 1588 (1988).

31. (a) L. M. Roth, Y. Huang, J. T. Schwedler, J. C. Cassady, D. Ben-Amotz, B. Kahn, and B. S. Freiser, *J. Am. Chem. Soc.* **113**, 6298 (1991); (b) Y. Huang and B. S. Freiser, *J. Am. Chem. Soc.* **113**, 8186 (1991); (c) Y. Huang and B. S. Freiser, *J. Am. Chem. Soc.* **113**, 9418 (1991).

32. H. W. Kroto, J. R. Heath, S. C. O'Brien, R. F. Curl, and R. E. Smalley, *Nature* **318**, 165 (1985).

33. J. R. Heath, S. C. O'Brien, Q. Zhang, Y. Liu, R. Curl, H. W. Kroto, F. K. Tittel, and R. E. Smalley, *J. Am. Chem. Soc.* **107**, 7779 (1985).

34. (a) E. A. Rohlfing, D. M. Cox, and A. Kaldor, *J. Chem. Phys.* **81**, 3322 (1984); (b) D. M. Cox, D. J. Trevor, K. C. Reichmann, and A. Kaldor, *J. Am. Chem. Soc.* **108**, 2457 (1986).

35. (a) F. D. Weiss, J. L. Elkind, S. C. O'Brien, R. F. Curl, and R. E. Smalley, *J. Am. Chem. Soc.* **110**, 4464 (1988); (b) Review: R. E. Smalley, ACS Symposium Series on *Large Carbon Clusters*, Eds. G. Hammond and V. Kuck, Washington, D.C. 1992, p. 141.

36. S. W. McElvany, *J. Phys. Chem.* **96**, 4935 (1992).

37. (a) K. Raghavachari, R. A. Whiteside, and J. A. Pople, *J. Chem. Phys.* **85**, 6623 (1986); (b) K. Raghavachari and J. S. Binkley, *J. Chem. Phys.* **87**, 2191 (1987); (c) Review: W. Weltner, Jr. and R. J. Van Zee, *Chem. Rev.* **89**, 1713; d) R. E. Stanton, *J. Phys. Chem.* **96**, 111 (1992).

38. (a) R. F. Curl and R. E. Smalley, *Scientif. Amer.* 32 (1991); (b) Ref. [35b].

39. A reevaluation of the experiment using a better data set resulted in a binding energy of ~5.3 eV: (a) C. Lifshitz, private communication, March, 1992, (b) P. Sandler, T. Peres, G. Weissman, C. Lifshitz, *Ber. Bunsenges. Phys. Chem.* **96**, 1195 (1992); (c) P. Sandler, C. Lifshitz, C. E. Klots, *Chem. Phys. Lett.*, in press.

40. (a) J. H. Weaver, Y. Chai, G. H. Kroll, C. Jin, T. R. Ohno, R. E. Haufler, T. Guo, J. M. Alford, J. Conceicao, L. P. F. Chibante, A. Jain, G. Palmer, and R. E. Smalley, *Chem. Phys. Lett.* **190**, 460 (1992); (b) M. M. Alvarez, E. G. Gillan, K. Holczer, R. B. Kaner, K. S. Min, and R. L. Whetten, *J. Phys. Chem.* **95**, 10561 (1991); (c) R. D. Johnson, M. S. de Vries, J. Salem, D. S. Bethune, and C. S. Yannoni, *Nature* **355**, 239 (1992); (d) H. Shinohara, H. Sato, M. Ohkochi, Y. Ando, T. Kodama, T. Shida, T. Kato, and Y. Saito, *Nature* **357**, 52 (1992); (e) T. Guo, M. D. Diener, Y. Chai, M. J. Alford, R. E. Hanfler, S. M. McClure, T. Ohrer, J. H. Weaver, G. E. Scuseria, and R. E. Smalley, *Science,* submitted; (f) L. Soderholm, P. Wurz, K. R. Lykke, D. H. Parker, F. W. Lytle, *J. Phys. Chem.* **96**, 7153 (1992); (g) S. Suzuki, Kawata, H. Shiromaru, K. Yamauchi, K. Kikuchi, T. Kato, Y. Achiba, *J. Phys. Chem.* **96**, 7159 (1992); (h) C. S. Yannani, M. Hoinkis, M. J. de Vries, D. S. Behune, J. R. Salem, M. S. Crowder, R. D. Johnson, *Science,* **256**, 1191 (1992); (i) E. G. Gillan, C. Yeretzion, K. S. Min, M. M. Alvarez, R. L. Whetten, *J. Phys. Chem.* **96**, 6869 (1992).

41. (a) T. Pradeep, G. U. Kulkarni, K. R. Konnan, T. N. Guru Row, and C. N. R. Rao, *J. Am. Chem. Soc.* **114,** 2272 (1992); (b) R. E. Smalley, private communication, May, 1992.

42. (a) D. K. Böhme, T. Weiske, J. Hrušák, W. Krätschmer, and H. Schwarz, *39th Ann. Confer. Mass Spectrom. Allied Top.* (ASMS), Nashville, TN, 21 May 1991; (b) T. Weiske, D. K. Böhme, J. Hrušák, W. Krätschmer, and H. Schwarz, *Angew. Chem. Int. Ed. Engl.* **30,** 884 (1991); (c) T. Weiske, D. K. Bohme, and H. Schwarz, *J. Phys. Chem.* **95,** 8451 (1991); (d) T. Weiske, J. Hrušák, D. K. Böhme, and H. Schwarz, *Chem. Phys. Lett.* **186,** 459 (1991); (e) T. Weiske, J. Hrušák, D. K. Bohme, and J. Schwarz, *Helv. Chim. Acta* **75,** 79 (1992); (f) T. Weiske and H. Schwarz, *Angew. Chem. Int. Ed. Engl.* **31,** 605 (1992).

43. (a) M. M. Ross and J. H. Callahan, *J. Phys. Chem.* **95,** 5720 (1991); (b) K. A. Caldwell, D. E. Giblin, C. S. Hsu, D. Cox, and M. L. Gross, *J. Am. Chem. Soc.* **113,** 8519 (1991); (c) Z. Wan, J. F. Christian, and S. C. Anderson, *J. Phys. Chem.* **96,** 000; (d) R. C. Mowrey, M. M. Ross, J. H. Callahan, *J. Phys. Chem.*, in press; (e) E. E. B. Campbell, R. Ehrlich, A. Heilscher, J. M. A. Frazav, and I. V. Hertel, *Z. Phys. D.* **23,** 1 (1992).

44. (a) R. Srinivas, D. Sülzle, T. Weiske, and H. Schwarz, *Int. J. Mass Spectrom. Ion Processes* **107,** 369 (1991); (b) R. Srinivas, D. Sülzle, W. Koch, C. H. DePuy, and H. Schwarz, *J. Am. Chem. Soc.* **113,** 5970 (1991).

45. (a) C. G. Macdonald and M. J. Lacey, *Org. Mass Spectrom.* **19,** 55 (1984); (b) J. R. Chapman, *Practical Organic Mass Spectrometry,* Wiley, Chichester, 1985, Chapter 6.

46. R. E. Smalley, *Science* (N.Y.) **31,** 22 (1991).

47. R. G. Cooks, (Ed.), *Collision Spectroscopy,* Plenum Press: New York, 1978.

48. J. Hrušák, D. K. Bohme, T. Weiske, and H. Schwarz, *Chem. Phys. Lett.* **193,** 97 (1992).

49. T. Weiske, T. Wong, W. Krätschmer, J. K. Terlouw, and H. Schwarz, *Angew. Chem. Int. Ed. Engl.* **31,** 183 (1992).

50. Reviews: (a) C. Wesdemiotis and F. W. McLafferty, *Chem. Rev.* **87,** 485 (1987); (b) J. K. Terlouw and H. Schwarz, *Angew. Chem. Int. Ed. Engl.* **26,** 805 (1987); (c) H. Schwarz, *Pure Appl. Chem.* **61,** 685 (1989); (d) J. L. Holmes, *Adv. Mass Spectrom.* **11,** 53 (1989); (e) J. K. Terlouw, *Adv. Mass Spectrom.* **11,** 984 (1989); (f) J. L. Holmes, *Mass Spectrom. Rev.* **8,** 513 (1989); (g) F. W. McLafferty, *Science* **247,** 990 (1990); (h) F. W. McLafferty, *Int. J. Mass Spectrom. Ion Processes* **118/119,** 221 (1992).

51. For speculations and "experiments" regarding carbon clusters of unspecified structure as containers for noble gas atoms and their presence in interstellar space: (a) R. S. Lewis, B. Srinivasan, and E. Anders, *Science* **190,** 1251 (1975); (b) S. Niemeyer and K. Marti, *Proc. Lunar Planet. Sci.* **12B,** 1177 (1981); (c) D. Heymann, *J. Geophys. Res.* **B91,** E135 (1986).

52. S. G. Lias, J. E. Bartmess, J. F. Liebman, J. L. Holmes, R. D. Levin, and W. G. Mallard, *J. Phys. Chem. Ref. Data* Supplement 1 (1988).

53. G. Frenking and D. Cremer, *Structure and Bonding* **73,** 17 (1990), and references cited therein.

54. (a) P. Fournier, J. Appell, F. C. Fehsenfeld, and J. Durup, *J. Phys. B* **5,** L58 (1972); (b) F. C. Fehsenfeld, J. Appell, P. Fournier, and J. Durup, *J. Phys. B.* **6,** L268 (1973); (c) J. C. Lorquet, B. Ley-Nihant, F. W. McLafferty, *Int. J. Mass Spectrom. Ion Processes* **100,** 465 (1990).

55. Z. Wan, J. F. Christian, and S. L. Anderson, *J. Phys. Chem.* **96,** 3344 (1992).

56. J. H. Callahan, S. W. McElvany, and M. M. Ross, *40th Ann. Confer. Mass Spectrom. Allied Top.* (ASMS), Washington, D.C., May 31–June 5, 1992.

57. J. H. Callahan, M. M. Ross, T. Weiske, and H. Schwarz, *J. Phys. Chem.*, submitted.

58. T. Weiske, H. Schwarz, A. Hirsch, and T. Grösser, *Chem. Phys. Lett.*, **199,** 640 (1992).

CHAPTER

11

Fullerene Electrochemistry: Detection, Generation, and Study of Fulleronium and Fulleride Ions in Solution

Lon J. Wilson, Scott Flanagan, L. P. F. Chibante, and J. M. Alford

Since our initial report [1] in November of 1990 that electrochemical reduction of C_{60} proceeds smoothly to the C_{60}^- and C_{60}^{2-} ions in solution, fewer than 15 additional papers describing the electrochemical properties of the fullerenes have appeared in the literature. Together, these papers constitute only 3% of the more than 500 fullerene papers submitted for publication until the present time [2]. While few in number, these electrochemical papers bear a disproportionate importance to the field since they (1) first indicated that C_{60} and C_{70} would exhibit the chemistry of electron-deficient molecules, (2) first provided an experimental verification of initial molecular-orbital (MO) calculations [3] on C_{60}, and (3) first heralded (unknowingly at the time) the discovery of C_{60}-based superconducting materials [4]. Present work indicates that most, if not all, members of the fullerene family of molecules will exhibit a rich redox chemistry.

This chapter summarizes current general knowledge of the redox properties of fullerene cations (fulleronium ions) and anions (fulleride ions) in solution as revealed by electrochemical methodologies. While discussion focuses primarily on solution-state chemistry, a mention of the solid state is also included because of the inordinate importance of the C_{60}^{3-} anion to the solid-state field of superconductivity. Most of the information presented comes from the current literature, but we also present the first electrochemical data to appear for C_{84}. The chapter develops relevant concepts from a historical standpoint to emphasize the interplay between theoretical and experimental chemistry that has led to the current level of understanding of the reduced and oxidized states of C_{60} and other fullerenes. Finally, the chapter concludes with some projected glimpses

285

into the future of fullerene and metallofullerene redox chemistry. In such a rapidly moving field, this future undoubtedly will arrive sooner than later.

11.1 Electrochemistry of C_{60} and C_{70}

11.1.1 C_{60} Reduction Processes

The electrochemistry of C_{60} reveals the existence of at least eight oxidation states (C_{60}^n: $n = 0$, $1-$, $2-$, $3-$, $4-$, $5-$, $6-$, and $x+$, where x has yet to be definitely established) by use of cyclic voltammetry (CV) and differential pulse polography (DPP). The six known reductions of C_{60} have all been found to be reversible by cyclic voltammetry at $-10°C$, as shown in Figure 11.1. Historically, these six reduction processes for C_{60} were uncovered literally "one electron at a time," with reports of the two- [1], three- [5], four- and five- [6], and finally six-electron processes [7] appearing sequentially in the literature. All six electrochemical reductions of Figure 11.1 are reversible at a scan rate of 100 mV s^{-1} in a CH_3CN/toluene solvent system at $-10°C$. Even more remarkable, the reductions occur at fairly evenly spaced potentials ($\sim\Delta$ 400 mV). Initial MO calculations on C_{60} predict a triply degenerate LUMO (lowest unoccupied molecular orbital) of t_{1u} symmetry [3], and the observed reduction of C_{60} to the C_{60}^{6-} ion supports this view.

Other solvents such as THF or CH_2Cl_2 produce less reversible electrochemical behavior, especially for the more negative reduction processes [5]. The first four reductions of C_{60} have been verified as one-electron transfers by bulk electrolysis in benzonitrile [6c]. The C_{60}^-, C_{60}^{2-}, and C_{60}^{3-} fulleride ions remain stable in benzonitrile for up to several days, and neutral C_{60} can be quantitatively recovered from these solutions upon reoxidation. C_{60}^{4-} has so far only been produced by bulk electrolysis in either benzonitrile or pyridine using the wide reduction windows of these solvents; however, quantitative recovery of C_{60} was not achieved upon reoxidation of the C_{60}^{4-} solutions [8a]. The discovery of conditions under which all six reductions of C_{60} are reversible at slow scan rates indicates that it may be possible to generate the $C_{60}^- \rightarrow C_{60}^{6-}$ anions by bulk electrolysis, although no reports of such experiments have yet appeared [7].

Table 11.1 displays $E_{1/2}$ values for the C_{60} reduction processes for selected solvent/electrolyte/temperature conditions. The half-wave potentials for the electrode reactions of C_{60} vary dramatically from one medium to another, with solvent and supporting electrolyte effects in some cases as large as 600 mV. The $E_{1/2}$ values have been found to correlate with solvent donor–acceptor properties, as well as with the size and nature of the supporting electrolyte cation; the electrolyte anion has little effect [8a]. Solvents with high Gutmann donor numbers facilitate the first reduction process. Conversely, the third reduction correlates best with the Gutmann acceptor number. The second reduction does not correlate particularly well with either the acceptor or donor property of the

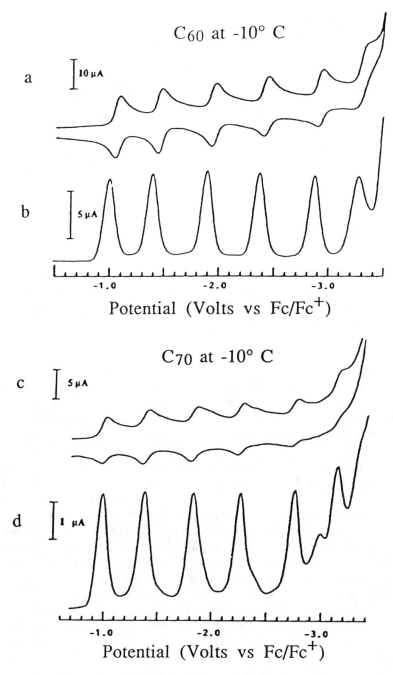

Figure 11.1 Reduction of C_{60} and C_{70} in CH_3CN/toluene at $-10°C$ using (a) and (c) cyclic voltammetry at a 100 mV/s scan rate and (b) and (d) differential pulse voltammetry (3 mm diameter glassy carbon working electrode, 50-mV pulse, 50-ms pulse width, 300-ms period, 25 mV/s scan rate). Reprinted with permission from Ref. [7].

Table 11.1 $E_{1/2}$ values (V vs. SCE) of C_{60} reductions for selected solvent/electro-lyte/temperature conditions

	benzonitrile/ TPAClO$_4$, RT[a]	CH$_2$Cl$_2$/ TBABF$_4$, RT[b]	THF/ TBAClO$_4$, RT[c]	benzene/ THAClO$_4$, RT[d]	(CH$_3$CN/toluene)/ TBAPF$_6$, $-10°C$[e] (vs. Fc/Fc$^+$)
$C_{60} \rightleftharpoons C_{60}^-$	-0.43	-0.44	-0.33	-0.36	-0.98
$C_{60}^- \rightleftharpoons C_{60}^{2-}$	-0.85	-0.82	-0.92	-0.83	-1.37
$C_{60}^{2-} \rightleftharpoons C_{60}^{3-}$	-1.32	-1.25	-1.49	-1.42	-1.87
$C_{60}^{3-} \rightleftharpoons C_{60}^{4-}$	-1.78	-1.72	-1.99	-2.01	-2.35
$C_{60}^{4-} \rightleftharpoons C_{60}^{5-}$	—	—	—	-2.60	-2.85
$C_{60}^{5-} \rightleftharpoons C_{60}^{6-}$	—	—	—	—	-3.26
$Fc^+ \rightleftharpoons Fc$	0.50	—	0.57	0.47	0.00

[a-c] Ref. [8a].
[d] Ref. [6c].
[e] Ref. [7].

solvent. The $E_{1/2}$ values have also been measured as a function of temperature in various media and the measured $\Delta E_{1/2}/\Delta T$ values used to calculate ΔS_{et} for each reduction step [8a].

11.1.2 C_{70} Reduction Processes

C_{70} also exhibits at least eight oxidation states (C_{70}^n: $n = 0, 1-, 2-, 3-, 4-, 5-, 6-, x+$, with x not yet determined), as shown in Figure 11.1. The elec-trochemical properties of C_{70} have become known almost in parallel to those of C_{60} due to C_{70}'s status as the second most abundant fullerene. Initially, Al-lemand et al. reported that C_{70} had the same electrochemistry as C_{60} [5a], but further study confirms this behavior for only the first three reduction potentials; the $E_{1/2}$ values for the $4 \rightarrow 6$ reduction processes differ substantially from those of C_{60} [6, 7]. Table 11.2 compares the six $E_{1/2}$ reduction potentials for C_{60} and C_{70}. Coulometry establishes one-electron transfers for the first three reductions of C_{70} [6c], and electrolyzed solutions of the C_{70}^- and C_{70}^{2-} fullerides exhibit essentially the same stabilities as their C_{60}^- and C_{60}^{2-} counterparts. Paralleling the case for C_{60}, conditions have now been found under which all six C_{70} anions can be observed at slow scan rates [7]. Thus, generation of all six anions may be possible by bulk electrolysis, but such experiments have yet to be reported for C_{70}^{3-} thru C_{70}^{6-}. As seen from the table, the $4 \rightarrow 6$ reductions for C_{70} occur at more positive potentials than the corresponding C_{60} reductions. Since the C_{70} molecule is larger than C_{60}, this observation is consistent with a simple charge-separation–delocalization model. However, this view may be naive since the LUMO of C_{70} is only doubly degenerate (e_{1u}'' symmetry) and can therefore ac-cept only four electrons; an additional low-lying orbital of a_1'' symmetry ac-counts for the two additional levels of reduction for the C_{70}^{5-} and C_{70}^{6-} ions [9]. The difference in the MO bonding scheme between C_{60} (LUMO of t_{1u}) and

Table 11.2 Comparison of $E_{1/2}$ values (V vs. Fc/Fc^+) for the six reductions of C_{60} and C_{70} under identical conditions[a,b]

	C_{60}	C_{70}
1st red.	−0.98	−0.97
2nd red.	−1.37	−1.34
3rd red.	−1.87	−1.78
4th red.	−2.35	−2.21
5th red.	−2.85	−2.70
6th red.	−3.26	−3.07

[a]Taken from Ref. [7].
[b]Conditions as given in Figure 11.1.

C_{70} (LUMO of e_1'' + LUMO + 1 of a_1'') could lead to interesting differences in the electronic ground states, and hence magnetic properties, of corresponding $C_{60}{}^{n-}$ and $C_{70}{}^{n-}$ ions for $n > 2$. A thorough study of solvent, electrolyte, and temperature effects has not been undertaken for C_{70}.

11.1.3 C_{60} and C_{70} Oxidation Processes

Early interest in fulleronium ions centered around $C_{60}{}^+$ in the gaseous state as a possible source of the diffuse interstellar lines [10]. In benzonitrile, both C_{60} and C_{70} oxidize electrochemically only with difficulty at +1.76 V (vs. SCE) [6c]. For C_{60} and C_{70} in benzonitrile, the $E_{1/2} = 1.76$ V oxidation process is irreversible at scan rates of up to 50 V s^{-1} and temperatures down to −15°C. Controlled-potential electrolysis of either C_{60} or C_{70} at +1.90 V indicates that four electrons are abstracted. The oxidized solutions are orange, air stable, and EPR silent at −150°C. Taken together, these data indicate that the first oxidation steps of C_{60} and C_{70} proceed via overall four-electron transfers that are followed or accompanied by one or more chemical reactions to render the overall electrochemical oxidations irreversible. The solvent window of nitrobenzene is wider than for benzonitrile, allowing observation of two irreversible oxidative processes for C_{60} at E_{ox} of +1.90 and +2.59 V (vs. Ag/AgCl), as seen in Figure 11.2. This second oxidative process, reported here for the first time, awaits more detailed characterization.

C_{70} exhibits a similar behavior to C_{60} showing E_{ox} at +1.97 and +2.74 (vs. Ag/AgCl), with both potentials shifted slightly positive relative to C_{60}, in contrast to the case for benzonitrile, where the first oxidation is coincident for the two fullerenes.

The question naturally arises as to whether the molecular integrity of C_{60} is maintained upon electrochemical oxidation. We continue to pursue this question via electrosynthetic methods for C_{60}.

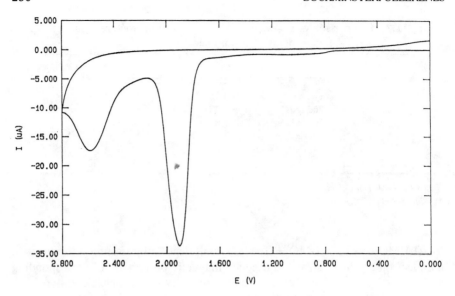

Figure 11.2 Oxidation of C_{60} in nitrobenzene ($0.1M$ TBABF$_4$, saturated C_{60}, 25°C) using cyclic voltammetry with a glassy carbon working electrode and a Ag/AgCl reference electrode at a 200 mV/s scan rate. Both peaks are absent when no C_{60} is present.

11.1.4   Spectroelectrochemical Studies of C_{60} and C_{70}

UV–Vis, EPR, and near-ir spectra have been obtained for a number of fulleride ions using spectroelectrochemical methods. Figure 11.3 depicts the time-dependent electronic absorption spectrum for reduction of C_{60} to C_{60}^- in CH_2Cl_2. Thin-layer controlled potential electrolysis generates C_{60}^- during the experiment, while depleting C_{60}. Similar data have also been gathered for the C_{60}^{2-}, C_{70}^{2-} fulleride ions in CH_2Cl_2 [6a], and the relevant spectral band positions are summarized in Table 11.3. In general, upon each reduction, the bands become broader and more red shifted. In the visible region, these spectral changes produce the colors of the different fulleride ions: C_{60}^-, dark red-purple; C_{60}^{2-}, red-orange; C_{60}^{3-} dark red-brown.

The electronic spectra of electrochemically generated C_{60}^- and C_{70}^- have also been obtained in the near-ir region and found to differ from one another, with the C_{60}^- spectrum displaying an intense band at 1064 nm ($\varepsilon \cong 1.2 \times 10^4$ mol^{-1} L^{-1}) accompanied by two weaker bands at 917 nm ($\varepsilon \cong 4.6 \times 10^3$ mol^{-1} L^{-1}) and 995 nm ($\varepsilon \cong 5.1 \times 10^3$ mol^{-1} L^{-1}), while the spectrum of C_{70}^- has no such bands [11d]. Figure 11.4 displays this spectral region for C_{60}^-. The source of these near-ir transitions for C_{60}^- have been interpreted on the basis of emerging MO theory for C_{60} and C_{70}. For C_{60}, the LUMO is t_{1u}, whereas for C_{60}^- the HOMO is the same t_{1u} orbital, but now singularly occupied. Thus, new symmetry-allowed transitions for C_{60}^- [t_{1u} (HOMO) $\rightarrow t_{1g}$ (LUMO)] become feasible at low energy, and the band positions and intensities of Figure

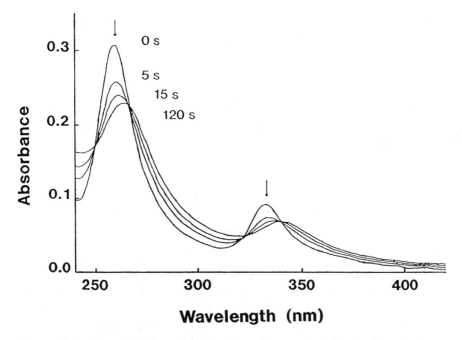

Figure 11.3 Time-dependent electronic absorption spectrum for the C_{60} reduction to C_{60}^- in CH_2Cl_2, $0.05M$ TBABF$_4$ by thin-layer controlled-potential electrolysis at -0.60 V versus SCE. Reprinted with permission from Ref. [6a].

11.4 agree with this assignment. The absence of such bands for C_{70}^- has been rationalized on the basis of recent ab initio SCF–HF calculations [9].

   EPR spectra of C_{60}^-, C_{60}^{2-}, C_{60}^{3-}, C_{70}^-, and C_{70}^{2-} have also been obtained via electrochemical generation of the appropriate fulleride ion in solution. The spectra of C_{60}^-, C_{60}^{2-}, C_{70}^-, and C_{70}^{2-} in frozen solution appear in Figure 11.5. The spectrum of C_{60}^- has been reported by several workers [6a, 8b, 11]. In general, these results now agree; thus, the g value of C_{60}^- lies between 1.997 and 2.000. These results are consistent with an $S = \frac{1}{2}$ ground state for C_{60}^-, with the negative shift of the g value relative to the 2.0023 free-electron value probably due to spin-orbit coupling [8b, 11e]. The g value of C_{60}^- displays a temperature dependence, increasing as the temperature rises. The linewidth also displays temperature dependence, broadening as the temperature rises. Based on this line-broadening behavior, Dubois et al. propose that the ground state of C_{60}^- thermally averages over two electronic states [8b]. The C_{60}^- ion apparently undergoes electron self-exchange with C_{60} on the EPR time scale, to cause line broadening of the C_{60}^- EPR signal [11d]. The greater anisotropy in the $S = \frac{1}{2}$ spectrum of C_{70}^- in Figure 11.5(c) seems reasonable in view of the fact that the symmetry of C_{70} is lower (D_{5h}) than that of C_{60} (I_h).

   Figure 11.5(b) displays the EPR spectrum of the C_{60}^{2-} ion. The complicated spectrum appears to consist of two sets of superimposed resonances, with the

Table 11.3 UV–VIS electronic absorption spectral data (λ_{max} values in nm) for C_{60}, C_{60}^-, C_{60}^{2-}, C_{70}, C_{70}^-, and C_{70}^{2-} in $CH_2Cl_2{}^a$

C_{60}	257	330				
C_{60}^-	262	339				
C_{60}^{2-}	263	340				
C_{70}	333	362	381	466		
C_{70}^-	340		386	483		
C_{70}^{2-}					609	636

^{a}Conditions are as given in Ref. [6a].

first set being a narrow line and the second set being similar to the two pairs of spectral lines anticipated for a randomly oriented solid solution of a triplet state radical with $E = 0$ [8b]. The intensity of the first resonance decreases at lower temperatures, while the intensity of the second set of resonances increases at lower temperatures. Such temperature dependency is consistent with a triplet ($S = 1$) ground state for the C_{60}^{2-} ion. Furthermore, with $D = 1.31 \times 10^{-3}$ cm^{-1} and $E = 0$, a distance of ~12 Å is calculated for the mean electron–electron distance in triplet C_{60}^{2-}.

The EPR spectrum of C_{60}^{3-} has features and a temperature dependency resembling that of C_{60}^- [8b]. This supports an $S = \frac{1}{2}$ magnetic ground state for the ion. No evidence for the alternative $S = \frac{3}{2}$ state has been detected.

A systematic spectroelectrochemical study of the more highly reduced C_{60}^{n-} and C_{70}^{n-} ions ($n > 3$) has yet to be undertaken.

11.2 Electrochemistry of the Higher Fullerenes

Little has appeared in the literature to date concerning the electrochemistry of the higher fullerenes ($>C_{70}$), largely due to the small quantities available and the difficulty of separating multiple isomers of the molecules [12]. The availability problem arises from the fact that the sum total of the higher fullerenes represents <5% of the crude soot extract generated by the carbon-arc heating method.

Figure 11.6 presents electrochemical data for C_{84}. HPLC of a mixture of higher fullerenes, eluted with a toluene/acetonitrile solvent mixture, provided about 3 mg of C_{84}. FABS mass spectrometry displayed peaks for C_{84}, C_{82}, and C_{80}, with the latter two probably due to successive C_2 loss in the gas phase. CV and DPV provided data in both benzonitrile and nitrobenzene. Reductive scans carried out in benzonitrile revealed four major, reversible reductions at -0.44, -0.75, -1.08, and -1.45 V versus Fc/Fc$^+$ by CV at 200 mV/s. These reductions occur at an average spacing of 0.34 V, and thus the region between -1.50 and -2.20 V appears conspicuous due to the absence of any additional major reductions at ~ -1.79 and -2.13 V vs. Fc/Fc$^+$. These data

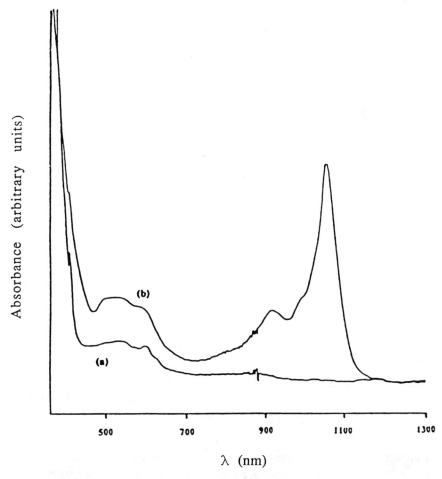

Figure 11.4 Room-temperature electronic spectrum of C_{60} and C_{60}^-. (a) C_{60} in $CH_2Cl_2/$ toluene before reduction; (b) C_{60}^- (end of the reduction). Reprinted with permission from Ref. [11d].

could indicate that the major isomer of C_{84} has a doubly degenerate LUMO, with no close-lying LUMO + 1, as with C_{70}. Theoretical studies indicate doubly degenerate LUMOs for the two most stable isomers of C_{84}, D_2 and D_{2d} [13]. The more revealing technique of DPV demonstrates that the "silent" region between −1.50 and −2.20 V vs. Fc/Fc^+ actually contains two minor peaks at −1.64 and −1.85 V versus Fc/Fc^+, as well as a shoulder ~0.1 V positive of the fourth major reduction, unobserved by CV. These smaller peaks present evidence for a second isomer of C_{84} with electronic properties different from those of the isomer responsible for the major peaks. The lower current

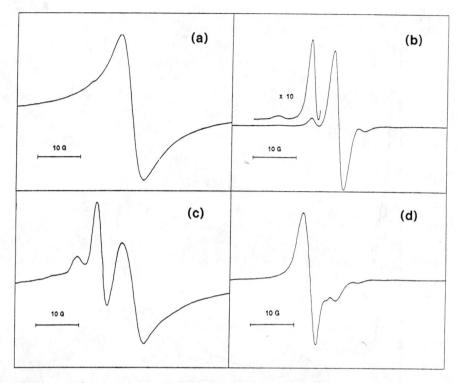

Figure 11.5 EPR spectrum at 120 K of a frozen CH_2Cl_2 glass, containing electrochemically generated (a) C_{60}^-, (b) C_{60}^{2-}, (c) C_{70}^-, and (d) C_{70}^{2-}. Reprinted with permission from Ref. [6a].

of these minor reductions may be due to a low abundance of a second isomer or to a lower solubility of a second isomer in the electroanalyzed solution. An alternative possibility would be the presence of minor amounts of C_{82} and C_{80}.

Oxidative scans of C_{84} in nitrobenzene yield data similar to those obtained for C_{60} and C_{70}. Two irreversible oxidative processes are observed at E_{ox} = +1.85 and +2.55 V versus Ag/AgCl at 200 mV/s by CV (Fc^+/Fc, +0.56 V vs. Ag/AgCl). As for C_{60} and C_{70}, the first oxidative process appears to modify the electrode surface, and successive scans produce lower current.

One electrochemistry report has mentioned finding two reversible oxidations for C_{76} as well as four reversible reductions for C_{76} and C_{78} (C_{2v} isomer) [12]; however, no $E_{1/2}$ values or spectroelectrochemistry studies have been published.

11.3 Electrochemistry of Surface-Derivatized C_{60}

As well-characterized surface derivatives of C_{60} become available, trends in the electrochemistry of the derivatives begin to become evident. (η^5-C_9H_7)Ir(CO)(η^2-

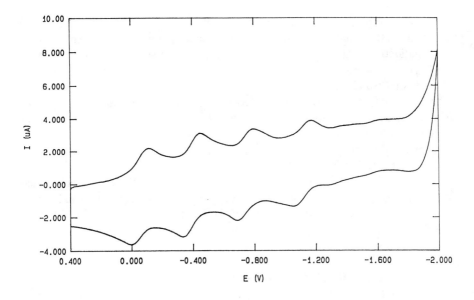

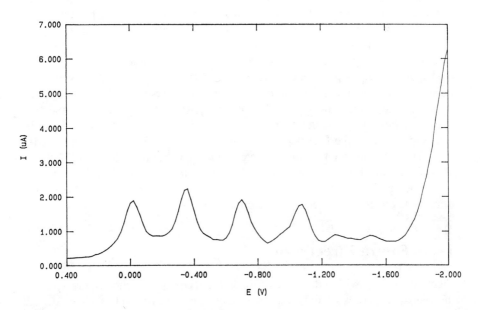

Figure 11.6 (a) Reduction of C_{84} by cyclic voltammetry in benzonitrile with a glassy carbon working electrode and a Ag/AgCl reference electrode at a scan rate of 200 mV/s (0.1M TBABF$_4$, saturated C_{84}, 25°C). (b) Reduction of C_{84}, under the same conditions as (a), using differential pulse voltammetry (20 mV scan increment, 200 mV/s scan rate, 50 mV pulse height, 50 mV pulse width). (a), (b) All peaks are absent when no C_{84} is present. Minor differences in $E_{1/2}$ values between the two figures are due to evaporation of benzonitrile in the dry Ar stream between runs. For this reason potentials have been reported relative to the internal Fc/Fc$^+$ standard.

C_{60}), for example, has been reported to exhibit two quasireversible reductions at potentials 0.12 and 0.07 V more negative than C_{60} under the same conditions [14], indicating that the metal center has transferred some electron density to the C_{60} cage. Likewise, the monosubstituted $(Et_3P)_2MC_{60}$ derivatives with $M =$ Ni, Pd, or Pt undergo the first reduction at potentials 0.34 V more negative than C_{60} [15]. The hexasubstituted derivatives reduce at proportionately lower potentials, 2.7–3.0 V more negative than C_{60} for a shift of ~ -0.47 V per added metal center. Strictly organic derivatives of C_{60} have also begun to be characterized electrochemically. For example, $E_{1/2}$ values for the first four reductions of a series of diphenyl fulleroids, $C_{60}(CPh_2)_x$, with $x = 1$ to 6, have been reported [16a]. All potentials shift toward the negative with each added $>CPh_2$ group. The average shift per group is ~ -0.11 V for the first reduction, ~ -0.09 V for the second, ~ -0.07 V for the third, and ~ -0.04 V for the fourth.

11.4   Electrochemistry of Solid Films

C_{60} has proved amenable to solid-state electrochemical studies by means of films deposited on the surface of electrodes. Jehoulet et al. [6b], have reported the cyclic voltammetry of the first four reductions and the first oxidation of a solid C_{60} film immersed in $CH_3CN/TBAAsF_6$ (Fig. 11.7). The cathodic peak potentials for the four reductions are close to the cathodic peak potentials of the same reductions in solution (-1.17, -1.39, -1.88, and -2.24 V vs. Fc/Fc$^+$). The irreversible oxidation occurs at $+1.6$ V versus Fc/Fc$^+$, close to the solution-state value reported for benzonitrile. Reoxidation waves of the first two reductions were found to be shifted ~0.5 and ~0.2 V positive of the cathodic peaks for the first two reductions. The authors interpreted the data for the first reduction in terms of a structural reorganization of the film. Miller et al. [17], used the solid film methodology to determine that C_{60} films behave as n-type semiconductors. Finally, an all-solid-state technique has been used to explore the intercalation of Li into C_{60} [18].

11.5   Future Directions

Electrochemistry will undoubtedly remain the technique of choice for exploring the redox properties of the fullerenes. Until now, most studies have focused on C_{60}, and to a lesser extent C_{70}, because of their relative abundances. Interest has centered around establishing the parameters of analytical electrochemistry such as the nature of the redox processes, the half-wave potentials, and some spectroelectrochemical characterizations. As the rarer members of the fullerene family become available in milligram quantities, they too will become as well characterized as their soccerball parent molecule. The dominance of the fullerenes as the "electron sponge" of molecules will likely remain a recurring

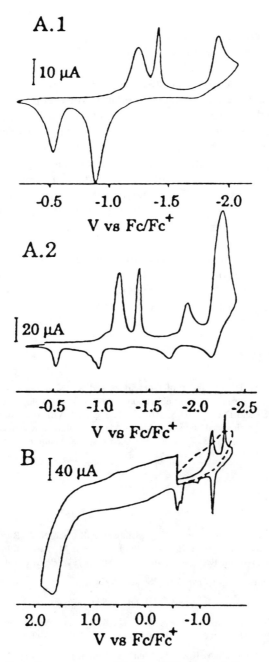

Figure 11.7 Cyclic voltammograms of a film of C_{60} on a platinum electrode (1 mm diameter) in a solution of MeCN [supporting electrolyte: $0.1M$ (TBA)BF$_4$] showing (A.1) third reduction, (A.2) fourth reduction, and (B) oxidation processes. Solid curve, first scan; dashed curve, second scan, $v = 200$ mV/s. Reprinted with permission from Ref. [6b].

theme in this research. The emergence of the C_{60}^{3-} anion as the crucial component of the new superconducting fullerides is a resounding consequence of this "electron-sponge" property. What other new electronic materials lie ahead?

More surface-derived fullerene molecules will emerge, and their redox chemistries will remain of interest, especially for those species containing redox- and/or photoactive metal centers intimately coupled to the fullerene framework. Perhaps even more exciting in this regard is the emerging family of $M@C_{60}$, etc., molecules with various metal atoms/ions encapsulated within the fullerene cage [19]. Metal atoms imprisoned by an electron-deficient spheroid cage promise a fascinating variety of intramolecular [metal $\xrightarrow{e^-}$ cage] redox reactions and a host of potential materials based on such internal electron transfers. Finally, as these and other redox-derived materials become desirable in bulk, electrosynthetic chemistry is likely to become the synthetic method of choice when careful control of the electrochemical potential is of paramount importance.

Without question, both electroanalytical and electrosynthetic chemistry will remain near the heart of developing fullerene chemistry, particularly when one remembers that whither goes a fullerene molecule, a rich redox chemistry surely follows.

Acknowledgment

We thank Dominique Dubois and Professors Jones and Kadish for insightful discussions and preprints of two manuscripts prior to publication. The Robert A. Welch Foundation under grant C-0627 and C-0689 and DARPA (U.S. Army) under grant DAAL-03-92-G-0186 are gratefully acknowledged for the financial support, which made our fullerene research and the writing of this chapter possible. Finally, S. F. thanks the U.S. National Institutes of Health for a NIGMS Training Grant (GM-08362) at Rice University.

References

1. R. E. Haufler, J. Conceicao, L. P. F. Chibante, Y. Chai, N. E. Byrne, S. Flanagan, M. M. Haley, S. C. O'Brien, C. Pan, Z. Xiao, W. E. Billups, M. A. Ciufolini, R. H. Hauge, J. L. Margrave, L. J. Wilson, R. F. Curl, and R. E. Smalley, *J. Phys. Chem.* **94**, 8634 (1990).

2. R. E. Smalley, *Acc. Chem. Res.* **25**, 98 (1992).

3. R. C. Haddon, L. E. Brus, and K. Raghavachari, *Chem. Phys. Lett.* **125**, 459 (1986).

4. A. F. Hebard, M. J. Rosseinsky, R. C. Haddon, D. W. Murphy, S. H. Glarum, T. T. M. Palstra, A. P. Ramirez, and A. R. Kortan, *Nature* **350**, 600 (1991).

5. (a) P.-M. Allemand, A. Koch, F. Wudl, Y. Rubin, F. Diederich, M. M. Alvarez, S. J. Anz, and R. L. Whetten, *J. Am. Chem. Soc.* **113**, 1051 (1991); (b) D. M. Cox, S. Behal, M. Disko, S. M. Gorun, M. Greaney, C. S. Hsu, E. B. Kollin, J. Millar, J. Robbins, W. Robbins, R. D. Sherwood, and P. Tindall, *J. Am. Chem. Soc.* **113**, 2940 (1991); (c) S. M. Gorun, M. A. Greaney, D. M. Cox, R. Sherwood, C. Day, V. Day, and R. Upton, *Proc. Mater. Res. Soc. Boston* **206**, 659 (1991).

6. (a) D. Dubois, K. M. Kadish, S. Flanagan, R. E. Haufler, L. P. F. Chibante, and L. J. Wilson, *J. Am. Chem. Soc.* **113,** 4364 (1991); (b) C. Jehoulet, A. J. Bard, and F. Wudl, *J. Am. Chem. Soc.* **113,** 5456 (1991); (c) D. Dubois, K. M. Kadish, S. Flanagan, and L. J. Wilson, *J. Am. Chem. Soc.* **113,** 7773 (1991).

7. Q. Xie, E. Pérez-Cordero, and L. Echegoyen, *J. Am. Chem. Soc.* **114,** 3978 (1992).

8. (a) D. Dubois, G. Moninot, W. Kutner, M. T. Jones, and K. M. Kadish, *J. Phys. Chem.* **96,** 7137 (1992). (b) D. Dubois, M. T. Jones, and K. M. Kadish, *J. Am. Chem. Soc.,* **114,** 6446 (1992).

9. G. E. Scuseria, *Chem. Phys. Lett.* **180***(5),* 451 (1991).

10. (a) H. W. Kroto, J. R. Heath, S. C. O'Brien, R. F. Curl, and R. E. Smalley, *Nature* **318,** 162 (1985); (b) *Polycyclic Aromatic Hydrocarbons and Astrophysics,* A. Léger, L. d'Hendecourt, and N. Boccara, Eds., D. Reidel Publishing Company, Boston, 1987, and references therein.

11. (a) P.-M. Allemand, G. Srdanov, A. Koch, K. Khemani, F. Wudl, Y. Rubin, F. Diederich, M. M. Alvarez, S. J. Anz, and R. L. Whetten, *J. Am. Chem. Soc.* **113,** 2780 (1991); (b) P. J. Krusic, E. Wasserman, B. A. Parkinson, B. Malone, E. R. Holler, Jr., P. N. Keizer, J. R. Morton, and K. F. Preston, *J. Am. Chem. Soc.* **113,** 6274 (1991); (c) J. A. Howard, M. Tomietto, and D. A. Wilkinson, *J. Am. Chem. Soc.* **113,** 7870 (1991); (d) M. A. Greaney, and S. M. Gorun, *J. Phys. Chem.* **95,** 7142 (1991); (e) P. N. Keizer, J. R. Morton, K. F. Preston, and A. K. Sugden, *J. Phys. Chem.* **95,** 7117 (1991).

12. F. Diederich and R. L. Whetten, *Acc. Chem. Res.* **25,** 119 (1992).

13. (a) B. L. Zhang, C. Z. Wang, and K. M. Ho, *J. Chem. Phys.* **96,** 7183 (1992); (b) K. Raghavachari, *Chem. Phys. Lett.* **190,** 397 (1992); (c) K. Raghavachari, personal communication.

14. R. S. Koefod, M. F. Hudgens, and J. R. Shapley, *J. Am. Chem. Soc.* **113,** 8957 (1991).

15. P. J. Fagan, J. C. Calabrese, and B. Malone, *Acc. Chem. Res.* **25,** 134 (1992).

16. (a) T. Suzuki, Q. Li, K. C. Khemani, F. Wudl, and O. Almarsson, *Science* **254,** 1186 (1991); (b) F. Wudl, *Acc. Chem. Res.* **25,** 157 (1992).

17. B. Miller, J. M. Rosamilia, G. Dabbagh, R. Tycko, R. C. Haddon, A. J. Muller, W. Wilson, D. W. Murphy, and A. F. Hebard, *J. Am. Chem. Soc.* **113,** 6291 (1991).

18. Y. Chabre, D. Djurado, M. Armand, W. F. Romanow, N. Coustel, J. P. McCauley, Jr., J. E. Fischer, A. B. Smith III, *J. Am. Chem. Soc.* **114,** 764 (1992).

19. (a) Y. Chai, T. Guo, C. Jin, R. E. Haufler, L. P. F. Chibante, J. Fure, L. Wang, M. J. Alford, and R. E. Smalley, *J. Phys. Chem.* **95,** 7564 (1991); (b) J. H. Weaver, Y. Chai, G. H. Kroll, C. Jin, T. R. Ohno, R. E. Haufler, T. Guo, M. J. Alford, J. Conceicao, L. P. F. Chibante, A. Jain, G. Palmer, and R. E. Smalley, *Chem. Phys. Lett.* **190,** 460 (1991); (c) R. D. Johnson, M. S. de Vries, J. R. Salem, D. S. Bethune, and C. S. Yannoni, *Nature* **355,** 239 (1992).

CHAPTER

12

Improved Preparation, Chemical Reactivity, and Functionalization of C_{60} and C_{70} Fullerenes

G. K. Surya Prakash, Imre Bucsi, Robert Aniszfeld, and George A. Olah

In 1985 it was discovered that laser-induced vaporization of graphite produces remarkably stable molecular allotropic forms of carbon C_{60} cluster **1** and to a lesser extent a stable C_{70} cluster **2**, as was shown by mass spectrometry [1]. Kroto et al. proposed [2] the structure for the sixty-carbon cluster to be a truncated icosahedron composed of 32 faces of which 12 are pentagonal and 20 are hexagonal, a structure analogous to a soccer ball and reminiscent of the geodesic domes of Buckminister Fuller. Thus, C_{60} is commonly referred to as ''buckministerfullerene.''

Initial experimental support for the spherical structure of C_{60} included mass spectrometric studies using lanthanum-impregnated graphite, which showed intense $C_{60}La$ peaks [3]. A number of theoretical studies [4] also indicated the truncated icosahedral structure **1** for C_{60} to be a stable closed-shell system and aromatic. In early 1990, Krätschmer and co-workers obtained spectroscopic evidence for C_{60} in smoke from thermal evaporation of graphite rods [5a] by observing four weak but distinct ir bands that were consistent with theory [4h–k]. Later Krätschmer et al. reported isolation and characterization of pure C_{60} and C_{70} using a chemical extraction technique [5b]. At the same time Kroto and co-workers [6] and Johnson et al. [7] have prepared a mixture of C_{60} and C_{70} using similar techniques. The fullerenes were separated by Kroto [6] via column chromatography on neutral alumina and their structure characterized by ^{13}C NMR spectroscopy. The ^{13}C NMR spectrum of C_{60} in benzene consists of a single line at 142.7 ppm, confirming the icosahedral structure. The ^{13}C NMR spectrum of C_{70} in benzene consists of 5 lines (150.7, 148.1, 147.4, 145.5, and 130.9 ppm in $1:2:1:2:1$ ratio, respectively, confirming a highly symmetrical egg-shaped structure [8] **2** (C_{5h} symmetry). The structure of C_{70} has been recently confirmed by 2D INADEQUATE NMR spectroscopy [9]. The C–C bond lengths in C_{60} has been determined by solid-state ^{13}C NMR spectroscopy [10]. Furthermore, x-ray crystal structures of several C_{60} derivatives have also been determined [11–13].

C_{60} and C_{70} are now being routinely prepared in gram quantities using benchtop reactors [14]. Separation of C_{60} and C_{70} and related higher fullerenes from fullerene mixtures is also relatively well established [6, 11, 15] using chromatographic techniques. In an elegant work Diederich et al. [15b, c] have isolated and characterized two isomers of C_{78} and one distinct chiral C_{76} allotrope. Other higher fullerenes have also been isolated [15].

Most of the chemical reactivity and functionalization studies have been centered on C_{60} because of its relative abundance in the fullerene extracts compared to other higher fullerenes [16, 17]. C_{60} is a fully delocalized closed-shell system showing "ambiguous" aromatic character [18]. The relatively high electronegativity of C_{60} has been proposed to be due to pyracyclene character of certain five-membered ring bonds (fulvalenelike character) (see structure **3**, and this is responsible for its facile reaction with nucleophiles.

This chapter discusses our work on the improved preparation of fullerenes using a nine-rod gravity-fed reactor allowing continuous burning of graphite electrodes without the need of opening the reactor after burning of each rod. We also report detailed studies on reduction, halogenation, methylation, and arylations.

12.1   Improved Preparation of Fullerenes

Several different reactors for the generation of fullerenes have been described in the literature [14]. Mixtures of fullerenes are now also available commercially but at relatively high cost. There is hope that commercial application of fullerenes will make it feasible to generate them on a bulk scale with a modest price tag. However, at present there is a need for improved laboratory preparation of fullerenes.

We have developed a laboratory reactor modeled after the established technique of resistive heating of graphite electrodes in helium atmosphere using gravity feed [14]. The reactor is capable of producing about 20 to 30 grams of soot per working day in which the C_{60}/C_{70} content is 8–11%. The resulting soot is extracted with toluene, and the extract is further purified by chromatographic techniques. The design simplifies the procedure and greatly reduces the effort required for preparation of fullerenes compared to other published methods [14]. The reactor is capable of "burning" nine 12-in.-long 1/8-in.-diameter high-purity graphite rods by consecutively switching from one electrode to the next without breaking vacuum, dis- and reassembling the reactor. Optimum yield for C_{60} is achieved when 70–75 A of current at 50 V is used in 100 Hg mm of helium atmosphere to heat high-purity graphite resistively.

Under these conditions each rod is consumed in approximately 9 min. The produced soot is collected using a vacuum cleaner equipped with a paper dust bag.

The nine-rod gravity-fed reactor is constructed from a 316 stainless steel base, a glass bell jar, OFHC copper electrode contacts and guiders, high-purity graphite electrodes, Macor insulators, Varian OFHC copper high-current feedthroughs, and high vacuum valves and fittings. A schematic representation of the reactor is depicted in Figure 12.1. In the process of resistive heating of graphite, substantial heat is generated. The size of the reactor is commensurate for the efficient dissipation of heat. Operation starts with evacuation and purging of the reactor with helium. After three evacuation–purge cycles the helium pressure is set to 100 Hg mm. Next the reactor's stainless steel base is cooled by circulating dry methanol at $-40°C$ in silicone rubber tubing wrapped around the base. By monitoring pressure changes the reactor is safeguarded against excessive temperature surges. Each 1/8-in. graphite rod is "ignited" by passing 150 A of current for 15 sec. The current is then dropped to 70–75 A and maintained at this value throughout the run. After about 9 min the burning is complete and the power disconnected. The reactor is cooled for 10 min. The process is repeated for each of the nine rods.

12.2   Reduction Studies

Theory predicts an extremely high electron affinity (facile reduction) for both C_{60} and C_{70} fullerenes [4a, k]. Initial experimental support for the ease of reduction of C_{60} was the formation of $C_{60}H_{36}$ via a Birch reduction (Li in liquid ammonia) of C_{60} by Smalley and co-workers [19b]. They successfully dehydrogenated $C_{60}H_{36}$ back to C_{60} by treatment with DDQ reagent and have also carried out cyclic voltammetry studies that indicated that C_{60} undergoes reversible two-electron reduction. Wudl, Diederich, and co-workers [19a] carried out cyclic voltammetry studies on chromatographically pure samples of C_{60} and C_{70}, which showed that each fullerene undergoes reversible three-electron reduction (down to -1.5 V vs. Ag/AgCl electrode). More recently, even reversible four-, five-, and six-electron reductions of C_{60} have been carried out [20, 21].

C_{60} and C_{70} fullerene mixtures (in approximate 85:15 ratio) undergo reduction with excess Li metal (reduction potential of $Li°$ ~ -3.0 V in THF-d_8) under ultrasound irradiation [22a]. The C_{60} and C_{70} fullerenes are only slightly soluble in THF. However, the reduced fullerenes are highly soluble and gave a deep red-brown solution after sonication. The ^{13}C NMR spectrum [Fig. 12.2(a)] at room temperature shows a single resonance at $\delta^{13}C$ 156.7 for the reduced C_{60}. The deshielding of 14 ppm per carbon atom is remarkable because generally carbanionic carbons are shielded compared to their neutral precursors. Such deshielding in the case of C_{60} polyanion may be rationalized by populating antibonding LUMO [22b] orbitals (paratropic deshielding). The ^{13}C NMR spec-

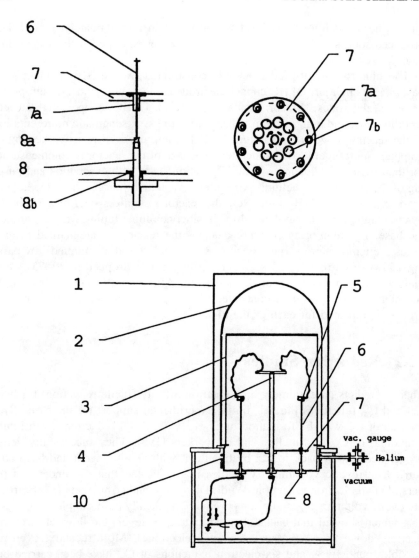

Figure 12.1 Schematic layout of the "nine barrel" fullerene generator. (1) Protective shield; (2) 18 × 30-in. diameter Pyrex brand bell jar; (3) 12-in. diameter inner glass jar for soot collection; (4) 1-in. copper center pole to support guide plate and high-current wire; (5) OFHC copper contact [22–24 g ea.] with attached 1/2-in. wide braided copper cable (6) 1/8-in.-diameter graphite rod; (7) 10.5-in.-diameter 3/8-in.-thick OFHC copper plate to support the nine guides; (7a) OFHC copper guide with a bore to allow "slip-fit" for 1/8-in.-diameter graphite rod; (7b) nine 1-in. holes on a 4.5-in.-diameter for better heat distribution and visibility; (8) 5/8-in.-diameter OFHC copper high-current feedthrough with bored tip to accommodate 3/8-in.-diameter graphite electrode; (8a) 5/8-in.-diameter 7/8-in. long graphite electrode; (8b) Macor insulator; (9) Sears–Craftsman 230/140 AC/DC Infinite Amp Arc Welder as the power supply; (10) silicone rubber tubing was wrapped around the stainless steel reactor body in which −40°C abs. methanol was circulated for cooling.

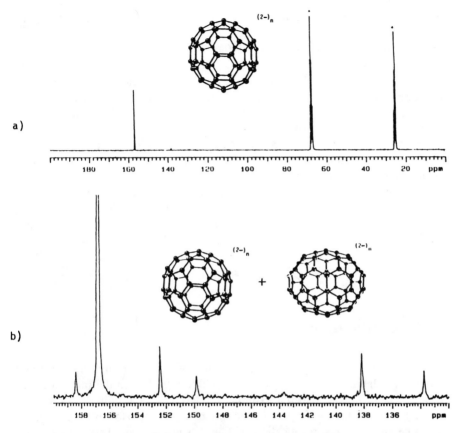

Figure 12.2 (a) 75 MHz ^{13}C NMR spectrum at room temperature of C_{60} polyanion in THF-d_8 (*–peaks due to solvent; (b) 75 MHz ^{13}C NMR spectrum of a mixture of C_{60} and C_{70} polyanions in THF-d_8 at $-80°$C.

trum, which indicates the presence of reduced C_{70} [see Fig. 12.2(b)] was obtained at $-80°$C to improve the signal-to-noise ratio. Five resonances were observed at δ^{13}C 158.3, 152.3, 149.6, 137.9, and 133.7 in a $1:2:1:2:1$ ratio, respectively, showing a slight overall deshielding compared to neutral C_{70} by about 0.9 ppm per carbon atom. Reduction of a chromatographically purified (alumina and hexane/toluene as eluent) sample of C_{60} confirmed the spectral assignments of reduced C_{60}. Further, the ^{7}Li NMR spectrum of the reduced C_{60}/C_{70} solution at $-80°$C showed a fairly sharp resonance at $+1.6$ ppm (vs. $1M$ LiCl in THF). The ^{7}Li NMR spectrum at room temperature was extremely broad, indicating a solvent separated ion pair/contact-ion pair equilibrium in the temperature range studied [23].

The polyanion obtained contains an even number of electrons judging from the sharpness of ^{13}C NMR signals, indicating a diamagnetic species. At early stages of sonication, cloudy green-colored solutions were obtained and no ^{13}C

NMR signals for the solution could be detected. This solution was ESR active and showed a strong signal at g value close to that of a free electron. Since previous cyclic voltammetry studies [19a, b, 20, 21] indicated reversible four-, five-, and six-electron reductions for each fullerene, it seems most likely under the conditions used for the reduction C_{60} and C_{70} that each has accepted four or more electrons. Theoretical calculations [4a, b, d–g, m] indicate a triply degenerate LUMO for C_{60}, making it probable that the hexanion, C_{60}^{6-}, is generated. A similar situation exists for C_{70} in which the LUMO and doubly degenerate LUMO +1 are closely spaced [4d]. Reversible six-electron electrochemical reductions of C_{70} have also been achieved [21b]. Attempts to determine the exact number of electrons added to fullerenes **1** and **2** by quenching the polyanions with D_2O were unsuccessful. The isolated mixture was shown by field ionization mass spectrometry (FIMS) to be a mixture of C_{60} and C_{70}. Presumably, the polyanions either transfer electrons to D_2O or the deuterated product mixture undergoes rapid oxidation to regenerate the more stable starting fullerenes in both cases. Similar observations have been made by Volhardt et al. [24] in the case of [3]phenylene dianion.

12.3   Oxidation of C_{60}

Contrary to reduction experiments, oxidation of C_{60} was found to be difficult, and electrochemical oxidations are irreversible [19–21]. In the mass spectrum of C_{60}, a peak at exact half mass (360 amu) was observed, indicating formation of C_{60}^{2+} under electron impact in the gas phase [15a]. However, attempts to generate the diamagnetic dication of C_{60} in solution have been unsuccessful [22a]. Attempted oxidation of C_{60} in a large excess of SbF_5/SO_2ClF gave a dark-green-colored solution that showed extremely broad ^{13}C NMR spectra at all temperatures employed ($-80°C$ to room temperature). Similar spectra were obtained using SbF_5 and Cl_2 as oxidants in SO_2ClF solution. It appears that radical cations have been generated and no diamagnetic di (or poly-) cations were formed. This is not surprising since electrochemical studies [19–21] and FT-ICR experiments [25] indicate a high oxidation potential for C_{60}. It is also in accord with theoretical studies [4] that predict the oxidation potential for C_{60} to be comparable to that of naphthalene, a molecule not oxidized to a stable dication under superacidic conditions [26].

C_{60} also dissolves in oleum to give a dark-green-colored solution. An electron spin resonance spectrum of the solution indicated formation of a radical cation [27]. Miller and co-workers [28a] have used Magic acid (FSO_3H-SbF_5) to oxidize C_{60} to its radical cation. The oxidized C_{60} has been trapped with nucleophiles to obtain nucleophilic addition products. Two, four, and six groups add symmetrically to the C_{60} radical cation. Methanol and butanol and even benzene (vide infra) was found to add to the C_{60} radical cation. Reaction of $C_{60}^{\cdot+}$ with 1,6-hexanediol produced 1,6-difulleroxyhexane. Christe recently [28b] reported preliminary results of the oxidation of C_{60} by XeF_2 to the fullerene radical cation $C_{60}^{\cdot +}$.

12.4   Alkylation and Related Reactions of Fullerene, C_{60}

Alkylation of C_{60} polyanion with excess methyl iodide yielded a light brown solid that field ionization mass spectrometry indicated to be a mixture of poly-methylated fullerenes [22a]. ^{1}H and ^{13}C NMR studies in benzene-d_6 indicate methyl absorptions at δ^1H 0.06 and δ^{13}C 1.0, respectively. The FIMS analysis clearly shows a range of methylated products from 1 all the way to 24 methyls.

$$C_{60}^{6-}6Li^+ \xrightarrow{\text{xs } CH_3I} C_{60}(CH_3)_n \quad n \leq 24$$

There is a preponderance of products with an even number of methyl groups (with 6 and 8 methyl groups predominating). The nominal masses of the products with odd number of methyl groups corresponds to the addition of a methyl group(s) and hydrogen atom(s). However, the exact mechanism of the observed alkylation is not yet clear, but possibly involves electron transfer to methylated fullerenes during quenching. The polymethylation of these fullerene polyanions represented the first functionalization of C_{60} via alkylative C–C bond formation [22a]. Attempts to obtain selective methylation have, however, been unsuccessful. Reaction of the polyanion with trimethylsilyl chloride did not result in a well-defined trimethylsilylated product.

Wudl and co-workers [29] have shown that C_{60} reacts with organometallic reagents such as alkyl/aryl lithiums and Grignard reagents. C_{60} stoichiometrically reacts with $LiB(Et)_3H$ to form the salt $C_{60}HLi \cdot 9H_2O$, which is indefinitely stable and does not react with methyl iodide or methanol. In contrast, organometallic adducts of C_{60} react with methyl iodide to give mixed alkyl/methyl and aryl/methyl adducts of C_{60}.

$$C_{60} + 60\ PhMgBr \xrightarrow[\text{MeI}]{\text{THF, RT}} C_{60}Ph_{10}Me_{10}$$

$$C_{60} + t\text{-BuLi} \xrightarrow[\text{MeI}]{\text{THF, RT}} C_{60}t\text{-BuMe}$$

$$C_{60} + t\text{-BuMgBr} \xrightarrow[\text{MeI}]{\text{THF, RT}} C_{60}t\text{-Bu}_{10}Me_{10}$$

Using equimolar amounts of C_{60} and t-BuLi, a mono t-BuLi adduct of C_{60} has also been isolated and characterized [30].

C_{60} also reacts with trimethylsilyl chloride in the presence of magnesium metal in THF under sonication [22a] (Barbier conditions) to give di- and tetra(trimethylsilyl) C_{60} adducts as determined by FIMS analysis.

Negative ion ionization of C_{60} in the mass spectrometer in the presence of methane gives methylene adducts [29]. The maximum number of carbons attached to C_{60} would seem to be 13 based on mass spectrometric analysis.

Several alkyl and benzyl radical reactions of C_{60} are also known [31, 32]. In fact C_{60} has been called a "radical sponge." Most recently, a t-butyl radical adduct of C_{60} has been studied [33] by ESR spectroscopy as a localized radical. The intriguing dimerization of the radical has also been explored.

12.5 Halogenation

The first halogenation carried out on C_{60} was fluorination. Smith et al. [34] have fluorinated C_{60} with elemental fluorine and obtained a mixture of poly-fluorinated products. The major component of the mixture was $C_{60}F_{36}$, analogous to the claimed Birch reduction product $C_{60}H_{36}$ [19b]. Subsequently, Holloway and co-workers [35a] have claimed the formation of $C_{60}F_{60}$, when 10 mg of C_{60} was exposed to fluorine gas over a period of 12 days. The product shows single broad fluorine absorption in the ^{19}F NMR spectrum. However, further studies indicate that the cage structure may have been lost during the fluorination reaction, and some kind of C_nF_n product resulted (fluorographites of such compositions are known) [35b]. Interestingly, nucleophilic substitution of fluorinated C_{60} has also been reported [36].

We have achieved both the first polychlorination as well as polybromination of C_{60} [37]. Treatment of C_{60}/C_{70} mixtures or pure C_{60} with neat chlorine or with chlorine in the presence of catalysts such as $AlCl_3$ or $AlBr_3$ in chloro organic solvents at various temperatures did not furnish any chloro products. However, in a hot glass tube under a slow stream of chlorine gas, both the C_{60}/C_{70} mixture and pure C_{60} react readily. At 250°C C_{60}/C_{70} were chlorinated (chlorine flow rate of 10 mL/min), and a maximum uptake of chlorine was achieved within 5 h. The weight increase of the yellowish-brown product indicated that on average 24 chlorine atoms (confirmed by elemental analysis) were added.

$$C_{60} \xrightarrow{Cl_2(g)} C_{60}Cl_n$$

At higher temperatures (300, 350, and 400°C) chlorination occurred more rapidly. With increasing temperature, however, the amount of chlorine added to the fullerene skeleton decreases (at 300, 350, and 400°C an average 24, 18, and 12 chlorine atoms were added to the C_{60} skeleton, respectively). Highly chlorinated fullerene is light orange in color and soluble in many organic solvents. No complete chlorination of C_{60} was achieved under any of the conditions employed. The ir spectrum of the chlorinated C_{60}/C_{70} shows strong C–Cl stretches at 750 to 950 cm^{-1} and is distinctly different from the simple ir spectrum of a C_{60}/C_{70} mixture. The solution 75 MHz ^{13}C NMR spectrum did not show any discernible peak even after 48 h of time averaging (due to chemical shift anisotropy and long carbon relaxation times). The 22.49 MHz solid-state ^{13}C NMR spectrum, however, showed broad absorptions at $\delta^{13}C$ 50 to 150. Efforts to obtain mass spectra of polychlorinated fullerenes by FAB (fast atom bombardment) or FIMS were unsuccessful. Even under these mild ionizing conditions, polychlorofullerenes lose chlorine (only peaks due to the C_{60} (M^+, 720) and C_{70} (M^+, 840) were observed). Indeed, polychlorofullerenes at 400°C under argon completely dechlorinate to the parent C_{60}/C_{70} fullerene mixture. The nature of polychlorinated C_{70} could not be ascertained since the percentage of C_{70} in the fullerene mixture was less than 10%.

Tebbe and co-workers [38] have also achieved multiple reversible chlorination of C_{60} with liquid chlorine near −35°C in the absence of light. The polychlorinated C_{60} in the presence of PPh_3 regenerates to C_{60} in 54–88% yield. The multielectron electrochemical reduction of polychlorinated C_{60} was found to occur at potentials significantly more anodic than reduction potentials of activated chlorocarbons such as benzylic and allylic chlorides.

Bromination of C_{60}/C_{70} mixture or pure C_{60} in neat bromine at 20 and 55°C gave evidence [37] for the addition of 2 to 4 bromine atoms based on the weight increase of the product. The FT–ir spectrum showed C–Br stretches at 515–545 cm^{-1}. Upon heating to 150°C, the brominated fullerenes lost their bromine readily.

$$C_{60} \xrightarrow[\text{RT or 50°C}]{\text{Br}_2\text{ (neat)}} C_{60}Br_n \quad n = 2 \text{ or } 4$$

The solution ^{13}C NMR spectrum again did not indicate discernible absorptions due to lack of symmetry. However, the solid-state ^{13}C NMR spectrum showed broad peaks between $\delta^{13}C$ 30 and 150 ppm. FAB mass spectra show a loss of bromine, giving mass peaks only due to C_{60}. Subsequently, higher polybromination of C_{60} was also achieved under prolonged exposure to Br_2 [27]. Similar studies have been reported by Rao et al. [39a].

More recently, Birkett, Kroto, et al. [39b] have obtained x-ray crystallographic evidence for the formation of $C_{60}Br_6$ and $C_{60}Br_8$. These molecular crystals contained occluded bromine in them. The authors were even able to characterize $C_{60}Br_{24}$ and propose a plausible structure for the sterically hindered molecule. The crystal structure of $C_{60}Br_{24}$ has also been obtained [40a]. Even molecular complex formation between C_{60} and I_2 has been reported [40b].

One could consider the possibility of adding hydrogen halides (HCl and HBr) across the double bonds of C_{60} as a method to obtain hydrohalogenated derivatives. Experiments were carried out in chlorinated organic solvents (CCl_4, $CHCl_3$) with or without catalysts such as $AlCl_3$ or $AlBr_3$ and in heterogeneous gas–solid phase in the 25–400°C temperature range with or without the influence of uv radiation. Addition of HCl and HBr was, however, not observed [27].

12.6 Friedel–Crafts Fullerylation of Aromatics

Polychlorinated fullerenes were found to fullerylate aromatics under Friedel–Crafts conditions [37]. Stirring a benzene solution of $C_{60}Cl_n$ with a catalytic amount of aluminum chloride at ambient temperature for two hours gave a dark reddish-brown solution that upon aqueous workup yielded a brown solid (soluble in organic solvents). The ^{1}H NMR spectrum (in CDCl$_3$) showed a broad absorption centered at δ^1H 7.2. The ^{13}C NMR spectrum also showed a broad absorption at δ^{13}C 128, indicating phenylation.

$$C_{60}Cl_n \xrightarrow[\text{AlCl}_3]{\text{benzene}} C_{60}(Ph)_n$$

The FAB mass spectrum indicates substitution of at least 22 phenyl groups (M^+, 2414), with no residual chlorine remaining in the product (by ir analysis). Further, consecutive loss of phenyl groups, demonstrating the fragmentation from $C_{60}(C_6H_5)_{22}$ to $C_{60}C_6H_5$, has also been observed. Fullerylation of toluene occurred similarly to provide poly para-substituted toluene adducts.

Brominated C_{60} also undergoes Friedel–Crafts-type fullerylation of benzene catalyzed by aluminum trichloride (or related catalysts). In addition to phenyl substitution, benzene addition (i.e., C_6H_5 and H) was also observed (vide infra). Taylor et al. [41] have observed formation of C_{60} Ph$_{12}$ by heating C_{60} in benzene with bromine and iron (III) chloride.

12.7   Methoxy and Hydroxyfullerenes

When a polychlorofullerene mixture was treated with excess methanol/KOH under reflux for 3 h, quenching by aqueous ammonium chloride and workup provided a yellowish material [37]. The proton NMR spectrum (in CDCl$_3$) indicated a broad envelope of methoxy groups centered at δ^1H 3.7. The FT–ir spectrum also shows a broad C–O stretching frequency at 1091 cm^{-1}. Further evidence for the replacement of the chlorine atoms by methoxy groups came from FAB mass spectrometry (polymethoxylation up to 26 methoxy groups M^+, 1526). Subsequential loss of methoxy groups was observed, indicating fragmentation from $C_{60}(OMe)_{26}$ all the way to $C_{60}(OMe)$. Mass spectrometric evidence for some hydroxylation was also obtained. However, the C_{60}/C_{70} mixture itself does not react with methanol/KOH. Observation of a product component with 26 methoxy groups implies that more than 24 chlorine atoms may have added to the fullerene skeleton. The substitution of chlorines by methoxy groups would probably need to be rationalized by an unprecedented frontside nucleophilic substitution process. Polychlorinated pure C_{60} behaved similarly.

$$C_{60}Cl_n \xrightarrow[\text{reflux}]{\text{MeOH, KOH}} C_{60}(OMe)_n$$

Chiang and co-workers [42] have obtained polyhydroxy C_{60} by treating C_{60} with an aqueous mixture of nitric acid and sulfuric acid. Similar products [43] were obtained by treating C_{60} with nitronium tetrafluoroborate in carboxylic acid solvents followed by hydrolysis.

12.8   Fullerenation of Aromatics

We have found that pure C_{60} and C_{60}/C_{70} mixture undergo $AlCl_3$- (or other strong Lewis acids such as $AlBr_3$, $FeBr_3$, $FeCl_3$, $GaCl_3$, $SbCl_5$) catalyzed reaction with aromatics such as benzene and toluene to polyarenefullerenes [44]. Since Ar and H are added across fullerene double bonds, the reaction can be considered as fullerenation of aromatics.

$$C_{60} \xrightarrow[\text{AlCl}_3]{\text{H–Ar (excess)}} C_{60}(\text{H–Ar})_n$$

Reaction of C_{60}/C_{70} mixture with aluminum trichloride in benzene at room temperature gave a dark reddish-brown homogeneous solution. Quenching the solution with water and usual workup gave upon evaporation of solvent a brown-colored solid. The FT–ir spectrum showed aromatic stretchings at 3056, 3022, 1598, 1492, 1445, and 694 cm^{-1}. The 300 MHz ^1H NMR spectrum (in $CDCl_3$) indicated not only broad aromatic absorption centered at δ^1H 7.4, but also a broad C–H absorption centered at δ^1H 4.5. Integration of the peaks showed about 5:1 relative intensities (implying monosubstitution on the phenyl ring). The 75 MHz ^{13}C NMR spectrum showed a broad absorption centered around $\delta^{13}C$ 128 with extremely broad absorptions centered at $\delta^{13}C$ 146 and 54. Most useful analytical information was obtained from FAB mass spectrometry. The FAB mass spectrum of the product showed a strong mass peak at M^+, 1656, indicating the formation of $C_{60}(C_6H_6)_{12}$. Some evidence for the formation of $C_{60}(C_6H_6)_{16}$ was also obtained in the mass spectrum (M^+, 1968, weak absorption). Other mass peaks corresponding to species containing decreasing amount of added benzene were also detected all the way to $C_{60}(C_6H_6)_6$. These peaks are indicative of consecutive C_6H_6 loss from higher phenylated species. The addition of $H–C_6H_5$ is also supported by the observation of the methine (C–H) hydrogens at δ^1H 4.5.

Friedel–Crafts fullerenation was also confirmed by carrying out the reaction with C_6D_6. Analysis of the product by FAB showed M^+ at 1728, which corresponds to $C_{60}(C_6D_6)_{12}$ and peaks due to progressive loss of mass 84 (C_6D_6). The FT–ir spectrum showed a characteristic C–D stretching frequency at 2271 cm^{-1}. In all deuterated and nondeuterated samples, peaks corresponding to C_{60} were also observed. However, no evidence for the phenylation of C_{70} in the fullerene mixture was observed. Similar results were also obtained with pure C_{60}. Related reaction with toluene gave toluene addition products. The FAB mass spectrum showed M^+ at 1824, indicating the formation of $C_{60}(C_6H_6–CH_3)_{12}$. In this case, too, consecutive loss of toluene (m/e 92) was observed all the way to $C_{60}(C_6H_6–CH_3)_{12}$. In the 300 MHz ^1H NMR spectrum (Figure 12.3), absorptions at δ^1H 7.2 (broad), 4.5 (extremely broad), and 2.35 (broad) were observed. The spectrum is in accord with the formation of polytoluene fullerenes. Relative integration of the proton signals at δ^1H 7.2 and 4.5 gave a ratio of 4:1, in accord with monosubstitution of toluene (in all probability in the *para* position according to the FT–ir spectrum).

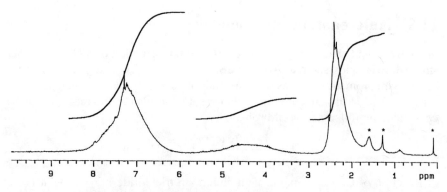

Figure 12.3   300 MHz ^{1}H NMR spectrum of fullerenated toluene [$C_{60}(C_6H_5-CH_3)_{12}$] in CDCl$_3$ solution at ambient temperature. *: peaks due to unidentified impurities.

The reaction takes place only with relatively strong Friedel–Crafts catalysts. Weak Lewis acids such as stannic chloride or titanium tetrachloride did not catalyze the reaction. The order of Lewis acid activity was found [27] to be AlBr$_3$ > AlCl$_3$ > FeCl$_3$ > FeBr$_3$ > GaCl$_3$ > SbCl$_5$. Inactive Lewis acids for the reaction included TiCl$_4$, SnCl$_4$, GeCl$_4$, BF$_3$, and BCl$_3$.

The reaction could be rationalized by initial protonation (by the inevitable moisture in AlCl$_3$) of fullerene to fullerene cation followed by electrophilic fullerenation of aromatics. The sequence repeats until on the average 12 Ar–H units are added. The reaction is similar to alkylation of arenes by alkenes (polyenes) under acidic catalysis. However, no Friedel–Crafts fullerenation of aromatics could be observed in the presence of such Bronsted acids as H$_2$SO$_4$, triflic acid, HBr, HBr–BBr$_3$. Thus the mechanism of the fullerenation is not yet clear. Redox processes, particularly with more easily reducible Lewis acids (FeCl$_3$, SbCl$_5$, etc.), can also be operative, involving radical cations as reported by Miller [28a] (single electron transfer, SET).

$$C_{60} \xrightarrow{\text{AlCl}_3} C_{60}H^+ \xrightarrow{\text{HC}_6\text{H}_5} \overset{\text{H}}{\underset{\text{H}}{C_{60}}} \diagdown \text{(+)} + \xrightarrow{-\text{H}^+}$$

$$C_{60}(H-C_6H_5) \longrightarrow \longrightarrow C_{60}(H-C_6H_5)_n$$

C$_{60}$/C$_{70}$ fullerene mixture was also found to undergo reaction with non-cross-linked polystyrene under aluminum trichloride catalysis in CS$_2$ solvent to give fullerenated polystyrene of highly cross-linked nature (highly insoluble).

Hooke et al. [45] have found similar arene adducts while refluxing C$_{60}$/C$_{70}$ mixture with toluene, p-xylene, and other aromatic compounds in the presence of FeCl$_3$. They report formation of benzene derivatives, however, only in the presence of bromine. They obtained not only 1:1 adducts but also multiple addition products and dehydrogenation products as determined by mass spectrometry. On the other hand, under similar conditions Taylor et al. have obtained fullerylation products [41] (vide infra).

12.9 Conclusions

This review covers various aspects of studies on reactivity and functionalization of C_{60} and C_{70} fullerenes. In this fast-developing area progress is so rapid that any account soon becomes obsolete. It can be said with some confidence, however, that further developments in fullerene chemistry will need to center on finding selective reactions giving well-defined products. Chemists are increasingly becoming successful in developing methods for obtaining selectively functionalized products, such as fullerene epoxide [46], metal complexes, etc. [47, 48]. Also, improved separation techniques will help in the isolation of remarkable products that can be obtained from the unique soccer-ball-shaped, closed-shell polycarbon supercage. Results obtained make it clear that it is increasingly becoming possible to attach a variety of functional groups to C_{60}, as well as to attach C_{60} to aromatics, polymers, etc. To what degree C_{60} and related fullerenes will live up to present-day expectations in practical applications (superconductivity, high-temperature lubricants, organic ferromagnets, optical devices, even effective biological and pharmaceutical products, etc.) remains to be seen. There is, however, continuing intense interest in the fascinating new chemistry of polycarbon supercage compounds and their derivatives.

Acknowledgment

Support of our work by the National Institutes of Health, National Science Foundation, and Office of Naval Research is gratefully acknowledged. Dr. R. Malhotra of SRI International is thanked for an initial sample of C_{60}. Names of other colleagues who worked with us on fullerenes are cited in references, and are thanked for their contributions.

References

1. *Considered Polycarbon Supercage Chemistry.* 5. Part 4: see G. A. Olah, I. Bucsi, R. Aniszfeld, and G. K. S. Prakash, *Carbon,* in press.

2. H. A. Kroto, J. R. Heath, S. C. O'Brien, R. F. Curl, and R. E. Smalley, *Nature (London)* **318,** 162 (1985). Also see: E. A. Rohlfing, D. M. Cox, and A. Kaldor, *J. Chem. Phys.* **81,** 322 (1984). For a recent historical account on fullerenes, see: H. W. Kroto, *Angew. Chemie* **104,** 113 (1992).

3. J. R. Heath, S. C. O'Brien, Q. Zhang, Y. Liu, R. F. Curl, H. W. Kroto, F. K. Tittle, and R. E. Smalley, *J. Am. Chem. Soc.* **107,** 7779 (1985).

4. (a) R. C. Haddon, L. E. Brus, and K. Raghavachari, *Chem. Phys. Lett.* **125,** 459 (1986);

(b) R. L. Disch and J. M. Schulman, *Chem. Phys. Lett.* **125,** 465 (1986); (c) R. C. Haddon, L. E. Brus, and K. Raghavachari, *Chem. Phys. Lett.* **131,** 165 (1986); (d) P. W. Fowler and J. Woolrich, *Chem. Phys. Lett.* **127,** 78 (1986); (e) A. J. Stone and D. J. Wales, *Chem. Phys. Lett.* **128,** 501 (1986); (f) M. D. Newton and R. E. Stanton, *J. Am. Chem. Soc.* **108,** 2469 (1986); (g) J. M. Rudzinski, Z. Slanina, M. Togasi, and E. Osawa, *Thermochim. Acta* **125,** 155 (1988); (h) M. D. Stanton and R. E. Newton, *J. Phys. Chem.* **92,** 2141 (1988); (i) Z. C. Wu, D. A. Jelski, and T. F. George, *Chem. Phys. Lett.* **137,** 291 (1988); (j) D. E. Weeks and W. G. Harter, *Chem. Phys. Lett.* **144,** 366 (1988); (k) D. E. Weeks and W. G. Harter, *J. Chem. Phys.* **90,** 4744 (1989); (l) Z. Slanina, J. M. Rudzinski, M. Togaso, and E. J. Osawa, *J. Mol. Struct.* **202,** 169 (1989); (m) P. W. Fowler, P. Lazzeretti, and R. Zanasi, *Chem. Phys. Lett.* **165,** 79 (1990).

5. (a) W. Krätschmer, K. Fostiropoulos, and D. R. Huffmann, *Chem. Phys. Lett.* **170,** 167 (1990); (b) W. Krätschmer, L. W. Lamb, K. Fostiropoulos, and D. R. Huffmann, *Nature (London)* **347,** 354 (1990).

6. R. Taylor, J. P. Hare, A. K. Abdul-sada, and H. W. Kroto, *J. Chem. Soc., Chem. Commun.* 1423 (1990).

7. R. D. Johnson, G. Meijer, and D. S. Bethune, *J. Am. Chem. Soc.* **112,** 8983 (1990).

8. (a) H. W. Kroto, *Nature (London)* **329,** 529 (1987); (b) T. G. Schmalz, W. A. Seitz, D. J. Klein, and G. E. Hite, *J. Am. Chem. Soc.* **110,** 1113 (1988).

9. R. D. Johnson, G. Meijer, J. R. Salem, and D. S. Bethune, *J. Am. Chem. Soc.* **113,** 3619 (1991).

10. C. S. Yannoni, P. P. Bernier, D. S. Bethune, G. Meijer, and J. R. Salem, *J. Am. Chem. Soc.* **113,** 3190 (1991). Also see R. D. Johnson, D. S. Bethune, and C. S. Yannoni, *Acc. Chem. Res.* **25,** 169 (1992).

11. J. M. Hawkins, A. Meyer, T. A. Lewis, S. Loren, and F. J. Hollander, *Science* **252,** 312 (1991).

12. P. J. Fagan, J. C. Calabrese, and B. Malone, *Science* **252,** 1160 (1991).

13. P. J. Fagan, J. C. Calabrese, and B. J. Malone, *J. Am. Chem. Soc.* **113** 9408 (1991).

14. A. S. Koch, K. C. Khemani, and F. Wudl, *J. Org. Chem.* **56,** 453 (1991) and references cited therein.

15. (a) H. Ajie, M. M. Alvarez, S. J. Anz, R. E. Beck, F. Diederich, K. Fostiropoulos, D. R. Huffmann, W. Krätschmer, Y. Rubin, K. E. Schriver, D. Senharman, and R. L. Whetten, *J. Phys. Chem.* **94,** 8630 (1990); D. H. Parker, P. Wurtz, J. Chatterjee, K. P. Lykke, J. E. Hunt, M. J. Pellin, J. Hemminger, D. M. Gyuen, and L. M. Stock, *J. Am. Chem. Soc.* **113,** 7499 (1991); E. A. Arafa, K. I. Kinnear, and J. E. Lockhart, *J. Chem. Soc., Chem. Commun.* 61 (1992), K. C. Khemani, M. Prato, and F. Wudl, unpublished results, M. S. Meier and J. P. Selegue, *J. Org. Chem.* **57,** 1924 (1992); (b) F. Diederich, R. L. Whetten, C. Thilgen, R. Ettl, I. Chao, and M. M. Alvarez, *Science* **254,** 1768 (1991); (c) F. Diederich and R. L. Whetten, *Nature (London)* **353,** 149 (1991); (d) F. Diederich, F. Wudl, K. Khemani, and A. Koch, *Science* **252,** 548 (1991); (e) F. Diederich and R. L. Whetten, *Acc. Chem. Res.* **25,** 119 (1992).

16. Y. Huang and B. S. Freiser, *J. Am. Chem. Soc.* **113,** 9418 (1991); L. Sunderlin, J. E. Paulino, J. Chow, B. Kahr, D. Ben-Amotz, and R. R. Sanires, *J. Am. Chem. Soc.* **113,** 5489 (1991); see also: H. Schwarz, Highlight, *Angew. Chem. Int. Ed. Engl.* **31,** 293 (1992).

17. (a) R. E. Smalley, in *Fullerenes,* ACS Symposium Series 481, G. S. Hammond and V. J. Kuck, Eds., American Chemical Society, Washington D.C., Chap. 10, 1992; (b) F. Diederich and R. L. Whetten, *Acc. Chem. Res.* **25,** 119 (1992).

18. R. C. Haddon, L. F. Schneemeyer, J. V. Waszczak, S. H. Glarum, R. Tycko, G. Dabbagh, A. R. Kortan, A. J. Muller, A. M. Mujsce, M. J. Rosseinksy, S. M. Zahurak, A. V.

Makhija, F. A. Thiel, K. Raghavachari, E. Cockayne, and V. Elser, *Nature (London)* **350**, 46 (1991).

19. (a) P. M. Allemand, A. Koch, F. Wudl, Y. Rubin, F. Diederich, M. M. Alvarez, S. J. Anz, and R. L. Whetten, *J. Am. Chem. Soc.* **113**, 1050 (1991); (b) R. E. Haufler, J. Conceicao, L. P. F. Chibante, Y. Chai, N. E. Byrne, S. Flanagan, M. M. Haley, S. C. O'Brien, C. Pan, Z. Xiao, W. E. Billups, M. A. Ciufolini, R. H. Hauge, J. L. Margrave, L. J. Wilson, R. F. Curl, and R. E. Smalley, *J. Phys. Chem.* **94**, 8634 (1990).

20. D. Dubois, K. M. Kadish, S. Flanagan, R. E. Haufler, L. P. F. Chibante, and L. J. Wilson, *J. Am. Chem. Soc.* **113**, 4634 (1991).

21. (a) D. Dubois, K. Kadish, S. Flanagan, and L. J. Wilson, *J. Am. Chem. Soc.* **113**, 7773 (1991); (b) Q. Xie, E. Perez-Cordero, and L. Echegoyen, *J. Am. Chem. Soc.* **114**, 3978 (1992); (c) Y. Ohsawa and T. Saji, *J. Chem. Soc., Chem. Commun.* 781 (1992).

22. (a) J. W. Bausch, G. K. S. Prakash, G. A. Olah, D. S. Tse, D. C. Lorents, Y. K. Bae, and R. Malhotra, *J. Am. Chem. Soc.* **113**, 3206 (1991); (b) According to Ref. [4k], C_{60} hexaanion should be more diamagnetic than neutral C_{60}. However, the paramagnetic contribution to the carbon chemical shift is greater in the hexaanion, resulting in net deshielding.

23. T. E. Hogen-Esch and J. Smid, *J. Am. Chem. Soc.* **89**, 2764 (1967).

24. B. C. Berris, G. H. Hovakeemian, Y. H. Lai, H. Mestdagh, and K. P. C. Vollhardt, *J. Am. Chem. Soc.* **107**, 5670 (1985).

25. J. Zimmermann, J. R. Eyler, S. B. H. Bach, S. W. McElvany, *J. Chem. Phys.* **94**, 3556 (1991). Also see: S. W. McElvany, M. M. Ross, and J. H. Callahan, *Acc. Chem. Res.* **25**, 162 (1992).

26. G. A. Olah and D. A. Forsyth, *J. Am. Chem. Soc.* **98**, 4086 (1976).

27. G. A. Olah, I. Bucsi, G. K. S. Prakash, and R. Aniszfeld, unpublished results.

28. G. P. Miller et al., unpublished results from Exxon Laboratory (see *Chem. & Eng. News,* Dec. 16 issue, 1991); (b) K. O. Christe and W. W. Wilson, paper presented at the Fluorine Division, 203rd American Chemical Society Meeting, San Francisco, CA, 1992, Paper No. 66.

29. F. Wudl, A. Hirsh, K. C. Khemani, T. Suzuki, P.-M. Allemand, A. Koch, H. Eckert, G. Srdanov, and H. M. Webb in *Fullerenes,* ACS Symposium Series 481, G. S. Hammond and V. J. Kuck, Eds., American Chemical Society, Washington D.C., Chap. 11, 1992. Also see: F. Wudl, *Acc. Chem. Res.* **25**, 157 (1992).

30. P. J. Fagan, P. J. Krusic, D. H. Evans, S. A. Lerke, and E. Johnston, *J. Am. Chem. Soc.,* submitted.

31. C. N. McEwen, R. G. McKay, and B. S. Larsen, *J. Am. Chem. Soc.* **114**, 4412 (1992).

32. D. A. Loy and R. A. Assink, *J. Am. Chem. Soc.* **114**, 3977 (1992).

33. J. R. Morton, K. F. Preston, P. J. Krusic, S. A. Hill, and E. Wasserman, *J. Am. Chem. Soc.* **114**, 5454 (1992).

34. H. Selig, C. Lifshitz, T. Peres, J. E. Fischer, A. R. McGhie, W. J. Romanow, J. P. McCauley, Jr., and A. B. Smith III, *J. Am. Chem. Soc.* **113**, 5475 (1991).

35. (a) J. H. Holloway, E. G. Hope, R. Taylor, G. J. Langley, A. G. Avent, J. H. Dennis, J. P. Hare, H. W. Kroto, and D. R. M. Walton, *J. Chem. Soc., Chem. Commun.* 966 (1991); (b) A. B. Smith III, personal communication.

36. R. Taylor, J. H. Holloway, E. G. Hope, A. G. Avent, G. J. Langley, J. J. Dennis, J. P. Hare, H. W. Kroto, and D. R. M. Walton, *J. Chem. Soc., Chem. Commun.,* 665 (1992).

37. G. A. Olah, I. Bucsi, C. Lambert, R. Aniszfeld, N. J. Trivedi, D. K. Sensharma, and G. K. S. Prakash, *J. Am. Chem. Soc.* **113,** 9385 (1991).

38. F. N. Tebbe, J. Y. Becker, D. B. Chase, L. E. Firment, E. R. Holler, B. S. Malone, P. J. Krusic, and E. Wasserman, *J. Am. Chem. Soc.,* **113,** 9900 (1991).

39. (a) R. Seshadri, A. Govindaraj, R. Nagarajan, T. Pradeep, and C. N. R. Rao, *Tetrahedron Lett.* 2069 (1992); (b) P. R. Birkett, P. B. Hitchcock, H. W. Kroto, R. Taylor, and D. R. Walton, *Nature (London)*, **357,** 479 (1992).

40. (a) F. N. Tebbe, R. L. Harlow, D. B. Chase, D. L. Thorn, G. C. Campbell, Jr., J. C. Calabrese, N. Herron, R. J. Young, Jr., and E. Wasserman, *Science* **256,** 822 (1992). (b) M. T. Beck, S. Keki, and E. Balazs, *Nature (London),* submitted.

41. R. Taylor, G. J. Langley, M. F. Meidine, J. P. Parsons, A. K. Abdul-sada, T. J. Dennis, J. P. Hare, H. W. Kroto, and D. R. M. Walton, *J. Chem. Soc., Chem. Commun.,* 667 (1992).

42. L. Y. Chiang, J. W. Swirczewski, C. S. Hsu, S. K. Chowdhury, S. Cameron, and K. Creegan, *J. Chem. Soc., Chem. Commun.* 1791 (1992).

43. L. Y. Chiang, R. B. Upasani, J. W. Swirczewski, and K. Creegan, *J. Am. Chem. Soc.* **114,** 10154 (1992).

44. G. A. Olah, I. Bucsi, C. Lambert, R. Aniszfeld, N. J. Trivedi, D. K. Sensharma, and G. K. S. Prakash, *J. Am. Chem. Soc.* **113,** 9387 (1991).

45. S. H. Hooke II, J. Molstad, G. L. Payne, B. Kahr, D. Ben-Amotz, and R. G. Cooks, *Rap. Commun. Mass. Spec.* **5,** 472 (1991).

46. K. M. Creegan, J. L. Robbins, W. K. Robbins, J. M. Millar, R. D. Sherwood, P. J. Tindall, D. M. Cox, A. B. Smith III, J. P. McCauley, Jr., D. R. Jones, and R. T. Ghallagher, *J. Am. Chem. Soc.* **114,** 1103 (1992).

47. J. W. Hawkins, T. A. Lewis, S. D. Loren, A. Meyer, J. R. Heath, Y. Shibato, and R. J. Saykelly, *J. Org. Chem.* **55,** 6259 (1990). Also see: J. W. Hawkins, *Acc. Chem. Res.* **25,** 150 (1992).

48. P. J. Fagan, J. C. Calabrese, and B. Malone, in *Fullerenes,* ACS Symposium Series 481, G. S. Hammond and V. J. Kuck, Eds., American Chemical Society, Washington D.C., Chap. 12, 1992. Also see: P. J. Fagan, J. C. Calabrase, and B. Malone, *Acc. Chem. Res.* **25,** 134 (1992).

13

Chemistry of Fullerenes

Fred Wudl

The discovery of buckminsterfullerene (C_{60}) and related substances by Smalley, Curl, Kroto, and collaborators in 1985 [1] stimulated enormous interest in the new carbon allotropes. These molecules were initially available only in vanishingly small amounts in the gas phase. While a number of important experimental studies appeared in the literature during the infancy of C_{60} and congeners [2], more traditional chemical studies had to await the advent of a method for the production of macroscopic quantities of the new substances.

Such a method was developed in mid-1990 by D. R. Huffmann and W. Krätschmer [3]. Their remarkably simple technique involves resistive heating and vaporization of inexpensive graphite electrodes in an atmosphere of helium. A black soot that thus forms contains about 5–10% of a mixture of fullerenes, consisting of approximately 80% of C_{60}, 15% of C_{70}, and 5% of higher fullerenes (C_{76}, C_{78}, C_{84}, etc.). Fullerenes, unlike the highly insoluble graphitic slag, dissolve easily in certain organic solvents and are readily extracted from the soot. We implemented this practical technique in a convenient benchtop reactor for fullerene synthesis [4].

The Huffman–Krätschmer results were easily duplicated by several research groups, most notably by those of Robert L. Whetten and François Diederich (UCLA) [5], Kroto (Sussex) [6], and Smalley and Curl (Rice) [7]. These research groups characterized the new all-carbon molecules by ^{13}C NMR, IR, UV–VIS, and mass spectrometry. In addition, the UCLA and Sussex teams described methodology for the chromatographic separation of individual fullerenes.

The Rice scientists take full credit for breaking ground on the study of the chemistry of fullerenes. In a seminal 1990 paper [7], they described the first cyclic voltammetric measurements on C_{60}/C_{70} mixtures and discovered that, contrary to intuition, fullerenes are excellent electron acceptors. Moreover, they described the first chemical reaction of a fullerene: the Birch reduction of C_{60} to the novel hydrocarbon $C_{60}H_{36}$, first achieved in the laboratory of one of the editors (MAC). The success of this transformation is a logical consequence of the electron-acceptor properties of C_{60}. Interestingly, $C_{60}H_{36}$ was observed to revert cleanly back to C_{60} upon treatment with DDQ, indicating that no alteration of the carbon cage had occurred during reduction.

These initial results stimulated a flurry of activity directed towards establishing a preliminary picture of fullerene reactivity. A reaction involving alkylation, instead of protonation, of a fullerene polyanion was described shortly thereafter. More than twenty methyl groups were attached to the cluster by quenching a reduction product of C_{60} resulting from sonication in the presence of a large excess of lithium metal with MeI [8]. A number of electrophilic additions to the cluster were also reported. Notable among these are Lewis acid–mediated arylation [9], halogenation [10], cation-mediated hydroxylation [11], and alkoxylation [12], as well as osmylation [13]. The latter reaction occupies a privileged position in the history of fullerenes because it provided the first derivative of C_{60} to be characterized by x-ray diffractometry. Radicals were also observed to react readily with C_{60} to give rather stable adducts [14]. Finally, some remarkable displacements of halide leading to aryl and alkoxy derivatives of the cluster from chlorofullerenes were described [15], and a number of important fullerene organometallic complexes were prepared and characterized [16]. Several of these general types of reactions are reviewed in detail by Olah and Kroto [17].

Soon, the cyclic voltammetry of fullerenes was thoroughly explored [18]. It became clear that the chemistry of fullerenes, especially C_{60}, would be dominated by their electrophilic character. This observation was in accord with the theoretical prediction of low-lying LUMO orbitals of C_{60} [19], but it did not provide a structure–property relationship picture of the origin of such electron affinity. This type of construct would be particularly appealing to the organic chemist seeking to develop a principle that might rationalize and predict the reactivity of the clusters.

Close examination of a molecular model revealed that C_{60} incorporates a number of pyracyclene units. Because pyracyclene is a $4n$-electron π system [20], it was anticipated that each unit would be able to accept up to two electrons from appropriate donors, either by sequential SET processes to give a $4n + 2$ π-electron dianion, or in the form of a lone pair to give an adduct wherein one of the five-membered rings of the former pyracyclene unit is in the form of a cyclopentadienide ion. Buckminsterfullerene was therefore assigned the symbolic representation shown throughout this article to emphasize the importance of the pyracyclene moieties vis-á-vis the chemical reactivity of the cluster. Similar conclusions about the reactivity of fullerenes were drawn independently by Fagan during his attempts to prepare transition-metal derivatives of C_{60} [21].

13.1   Selective Functionalization of Fullerenes

The foregoing considerations suggested that if chemical transformations of fullerenes were to be successful, they would probably involve addition to the pyracyloid double bonds. To test this hypothesis, we started by exploring the reactivity of C_{60} towards nucleophiles such as amines, phosphines and phosphites, Grignard and organolithium reagents, mercaptides, alkoxides, and phenoxides [22]. It was observed that, in general, a large number of products are formed when a particular reagent is employed. These are very difficult to separate and characterize, particularly since small amounts of reaction products were being studied due to the expense of C_{60}, and since a large excess of reagent was always used in order to convert all the available active sites on the fullerene. A number of research groups experienced similar difficulties [23, 24], and, as a result, the first several papers dealing with derivatization of fullerenes described product characterizations based solely on mass spectrometry, especially FAB-MS.

As can be seen from the results in the equations in Scheme 13.1 reaction of C_{60} with Grignard or organolithium reagents results in addition of as many as 10 carbanions. The products of such additions reverted back to C_{60} upon acid quenching of the reaction, but, fortunately, quenching the reaction mixtures with methyl iodide prevents the problem (cf. Ref. 8). The scheme also reveals that product distribution is sensitive to the nature of the Grignard reagent. We believe that the differences between commercial phenyl Grignard and that prepared in situ from phenyllithium is due to the presence of colloidal magnesium metal in the commercial reagent [25]. The magnesium, in competition with phenyl addition, acts as a reducing agent, and the reduced species is quenched with methyl iodide to give products in which the ratios of Me to Ph are > 1. As mentioned later, nucleophilic addition is not restricted to the condensed phase because the elements of negatively charged methane add to fullerene C_{60} in the gas phase.

The product of lithium methoxide addition proved to be a mixture of lithium fulleroxides. Apparently the initially formed methoxy fullerides were demethylated by excess reagent. Iodide and thiocyanate were found to be unreactive towards C_{60}. This observation is consonant with the redox properties of the cluster: Because I_2 and $(SCN)_2$ are better oxidizing agents than C_{60}, I^- and SCN^- are more stable than C_{60}^-. Preliminary data indicated that triethyl phosphite adds to C_{60} to produce an Arbuzov type of product, as determined by infrared spectroscopy (alkyl phosphonate bands). As many as six phosphonate groups are added (FABMS) [26]. Most other nucleophiles reacted to give intractable mixtures. Two notable exceptions were $Et_3BH^-Li^+$, which induced single hydride addition, and morpholine, which reacted with C_{60} to form a hexaadduct.

Addition of THF solution of $LiB(Et)_3H$ to a benzene solution of C_{60} resulted in immediate precipitation of a substance, which upon aqueous workup was identified as $C_{60}HLi \cdot 9H_2O$ [27]. This hydrofulleride salt does not react with

$$60 \text{ PhMgBr}^a + C_{60} \xrightarrow[\text{2) MeI}]{\text{THF, RT}} C_{60}Ph_{10}Me_{10}$$

$$60 \text{ PhLi} + C_{60} \xrightarrow[\text{2) MeI}]{\text{THF, RT}} C_{60}Ph_xMe_y$$

$$x = 10\text{-}0; \ y = 10\text{-}1$$

$$60 \text{ PhMgBr}^b + C_{60} \xrightarrow[\text{2) MeI}]{\text{THF, RT}} C_{60}Ph_xMe_y$$

$$x = 2, 1, 0; \ y = 3, 2, 1$$

$$60 \ t\text{-BuMgBr}^a + C_{60} \xrightarrow[\text{2) MeI}]{\text{THF, RT}} C_{60}t\text{-Bu}_{10}Me_{10}$$

$$60 \ t\text{-BuMgBr}^b + C_{60} \xrightarrow[\text{2) MeI}]{\text{THF, RT}} C_{60}t\text{-Bu}_{10}Me_{10}$$

$$60 \ t\text{-BuLi} + C_{60} \xrightarrow[\text{2) MeI}]{\text{THF, RT}} C_{60}t\text{-Bu}_xMe_y$$

$$x = 2, 1, 0; \ y = 2, 1$$

[a] From PhLi + $MgBr_2$

[b] Commercial reagent

Scheme 13.1

methyl iodide and is indefinitely stable in methanol solution at room temperature, implying that its conjugate acid $C_{60}H_2$ has $pK_a \ll 16$.

The most remarkable result is the addition of neutral nucleophiles such as amines. As many as 12 propylamine molecules can be added onto C_{60}. The ethylenediamine, propylamine, and morpholine adducts are soluble in dilute hydrochloric acid, while the dodecylamine adduct is not. Morpholine adds smoothly to afford a precipitate of C_{60} [$HN(C_2H_4)_2O$], as determined by fast atom bombardment mass spectrometry (FABMS), but C_{60} [$HN(C_2H_4)_2O]_6$, as determined by elemental analysis [28]. Interestingly, cluster-bound hydrogens of the morpholine hexaadduct are mobile at room temperature and were labeled as "globe-trotting hydrogens."

The mechanism of additions of nucleophiles to fullerenes appears to involve SET chemistry. With ethylenediamine and propylamine, electron transfer was detected spectroscopically and by electron resonance (ESR) spectroscopy. With both techniques, a signal due to radical anion formation is replaced by that of the adduct. The decay of the open-shell intermediate is slow for propylamine but rapid for phenyl Grignard addition (minutes). With tertbutylamine, the reaction occurred only in polar solvents such as dimethylformamide (DMF), and no intermediate radical species was observed, but a deep blue intermediate can be followed spectrophotometrically.

Fullerenes soon became readily available in our laboratory through the use of our benchtop reactor [4] and purification apparatus [29], as well as from commercial sources. This allowed us to explore some of our earlier reactions in greater detail. Significantly, Hirsch et al. found that it is possible to titrate one mole of C_{60} with t-butyllithium. The acid-quenched product was fully characterized, except for x-ray crystallography [30].

In order to support our fullerene reactivity hypothesis further, we probed the ability of the double bonds of pyracyclene subunits to function as dienophiles and dipolarophiles. Indeed, C_{60} gave adducts with anthracene, cyclopentadiene, furan, and isobenzofuran [12]. The 1,3 dipoles p-nitrophenyl azide, ethyl diazoacetate, and phenyl-diazomethane were also found to add to C_{60} [12]. The isobenzofuran adduct, an addition product with two types of trimethylenemethanes (nonpolar and polar), were isolated and characterized.

Independently, Kahr, Cooks, and collaborators discovered that benzyne, generated in situ from anthranilic acid and isoamyl nitrite, reacts with C_{60} to form compounds of formula [C_{60} $(C_6H_4)_n$], where $n = 1$–4 [31]. It proved possible to isolate the monoaddition product chomatographically and characterize it by ^{1}H-NMR and UV–VIS spectroscopy; moreover, a benzyne monoaddition product was also obtained from C_{70}.

The C_{60}–benzyne monoadduct is yellow in color, and its UV–VIS spectrum in n-hexane showed λ_{max} at 320 nm with a weak visible band at 428 nm. The precise structure of this material, however, is uncertain. Benzyne typically reacts with polycyclic aromatic compounds by adding across the 1,4 positions within a ring, thus participating in [2 + 4] cycloadditions [32], but in C_{60}, addition may also take place in a [2 + 2] sense across the "inner" pyracyclene bond. Either or both modes of addition may be observed with fullerenes, and, indeed, in one anomalous reaction of benzyne with buckminsterfullerene, a mixture of two monoadducts was obtained. Full characterization of these has not been reported as of this writing. Multiple isomers are certainly obtained for the benzyne–C_{60} diadduct. Presumably, this is due to the large number of regiochemical possibilities for addition of a second benzyne to a monoadduct.

Finally, oxygen atoms may be attached to the cage in a controlled fashion by the use of appropriate techniques [33], such as the reaction of fullerenes with dimethyldioxirane.

13.2   The Preparation of Fulleroids

As evidence regarding the electrophilic, dienophilic, and dipolarophilic properties of C_{60} accumulated, we became interested in developing the means for systematic functionalization of the cluster. During a study of the reaction of C_{60} with diazoalkanes, we observed that an intermediate 1-pyrazoline obtained from ethyl diazoacetate partitioned to the 2-isomer and to a bridged molecule, as shown schematically in the following. At about this time, Professor F. Diederich informed us that one of the components of a sample of higher fullerenes that we had sent him was actually $C_{70}O$. The UV–VIS spectrum of this molecule was essentially identical to that of C_{70}. In accord with precedent in bridged annulenes [34], $C_{70}O$ was assigned [35] a ring-opened structure: an oxofulleroid [36] rather than an epoxyfullerene [37]. This observation raised the possibility of specifically preparing functionalized C_{60} derivatives through a dipolar cycloaddition followed by ring opening according to the formulation of Scheme 13.2. The electronic properties of the fullerene would remain essentially unperturbed by the ring-opening step. The concept, as depicted below, was examined with $a = b = N$ and $c = CR_2$.

Diphenyldiazomethane was initially chosen as a convenient reagent that would avoid complications arising from possible prototropic shifts within the pryazoline adduct. It was gratifying to observe that the expected pyrazoline intermediate lost nitrogen spontaneously at room temperature, and the product of single addition was the least soluble component in the reaction mixture, and

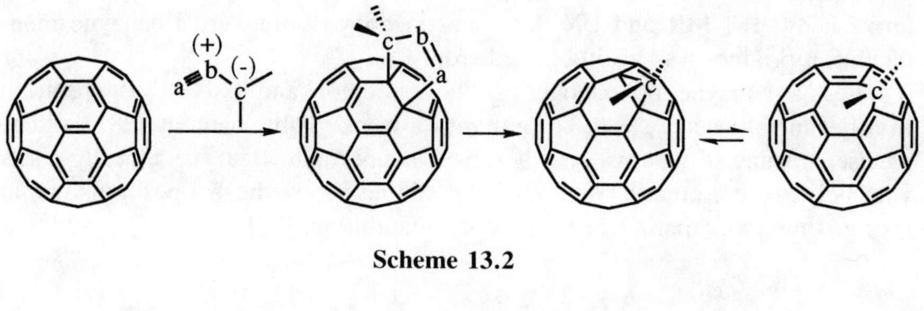

Scheme 13.2

was therefore readily isolated (>40% yield). Moreover, while as many as six cycles of this chemistry are possible with C_{60}, each incremental diphenyldiazomethane addition product caused a substantial decrease in reaction rate (10 days for hexaadduct formation), thus facilitating selective functionalization [38].

The monoadduct's UV–VIS spectrum was identical to that of C_{60} in the UV region but was different in the weakly absorbing visible region, and its cyclic voltammetry showed four reversible waves with a 0.08-V cathodic shift of the first reduction peak (−335, −920, −1470, and −1929 mV) relative to C_{60} (−228, −826, −1418, and −1916 mV, under the same conditions) [39]. Since the adduct resembled C_{60} in its electronic properties, we called it a *fulleroid*. Calculations carried out in collaboration with Professor T. C. Bruice's group revealed that the ring-opened form was indeed thermodynamically more stable than the cyclopropane isomer. On the basis of these data, we concluded that the monoadduct possessed the structure depicted for **1**. At the same time, we sought to produce definitive structural proof through x-ray crystallography. The impetus for this latter effort was provided by the observation that, much to our surprise, an isobenzofuran monoadduct of C_{60}, prepared in collaboration with Professor B. R. Rickborn's group, exhibited UV–VIS and cyclic voltammetry characteristics nearly identical to those of $C_{61}Ph_2$. Formation of an isobenzofuran monoadduct almost certainly did not induce rupture of a C–C bond within the cluster; therefore, our conclusion for the structure of the fulleroids was not on firm ground.

Suitable crystals of $C_{61}Ph_2$ were eventually obtained, but while a complete data set was collected in Professor Strouse's group at UCLA, the crystal structure could not be solved because of excessive disorder. Fortunately, we succeeded in preparing a sample of $C_{61}PhH$ where the methine carbon was 99% ^{13}C enriched. The chemical shift of the methine carbon was 53.6 ppm, much too low field for a cyclopropane, and the $^{13}C-^{1}H$ coupling constant was 140 ppm, in excellent agreement with the value J_{C-H} of 11-phenyl-1,6-methano[10]annulene (139 ppm) [40]. We were now confident that, at least for the monophenyl-substituted fulleroid, the transannular bond was open.

Additional evidence came from an x-ray study of C_{61} (p-$BrC_6H_4)_2$, which crystallized in beautiful hexagonal prisms. The structure was solved in collaboration with Dr. Saeed Khan (UCLA). The critical transannular bond length was found to be 1.84 Å, somewhat shorter [41] than that observed in most methanol[10]annulenes (2.1 Å) but still considerably longer than a fullerene bond (1.37–1.47 Å), proving, together with the ^{13}C NMR data, that the transannular bond was broken. The concept of fulleroid formation had been vindicated. However, ^{13}C NMR spectra of fulleroids (e.g., 4,4'-dimethoxydiphenylfulleroids–C_{61}) revealed that the chemical shift of the bridgehead carbons is only in the range of 90 ppm, instead of the typical fullerene resonances that appear at 143–156 ppm. This, together with the shorter than expected transannular distance of 1.84 Å mentioned and calculations by Haddon on more rigid C_{61} fulleroids (such as 9-fluorenyl–C_{61}), led us to conclude that some form of weak transannular bonding exists in disubstituted fulleroids.

13.3   Scope of the Fulleroid Synthesis

Several fulleroids were prepared and characterized in quick succession, and from these studies it became clear that the fulleroid synthesis has ample scope. In particular, p,p'-disubstituted-diphenyl fulleroids (1–6) are easily prepared from readily available, substituted benzophenones, which are readily converted to diazomethanes via the hydrazones [42].

A remarkable characteristic of the fulleroid formation reaction is that the product of single addition was a mixture of three isomers. For instance, the bridge carbon-^{13}C labeled diphenyl fulleroid 1 exhibited three bridge carbon resonances at 37.5, 57.9, and 64.8 ppm in the ^{13}C NMR spectrum, while bis(p-methylphenyl) fulleroid 4 showed three methyl singlets and a complex pattern in the aromatic region of its ^{1}H NMR spectrum. Later it was determined that 3–6 but not 7 nor 8, 10, 11 were also obtained as mixtures of at least two isomers [43]. It should be pointed out that all three isomers had identical TLC mobility. Remarkably, only the isomer with the 57.9 ppm resonance remained after overnight thermolysis of the isomer mixture of 1 in refluxing toluene solution. The same heat treatment caused the isomers of 2, 3, and 6 to converge to a single product (NMR), and induced considerable simplification and line narrowing in the NMR spectra of higher fulleroids. These observations imply that one of the fulleroid isomers must be appreciably more thermodynamically stable than the others.

Even the parent fulleroid, $C_{60}CH_2$, is readily obtained from C_{60} and diazo-

methane [44]. In this case, contrary to other systems, the intermediate pyrazoline can be isolated and characterized. This allowed us to determine that the
first step of the fulleroid synthesis is indeed a dipolar addition across the reactive 6-ring–6-ring (pyracylene) junction, and that thermal nitrogen loss occurs concomitantly with rearrangement (1,5-shift?). The fulleroid structure of
the compound is strongly supported by the ^{13}C NMR spectrum of a sample
prepared from 99%-^{13}C diazomethane. The bridge carbon appeared as a single
peak at 38.85 ppm, in excellent agreement with the case of methano[10]annulene (38.4 ppm) [45]. Additional evidence is provided by the coupling constant of the methylene hydrogens (J = 9.7 Hz), which may be correlated with the C–CH$_2$–C angle [46], and by the coupling constants between
the methylene carbon and its (nonequivalent) hydrogens, which are 145.0 and
147.8 Hz, respectively, as typical for methano[10]annulenes [47].

All fulleroids **1–7** have very similar UV–VIS spectra and cyclic voltammograms. In addition to the fulleroid CV waves, the anthrone-derived fulleroid
8 shows a wave due to the carbonyl redox system that almost overlaps the third
wave of the fulleroid (−1300 mV). The chemical shift of the aromatic ring
hydrogen atoms closest to the spheroid provide additional clues regarding the
electronic properties of fulleroids. For instance, the ortho hydrogens of diphenyl fulleroid **1** are the closest atoms to the surface of the sphere. These
appear at 8.02 ppm, considerably deshielded relative to 2,2-diphenylpropane.
Additional chemical shifts of ortho-hydrogens in fulleroids **1, 10, 11, 8** appear
in Table 13.1. It is apparent that as these hydrogens are pushed closer to the
fulleroid π system, the deshielding effect of the fulleroid becomes stronger,
and it is always in the opposite sense of the 1,6-methano[10]annulenes. A sim-
plistic interpretation of these results is that if ring currents exist in the fulleroids
(and by extension, in the fullerenes), they are paramagnetic, as expected from

Table 13.1   Table of Selected ^{1}H NMR Chemical Shifts

Compound	Chemical Shift of *ortho* Hydrogen Atom (ppm rel. to TMS)
1[a]	8.02
10[a]	8.76
11[a]	9.69
8[a]	8.40
2,2-diphenylpropane[b]	7.17
Flourene[c]	7.78
2-nitrofluorene[c]	8.30
Anthrone[d]	7.50

[a]The chemical shift of the hydrogen atoms *ortho* to the bridge carbon atom is reported.
[b]The chemical shift of the hydrogen atoms *ortho* to the dimethylbenzyl group is reported.
[c]The chemical shift of the C(1)-H is reported.
[d]The chemical shift of the the H *ortho* to the methylene is reported.

a nonaromatic or antiaromatic π system. However, an ^{1}H NMR study of $C_{60}CH_2$ suggests otherwise. Remarkably, the hydrogens of this fulleroid appear as *two* doublets at 2.87 (H_A) and 6.35 (H_B) ppm in a 1:1 ratio. Spin decoupling revealed that the two sets of doublets are coupled. The difference in chemical shift between the two hydrogens, $\Delta\delta = 3.48$ ppm, is unusually large. There are only a few cases reported where effects of such magnitude were observed in related systems. Of these, the most revealing is hydrocarbon 14 [48]. In the latter system, the relevant high-field and low-field protons appear at 1.2 and 4.7 ppm, respectively, that is, 1.67 and 1.65 ppm upfield from the corresponding fulleroid hydrogens. As discussed, in fulleroids where the aromatic protons are directly above the π system of the six-membered rings, drastic deshielding of those hydrogens is seen, but $C_{60}CH_2$ is the first example of a fulleroid with protons over a five-membered ring. Two possibilities emerge: Since aromatic ring currents cannot exist in 14 [49], and its $\Delta\delta$ is the same as in the fulleroid, one would conclude that there are no ring currents in the fulleroid spheroid, as is the case for C_{60} [50]. Alternatively, two types of paramagnetic ring currents may exist in fulleroids (and in buckminsterfullerene [51]). If that were the case, we would conclude that the ring currents of the six-membered rings are deshielding and those of the five-rings are shielding.

C_{61} Fulleroids

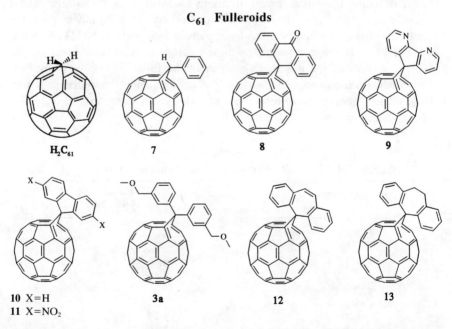

In summary, fulleroid preparation is facile and versatile. Furthermore, we have recently determined that it is possible to prepare azafulleroids by dipolar addition of an organic azide followed by nitrogen elimination [52]. The stage was set for the preparation of molecules that would have greater technological potential.

1 4

13.4 Bifulleroids, Pearl Necklaces, and Charm Bracelets

In principle one should be able to extend the fulleroid synthesis to the preparation of two types of polymers: one with fulleroids as part of the backbone, and the other with fulleroids as pendant groups. The former have been dubbed "pearl necklace" polymers (PNP), and the latter "charm bracelet" polymers (CBP) [53].

PNPs may conceivably be made by the use of bis(diazomethanes). Indeed, reaction of C_{60} with the diazomethanes derived from *m*-phenylenebis(benzoyl) and *p*-phenylenebis(benzoyl) provided compounds **15** and **16,** respectively. These are two-pearl sections of pearl necklace polymers. Again the "crude" compounds (single spot by TLC) produced very broad NMR spectra, implying again the presence of several isomers, but the heat treatment described for the simple fulleroids worked, and clean, interpretable NMR spectra were recorded. Also, in line with previous observations on **1–13,** the hydrogen atom sandwiched between two spheres in **15** appeared at 8.8 ppm, compared to the singlet at 8.19 ppm of the four hydrogen atoms of the *p*-phenylene moiety of **16.** The CV of these bifulleroids was essentially identical to those of diphenyl fulleroid ($C_{60}CPh_2$) [54].

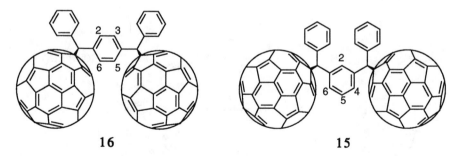

16 **15**

Unfortunately, these molecules were found to be quite insoluble and thus very likely not good candidates for further reaction to produce macromolecules. However, attachment of solubilizing groups on the phenylene moiety will undoubtedly render the polymers more tractable, by analogy with the well-precedented case of conjugated polymers [55].

Fulleroids **3** and **6** represent excellent monomers for the preparation of CBPs, because a diphenol or a dicarboxylic acid obtained after cleavage of the methyl ether of the former and the benzyl ester of the latter would afford poly(esters) and poly(urethanes) upon reaction with commercially available bis(acyl halides) and diisocyanates [56]. This hypothesis has been reduced to practice with **3** [57]. Treatment of this fulleroid with BBr_3 induced clean demethylation (94% yield) to a diphenol. Copolymerization of **17** and sebacoyl chloride [58] produced poly(4,4'-diphenyl–C_{61} sebacate), **18**, as a brown powder that was soluble in nitrobenzene and benzonitrile, but sparingly soluble in THF [59]. Thermogravimetric analysis (TGA) revealed that a weight loss of 9% of the original mass occurs between 360 and 480°C. By contrast, the parent fulleroid **17** is less thermally stable: It gradually loses weight upon heating, retaining ~90% of its original mass at 500°C and ~83% at 700°C [60]. Solution cyclic voltammetry of the polymer in THF showed the first three characteristic reduction waves of all diphenyl $C_{61}s$, a clear indication that the polymer retained the electronic properties of diphenyl fulleroids.

In a similar fashion, **17** copolymerized [61] with hexamethylene diisocyanate to furnish an insoluble brown powder to which structure **19** was assigned on the basis of the fact that its IR spectrum is essentially superimposable with that poly(bisphenol A hexamethyleneurethane) (PBAU). Comparative thermogravimetric analysis (TGA) results of PBAU and **19** shows that PBAU ex-

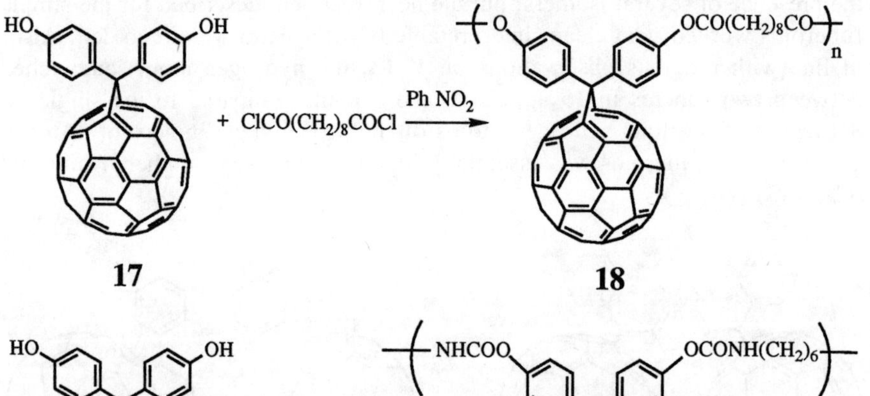

17   **18**

PHC$_{61}$U

19

hibits a rapid weight loss (90%) from 180 to 300°C [62]. Likewise, **19** experienced weight loss between 180 and 340°C, consistent with a urethane structure. The nonquantitative weight loss of about 9.5–10% (calculated for 100% loss of diisocyanate = 15.5%) for the unzipping of **19** is unexplained at this time, but it could be attributable to the thermal stability of the fulleroid moieties and their trapping of "nascent" diisocyanate. Finally, while PBAU displays an endothermal peak at about 88°C, which is believed to be the glass transition temperature, the differential scanning calorigram of **19** shows no transitions. This observation and the insolubility of the polymer could imply crosslinking during urethane formation, but in a control experiment, dimethoxyphenyl fulleroid and hexamethylene diisocyanate showed no reaction, indicating no crosslinking between C_{61} and the isocyanate function.

13.5 Outlook

The future for the chemistry of fullerenes and fulleroids is bright. The fullerene inflation reactions will continue to produce functionalized fulleroids that will find applications in polymer chemistry and materials science. Electronic modification to produce fulleroids that are actually electron donors by application of the concept of periconjugation further open the door to organic metal, ferromagnet, and superconductor research. In addition, more exotic fulleroids may have unusual materials properties. For example, alkylation of both nitrogens in **2** will lead to a fulleroid dication, which upon two-electron reduction will give a "self-doped" fulleroid, akin to the self-doped conducting polymers. The self-doped fulleroid would be expected to be a metal. It may turn out to be a ferromagnet as well, if the LUMO of the fulleroids is of the same character as that of C_{60}. Further doping with 1 equivalent of an alkali metal would produce a material that may undergo a transition to the superconducting state. Although the higher fulleroids were not discussed in this review, a C_{66} fulleroid might serve as a monomer for the preparation of very large dendrimers [63]. As more techniques for functionalization of fullerenes are developed, members of the chemical community will be provided with concepts that will allow the synthesis of wide range of novel molecules with intricate, spherical architecture and as yet unknown properties. While an organometallic complex of C_{70} was characterized, no other chemical derivatization of fullerenes other than C_{60} have appeared. Once larger quantities of the higher fullerenes, starting with C_{70}, are available, a large, exciting new area of organic chemistry will evolve naturally.

Acknowledgment

I am indebted to all my co-workers mentioned in the various references. I would also like to acknowledge Professors W. E. Billups and M. A. Ciufolini for help with the writing of the manuscript.

References

1. H. W. Kroto, J. R. Heath, S. C. O'Brien, R. F. Curl, and R. E. Smalley, *Nature* **318**, 162 (1985).

2. For a review see R. E. Smalley, *Acc. Chem. Res.* **25**, 98 (1992).

3. W. Krätschmer, L. D. Lamb, K. Fostiropoulous, and D. R. Huffmann, *Nature* **347**, 354 (1990).

4. A. S. Koch, K. C. Khemani, and F. Wudl, *J. Org. Chem.* **56**, 4543 (1991).

5. H. Ajie, M. M. Alvarez, S. J. Anz, R. D. Beck, F. Diederich, K. Fostiropoulous, D. R. Huffmann, W. Krätschmer, Y. Rubin, K. E. Schriver, D. Sensharma, and R. L. Whetten, *J. Phys. Chem.* **94**, 8630 (1990).

6. R. Taylor, J. P. Hare, A. K. Abdul-Sada, and H. W. Kroto, *J. Chem. Soc., Chem. Commun.* 1423 (1990).

7. R. E. Haufler, J. Conceicao, Y. Chai, L. P. F. Chibante, N. E. Byrne, S. Flanagan, M. M. Haley, S. C. O'Brien, C. Pan, Z. Xiao, W. E. Billups, M. A. Ciufolini, R. H. Hauge, J. L. Margrave, L. J. Wilson, R. F. Curl, and R. E. Smalley, *J. Phys. Chem.* **94**, 8634 (1990).

8. J. W. Bausch, G. K. Suria Prakash, G. A. Olah, D. S. Tse, D. C. Lorents, Y. K. Bae and R. Malhotra, *J. Am. Chem. Soc.* **113**, 3205 (1991).

9. G. A. Olah, I. Bucsi, C. Lambert, R. Aniszfeld, N. J. Trivedi, D. K. Sensharma, and G. K. Surya Prakash, *J. Am. Chem. Soc.* **113**, 9387 (1991).

10. Fluorination to $C_{60}F_{36}$: H. Selig, C. Lifschitz, T. Peres, J. E. Fischer, A. R. McGhie, J. W. Ronamow, J. P. McCauley, Jr., and A. B. Smith III *J. Am. Chem. Soc.* **113**, 5475 (1991); to $C_{60}F_{60}$: J. H. Holloway, E. G. Hope, R. Taylor, G. J. Langley, A. G. Avent, T. J. Dennis, J. P. Hare, H. W. Kroto, and D. R. M. Walton, *J. Chem. Soc., Chem. Commun.* 966 (1991). Chlorination and bromination: F. N. Tebbe, J. Y. Becker, D. B. Chase, L. E. Firment, E. R. Holler, B. S. Malone, P. J. Krusic, and E. Wasserman, *J. Am. Chem. Soc.* **113**, 9900 (1991); G. A. Olah, I. Bucsi, C. Lambert, R. Aniszfeld, N. J. Trivedi, D. K. Sensharma, and G. K. Surya Prakash, *J. Am. Chem. Soc.* **113**, 9385 (1991). For x-ray crystallography of bromine adducts: F. N. Tebbe, R. L. Harlow, D. B. Chase, D. L. Thorn, G. C. Campbell, Jr., J. C. Calabrese, N. Herron, R. Y. Young, E. Wasserman, *Science* **256**, 822 (1992), P. R. Birkett, P. B. Hitchcock, H. W. Kroto, R. Taylor, D. R. M. Walton, *Nature* **357**, 479 (1992).

11. L. Y. Chiang, R. Upasani, and J. W. Swirczewski. Presented at the Materials Research Society Meeting, Boston, MA, Dec. 1991.

12. G. P. Miller, C. S. Hsu, L. Y. Chiang, H. Thomann, and M. Bernardo, Presented at the Materials Research Society Meeting, Boston, MA, Dec. 1991.

13. J. Hawkins, *Acc. Chem. Res.* **25**, 150 (1992), and references cited therein.

14. P. J. Krusic, E. Wasserman, P. N. Keizer, J. M. Morton, and K. F. Preston, *Science* **254**, 1183 (1991).

15. G. A. Olah, I. Bucsi, C. Lambert, R. Aniszfeld, N. J. Trivedi, D. K. Sensharma, and G. K. Surya Prakash, *J. Am. Chem. Soc.* **113**, 9385, 9387 (1991). See also: S. H. Hoke II, J. Molstad, G. L. Payne, B. Kahr, D. Ben-Amotz, and R. G. Cooks, *Rapid Commun. Mass Spectrom.* **5**, 472 (1991).

16. See Ref. 13 as well as: P. J. Fagan, J. C. Calabrese, and B. Malone, *Science* **252**, 1160 (1991); A. L. Balch, V. J. Catalano, and J. W. Lee, *Inorg. Chem.* **30**, 3980 (1991); A. L. Balch, V. J. Catalano, J. W. Lee, M. M. Olmstead, and S. R. Parkin, *J. Am. Chem. Soc.* **113**, 8953

(1991); P. J. Fagan, J. C. Calabrese, and B. Malone, *J. Am. Chem. Soc.* **113**, 9408 (1991); R. S. Koefod, M. F. Hudgens, and J. R. Shapley, *J. Am. Chem. Soc.* **113**, 8957 (1991). See also: P. J. Fagan, J. C. Calabrese, and B. Malone, *Acc. Chem. Res.* **25**, 134 (1992).

17. Please see the contributions by G. A. Olah and H. W. Kroto in this volume.

18. P.-M. Allemand, A. Koch, F. Wudl, Y. Rubin, F. Diederich, M. M. Alvarez, S. J. Anz, and R. L. Whetten, *J. Am. Chem. Soc.* **113**, 1050 (1991); D. Dubois, K. M. Kadish, S. Flanagan, and L. J. Wilson, *J. Am. Chem. Soc.* **113**, 7773 (1991); Q. Xie, E. Perez-Cordero, L. Echegoyen, *J. Am. Chem. Soc.* **114**, 3978 (1992).

19. R. C. Haddon, *Acc. Chem. Res.* **25**, 127 (1992).

20. B. M. Trost, G. M. Bright, C. Frihart, and D. Brittelli, *J. Am. Chem. Soc.* **93**, 737 (1971).

21. P. J. Fagan, J. C. Calabrese, B. Malone, *Acc. Chem. Res.* **25**, 134 (1992).

22. A. Hirsch, A. Koch, S. Shi, T. Suzuki, T. White, B. R. Rickborn, and F. Wudl, unpublished results from these laboratories (1991).

23. Please see the contributions by G. A. Olah and H. W. Kroto in this volume, as well as Ref. 7.

24. M. A. Ciufolini, *Gendai Kagaku (Chemistry Today)* **253**(4), 12 (1992).

25. We thank Professor George A. Olah of the University of Southern California for the suggestion that colloidal Mg may be responsible for this strange reactivity.

26. A. Hirsch, and F. Wudl, unpublished data from these laboratories.

27. F. Wudl, A. Hirsch, K. C. Khemani, T. Suzuki, P.-M. Allemand, A. Koch, and G. Srdanov, presented at the 200th National Meeting of the American Chemical Society, Atlanta, GA, April 1991. Idem, in *Fullerenes: Synthesis, Properties and Chemistry of Large Carbon Clusters;* G. S. Hammond, V. J. Kuck, Eds.; American Chemical Society: Washington, DC 1992, p. 161.

28. A. Hirsch, Q. Li, and F. Wudl, *Angew. Chem.* **103**, 1339 (1991).

29. K. C. Khemani, M. Prato, and F. Wudl, *J. Org. Chem.* **57**, 3254 (1992).

30. A. Hirsch et al., *Angew. Chem.* **31**, 766 (1992).

31. S. H. Hoke II, J. Molstad, D. Dilettato, M. J. Jay, D. Carlson, B. Kahr, and R. G. Cooks, *J. Org. Chem.* **57**, 5069 (1992).

32. Cf. R. W. Hoffman, *Dehydrobenzene and Cycloalkynes,* Verlag Chemie; New York, N.Y., 1967, pp. 200–39.

33. J. M. Wood, B. Kahr, S. H. Hoke II, L. Dejarme, R. G. Cooks, and D. Ben-Amotz, *J. Am. Chem. Soc.* **113**, 5907 (1991), K. M. Creegan, J. L. Robbins, W. K. Robbins, J. M. Millar, R. D. Sherwood, P. J. Tindall, D. M. Cox, A. B. Smith III, J. P. McCauley, Jr., D. R. Jones, and R. T. Gallagher, *J. Am. Chem. Soc.* **114**, 1103 (1992); Y. Elemes, S. K. Silverman, C. Sheu, M. Kao, C. S. Foote, M. M. Alvarez, R. L. Whetten, *Angew. Chem. Int., Ed. Engl.* **31**, 351 (1992). See also: W. A. Kalsbeck, H. H. Thorp, *J. Electroanal. Chem.* **314**, 363 (1991).

34. E. Vogel, *Pure Appl. Chem.* **54**, 1015 (1982).

35. F. Diederich, R. Ettl, Y. Rubin, R. L. Whetten, R. D. Beck, M. M. Alvarez, S. J. Anz, D. Sensharma, F. Wudl, K. C. Khemani, and A. Koch, *Science* **252**, 548 (1991).

36. *Fulleroid* is a term introduced to describe an expanded fullerene where the electronic properties and number of π electrons remain unchanged with respect to the unexpanded fullerene.

37. A C_{60} oxide, where the structure assignment was of a closed epoxide rather than a fulleroid, has recently been prepared; see Ref. 34.

38. T. Suzuki, Q. Li, K. C. Khemani, F. Wudl, and O. Almarsson, *Science* **254,** 1186 (1991).

39. Pt working electrode and counterelectrode; Ag/AgCl/3 *M* NaCl, reference; ferrocene, internal reference (+620 mV); 0.1 *M* TBABF$_4$ (tetrabutylammonium tetrafluoroborate)/THF in a drybox.

40. Data provided kindly by Professor E. Vogel prior to publication.

41. There are a number of methano[10]annulenes that have bond lengths of that order and that were considered to be in the annulene rather than the norcaradiene form. However, this interpretation of crystallographic results has come into question since the advent of solid-state CPMAS ^{13}C NMR. See: R. C. Haddon, *Tetrahedron* **44,** 7611 (1988).

42. J. March, Advanced Organic Chemistry, 3rd Edition, John Wiley & Sons, New York, NY 1985, p. 1062.

43. At the time of this writing, we are certain that the thermodynamically most stable isomer is the one depicted. One could speculate that the structure of one other isomer has the bridge spanning a 6–5 ring juncture rather than the 6–6 juncture depicted in H_2C_{61}. The structure of the third isomer is uncertain, but it is apparent that isomers can be obtained by a series of allowed sigmatropic rearrangements of the bridge carbon.

44. T. Suzuki, Q. Li, K. C. Khemani, F. Wudl, *J. Am. Chem. Soc.* **114,** 7302 (1992).

45. H. Günther, H. Schmickler, W. Bremser, F. A. Straube, and E. Vogel, *Angew. Chem., Int. Ed. Engl.* **12,** 570 (1973).

46. E. Vogel, W. Wiedermann, H. D. Roth, J. Eimer, and H. Günther, *Justus Liebigs Ann. Chem.* **759,** 1 (1972).

47. R. Arnz, J. W. M. Carneiro, W. Klug, H. Schmickler, and E. Vogel, *Angew. Chem., Int. Ed. Engl.* **30,** 683 (1991).

48. E. Vogel, *Aromaticity,* O. D. Ollis, Ed.; The Chemical Society: London, 1967, Vol. 21, p. 142.

49. If there were a ring current in the "cyclohexatriene" moiety of **6**, then the proton oztho to the fulleroid would have a chemical shift of 0 to −1 ppm. However, Haddon has pointed out a possible danger of using model chemical shifts (MCS) in analyzing ring currents (private communication; also: R. C. Haddon, *Tetrahedron* **28,** 3613, 3635 (1972).

50. R. C. Haddon, L. F. Schneemeyer, J. V. Waszczak, S. H. Glarum, R. Tycko, G. Dabbadh, A. R. Kortan, A. J. Muller, A. M. Mujse, M. Rosseinsky, S. M. Zahurak, A. V. Makhija, F. A. Thiel, K. Raghavachari, E. Cockayne, and V. Elser, *Nature* **350,** 46 (1991) V. Elser and R. C. Haddon, *Nature* **325,** 792 (1987).

51. A. Pasquarello, M. Schluter, and R. C. Haddon, *Science* **257,** 1660 (1992).

52. M. Prato et al. *J. Am. Chem. Soc.,* in press.

53. Cf. I. Amato, *Science* **254,** 30 (1991).

54. T. Suzuki, Q. Li, K. C. Khemani, F. Wudl, and Ö. Almarsson, *J. Am. Chem. Soc.* **114,** 7300 (1992).

55. (a) *Handbook of Conducting Polymers,* T. A. Skotheim, Ed., Dekker: New York, 1986. (b) A. O. Putil, A. J. Heeger, F. Wudl, *Chem. Rev.* **88,** 183 (1988), and references within.

56. Polymers containing a C_{60} unit in the backbone of a poly(p-xylylene) have recently been

reported by D. A. Loy, and R. A. Assink, *J. Am. Chem. Soc.* **114,** 3977 (1992). A "fulleren-ated" poly(styrene) was prepared by Professor G. A. Olah (personal communication).

57. S. Shi, K. C. Khemani, Q. Li, and F. Wudl, *J. Am. Chem. Soc.*, in press.

58. The polymerization was conducted in dry nitrobenzene at 143° (22 h), using equimolar amounts of the monomers and no catalyst. HCl was removed by bubbling nitrogen through the reaction mixture.

59. This material showed ester IR absorptions (C=O at 1760, 1725 cm^{-1}) and ^{1}H-NMR signals (THF-d$_8$) at 10.83 ppm (OH end groups), 6.8–8.55 ppm (complex multiplet, phenylene protons) and 1.2–2.6 ppm (complex multiplet, hexamethylene protons). The UV–VIS spectrum of a THF solution is reminiscent of other C$_{61}$s^3 (bands at 690, 475, 430, 330, 275, and 250 nm).

60. A weight loss observed between 200 and 300°C is typical of diphenylfulleroids and has not been assigned to any particular fragmentation.

61. An equimolar mixture of **17** and the diisocyanate was heated in *o*-dichlorobenzene in the presence of a catalytic amount of 1,4-diaza-bicyclo[2,2,2]octane (DABCO).

62. The weight loss observed in almost all polyurethanes is attributed to "unzipping" to the original monomers.

63. G. Caminati, K. Gopidas, A. R. Leheny, N. J. Turro, and D. A. Tomalia, *Polym. Prepr. (Am. Chem. Soc., Div. Polym. Chem.)* **32,** 602 (1991), and references within. G. R. Newcome, and G. R. Baker, *Polym. Prepr. (Am. Chem. Soc., Div. Polym. Chem.)* **32,** 615 (1991), and references within. C. J. Hawker, K. L. Waoley, and J. M. Frechet, *J. Polym. Prepr. (Am. Chem. Soc., Div. Polym. Chem.)* **32,** 623 (1991), and references within.

INDEX

Angle defect, 90
Aromatic hydrocarbons, 8, 64, 98, 311, 326
Azafulleroids, 326

BCS Theory, 202, 205, 207, 226
Bifulleroids, 327
Buckminsterfullerene (C_{60}), 4, 21
  Addition mechanism, 321
  Adsorbed on metal surfaces, 247
    Binding energy, 247–52
    Electron transfer, 250
  Alkoxy, 306, 310, 318, 319
  Alkylation, 31, 307, 318–19
  Alternative structure (graphitene), 83
  Bond lengths, 105, 133–34, 137, 302
  Chemistry, 29, 127, 258, 285, 301, 317
  Collisions, at hypervelocities, 127, 147, 262
    Anisotropic compression during, simulated, 154
    Chemisorption during, simulated, 149
    Entrapment of atoms/molecules during, 148, 259, 262, 265
      efficiency/probability, 152, 267–68
    Molecular dynamics simulation of, 149
    Penetration barrier, 151, 259, 277
    With gases, 127, 147, 151, 262, 265
    With surfaces, 127, 148–49, 262
  Combustion, 127
  Conversion to diamond, 127, 148
  Conversion to graphite, 127, 148

Cyclic voltammetry, 69, 71, 286, 303, 306, 318
Cycloadditions, 321
Derivatives. *See also* specific derivatives
  Alkali metal, 40, 41, 45, 127, 201, 218, 221, 233, 303
  Chemical reactions, 303, 309, 310, 318
  Electrochemistry, 296, 323
  Endohedral, 148, 151, 217, 239, 257, 259
  Fulleroid, 134, 296, 322
  Globe-trotting hydrogens in, 321
  Semiconducting, 200–1, 296
  Spectra, 30, 32, 34, 47, 229, 305, 308, 311, 323–29
  Superconducting, 40, 45, 127, 199, 201, 203, 225, 285
  Transition metal, 296, 318
  X-ray structures, 33, 34, 302, 308, 318, 323
Electrochemistry, 69, 71, 285, 286
Electron affinity, 134, 138, 240, 257
Films, properties, 296–97
Fluorinated,
  Endo–exo isomerism in, 110
  F-F steric repulsion in, 110
  Physical properties, predicted, 147
  Possible structures, 109, 145
  Preparation, 29, 308
  Reactions, 308
Halogenation, 31, 29, 308, 318

With Br$_2$, 31, 309
With Cl$_2$, 308
With F$_2$, 29, 308
With XeF$_2$, 29
Hybridization, 192, 203
Hydrogenated, 107, 134, 144, 303
   C$_{60}$H$_{36}$, 108, 134, 144, 147, 303, 318
   C$_{60}$H$_{60}$, 108, 145
   Dehydrogenation, 303, 318
   Endo–exo isomerism in, 108, 147
   Physical properties, predicted, 145
   Possible structures, 108, 109, 144
Hydroxy, 310, 318, 319
In flames, 23
Ionization potential, 11, 136, 138, 257
IR spectrum, 35, 37, 136
Lubricant properties, predicted, 154
Matrix isolation spectrum, 15, 16, 35
Neutron,
   Diffraction, 40, 41
   Scattering, inelastic, 41
NMR studies, 23, 35, 43, 133, 192, 301
Osmylation, 134
Oxidation, 289, 306, 322
Penetration barrier, 151, 259, 277
Phase transitions, crystal, 35, 41, 219
Photoelectron spectra,
   Experimental, 133, 141
   Predicted, 140
Photoemission, 201
Physical measurements, 35, 43
Preparation, macroscopic quantities, 23, 25,
   59, 71, 301, 317
Proton affinity, 257
Protonation, 258, 271, 309, 312
Pyracyclene units within, 318
Pyramidalization angle, 193–94
Radical cations, 306, 312
Raman spectrum, 35, 38, 133
Reaction,
   Friedel–Crafts-type, 31, 309, 311
   Hydrogenation, 107, 134, 144, 303
   Oxidation, 289, 306, 322
   Reduction, electrochemical, 286
   Reduction, Birch, 107, 134, 144, 303, 318
   With alkoxides, 319
   With amines, 321
   With anthracene, 321
   With aromatic compounds and AlCl$_3$, 311–
   13
   With azides, 321, 326
   With benzyne, 321
   With Br$_2$, 31, 309
   With Br$_2$, FeCl$_3$, and benzene, 31, 310
   With *t*-BuLi, 307, 321
   With Cl$_2$, 308
   With cyclopentadiene, 321
   With diazo compounds, 322, 324, 325, 327
   With diazomethane, 325
   With ethyl diazoacetate, 321, 322

   With F$_2$, 29, 308
   With free radicals, 308, 318
   With furan, 321
   With Grignard reagents, 307, 319
   With HNO$_3$, 310
   With isobenzofuran, 321
   With Li in THF, 303, 307
   With LiB(Et)$_3$H, 307, 319
   With Mg, 319
   With NO$_2$BF$_4$, 310
   With organolithium reagents, 307, 319, 321
   With OsO$_4$, 134, 318
   With phenyldiazomethane, 321, 322
   With phosphites, 319
   With polystyrene and AlCl$_3$, 312
   With strong Brønsted acids, 306, 310
   With TMSCl and Mg, 307
   With XeF$_2$, 29, 307
Ring currents, 326
Selective functionalization, 323
Silylation, 307
Spectroelectrochemical studies, 290
STM imaging, 127, 133
Structure,
   By computational methods, 104, 135, 137,
   193
   By spectral methods, 134
   Electronic, 137, 185, 199, 217, 219
   Solid state, 42, 44, 134, 201, 219
Systematic functionalization, 322
Theoretical studies, 103, 125, 219, 306
Thermogravimetric measurements, 35
Triplet state, 14
UV–VIS spectrum, 14, 35, 36, 37, 70
van der Waals adducts, 14
X-ray diffraction, powder, 35, 39
Buckyonions, 111, 115
Buckytori, 95, 120
Buckytubes, 48, 50–52, 77, 96, 111, 115, 127,
   159
   Band gaps, predicted, 164
   Elastic modulus, 170
   Energetics, 115, 168
   HMO studies, 97
   Preparation, 175
   Properties, predicted, 159, 164
   Structures, 97, 111, 159

Carbon
   Allotropes, 148
   Clusters, 3, 23
   Fibers, 47, 111
   Microparticles, 47, 49
   Surfaces,
      genus, 84, 95, 97, 120
      with negative curvature, 97, 118
   Tubules. *See* Buckytubes
Charm Bracelets, 327
Circuits. *See* conjugated circuit theory
Clar Sextet Criterium, 94

Clusters
  Carbon, 3, 23
  Mass Spectrometry, 1
Conjugated Circuit Theory, 88
  vs. Hückel theory, fullerene structures, 89
Corannulene, 64, 98
Cyclic Voltammetry, 69, 71, 286, 318, 323,
    327, 328

Descartes Relation, 90
Dodecahedrane, 40, 117, 175

Electrochemical studies, 69, 71, 285
Electron affinity,
  C_{60}, 134, 138, 240, 257
  Metallofullerenes, 6
Eliahsberg equations, 205–7
Endohedral complexes, 6, 22, 92, 117, 127,
    148, 217, 239, 245, 257, 259
  Centered, 239
  Electronic structure, 217, 239, 245
  Evidence for, 259
  Neutralization-reionization, 275
  Off-centered, 242
  Preparation, 246, 259, 261
  Relativistic effects in, 130
  Solid state, 244
  Spectra, 7, 246, 257, 261
Euler's theorem, 61, 84, 90, 120, 185
Exohedral complexes, 134, 151, 257, 296, 318
Ethylene, model for fullerene structure, 188

Fibers, carbon, 47, 111
Fullerenes
  Adsorbed on metal surfaces, 247
  Anions or fullerides, 45
  Aromaticity of, 185
  Clar sextet, 94
  Chemistry, 29, 301
  Chromatography, 25, 26, 59, 62, 302, 317,
    321
  Complexed, 194
  Conjugation in, 186
  Crystal structures
    C_{60}, 42, 44
    C_{70}, 43
  Cyclic voltammetry, 69, 71, 286, 318
  Electrochemistry, 69, 71, 285, 286
  Definition, 83, 120
  Detection, 3, 4
  Doped. *See* Fullerides
  Fluorination, 29
  Formation
    Laser vaporization of graphite, 3
    Macroscopic quantities, 23, 25, 59, 71
    Possible mechanisms of, 98, 155, 173
      molecular dynamics simulation, 155
    Temperature dependence, 157
  Fractals of, 120
  Free, 194

Giant, 23, 25, 111, 135, 142
  STM imaging, 111, 127
  structures, ab initio, 110
Graphs, 85
Higher, 5, 59, 93, 127, 135, 142, 174
  Clar sextet, 94
  Cyclic voltammetry, 69, 71, 292
  Electrochemistry, 292
  HOMO–LUMO gap in, 69
  Isomers, HMO study, 61, 75
  NMR studies, 61–69
  Photoelectron spectra, predicted, 143
  Preparation, resistive vs. arc heating, 73
  Separation, 60
  Structural motifs in, 63, 64
  Structure, 61, 67, 75, 93, 106, 111
  UV–VIS spectra, 70
Hybridization, 185
Hyper, 120
In interstellar matter/space, 8, 15, 50, 289
Ionization potential, 11
Ion–molecule reactions, 257
Lower (or smaller), 23, 24, 35, 92, 135, 175
  Hydrogenated, 40, 41
Mass spectrometry, 1, 26, 72
Monolayers, 247
NMR studies, 23, 27, 28, 35, 61–69, 301
Oxidation, 289, 322
Oxides, 61, 322
Photofragmentation, 8, 98, 147, 257
  Possible mechanisms of, 75, 99
Preparation, Macroscopic Quantities, 23, 25,
    59, 71, 301, 317
Proton affinity, 257
Pyramidalization angles, 193
Reactor design, 302, 317
Resonance,
  And conjugated circuit theory, 88
  Stabilization, 87
  (Kekulé) structures, 5, 88, 93
Solid State Studies, 40
Spectroelectrochemical studies, 290
Spectroscopy. *See also* specific spectrosco-
    pies, 9, 133–34, 317
Stability, 26, 29, 154
Stone–Wales rearrangement in, 75, 99, 137
Strain, 89
Structure, 83, 103, 125, 185
  Angle defect, 90
  C_{28}, theory, 116
  C_{60}, experimental, 42, 44, 106
  C_{60}, theory, 105
  C_{70}, experimental, 43, 106, 121
  C_{70}, theory, 106
  C_{76}, experimental, 66, 67
  C_{78}, experimental, 66, 67
  C_{80}, theory, 111
  C_{84}, experimental, 68
  C_{120}, theory, 112–13
  C_{180}, theory, 114, 142

C_{240}, theory, 115, 142
Constraints, 6, 22, 61, 75, 84, 90, 91, 125, 185
Electronic, 103, 123, 185, 217
Prediction, 85, 92, 103, 125, 185
Theoretical Studies, 103, 125
Transient, 194
UV–VIS spectra, 14, 35, 36, 37, 69, 70
Fullerides
Alkali metal, 45, 199, 217, 303, 307
Electronic structure, experimental, 229
Electronic structure, theory, 218
Reaction with CH_3I, 307, 318
Solid state structure, 221, 226
Phonons, 202, 205, 207
Spectra, 47, 204, 207, 229, 291
Superconducting, 45, 127, 199, 225
Fullerite, 127, 148, 173
Fulleroids, 134, 296, 322
Aza, 326
Bi-, 327
Bond lengths, 323
Cyclic voltammetry, 323, 325, 327
Isomerization, 324, 327
Polymers, 327
Charm bracelets (CBP), 327–28
Cyclic voltammetry, 328
Dendrimers, 329
Pearl necklaces (PNP), 327
Polyesters, 328
Polyurethanes, 328
Thermogravimetric analysis (TGA), 328
Preparation, 322
Ring currents, 326
Spectra, 323–27
X-ray crystallography, 323
Fulleronium ions, 285, 289, 294, 306, 311

Gauss–Bonnet Theorem, 90
Graphitene, 83

Hubbard model, 212
Hückel Molecular Orbital Theory (HMO), Fullerene isomers, 61, 75, 86
Resonance stabilization, fullerenes, 87
vs. conjugated circuit theory, fullerene structures, 89
Hyperfullerenes, 120

Interstellar Bands, Diffuse (DIBs), 6, 50, 258, 289
Possible constituents of, 8, 258
Isolated Pentagon Rule (IPR), 22, 35, 61, 91, 112, 135, 158, 278
Isotope effects, 208, 211
Pressure dependence, 212

Kohn–Sham Equations, 218
Korringa relaxation mechanism, 235

Lonsdaleite, 148

Mass Spectrometry
Cluster beam, 1
Fullerenes, 1, 26, 72
Fullerene derivatives, 34, 41, 257, 301–13, 317–22
Lower fullerenes, 40
Metallofullerenes, 7, 257
Metallofullerenes
Electron affinity, 6
Formation, 6
Preparation, 261
Shrink-wrapping, 9, 126, 260
Spectra, 7, 246, 257, 261

Neutron
Diffraction, 40, 41
Scattering, inelastic, 41, 47
NMR Spectrometry
Buckminsterfullerene and derivatives, 23, 35, 228, 230, 235
Fluorinated C_{60}, 29, 30
Fullerenes, 23, 27, 28, 35, 61–69
Derivatives, 301–13, 317–29
Higher, 61–69

Pearl Necklaces, 327
Pentagon Isolation Rule. *See* Isolated Pentagon Rule (IPR)
Phonons, 202, 205
Anharmonic, 207
Photofragmentation
Fullerenes, energetics of, 8
Fullerenes, pattern of, 8, 98, 147
Mechanism, 10, 75, 99, 260
Plasma Frequency, 209, 228
Polarography, Differential Pulse (DPP), 286
Polyynes, 21
Pyracyclene, 318
Analogy, fullerene reactivity, 319, 321
Rearrangement. *See* Stone–Wales rearrangement

Reactor design, 302, 317
Resistivity, Temperature dependence, 209
Ring currents, 326

Scanning Tunneling Microscopy (STM), Of C_{60}, 127
Of giant fullerenes, 111, 127
Schlegel Diagrams, 31, 33, 53, 95
Schwartzons, 119
Shrink-wrapping, 9, 126, 175, 260, 271
Single Electron Transfer (SET), during addition reactions, 321
Spectroelectrochemistry, 290
Stone–Wales Rearrangement, 75, 99, 137, 174, 260

Superconductivity, 40, 45, 127, 199, 204, 217
  BSC–Eliahsberg model, 205, 226
  Electron–phonon interaction, 204, 226
  Hubbard model, 212
  Isotope effects, 208, 211
  Strong electron correlation, 212

Thermogravimetric measurements, 35, 328
Theoretical Studies, 103, 125, 185, 199, 217, 239
  Electronic structure, 131, 185, 199, 217, 239
  Endohedral complexes, 239, 245
  Molecular dynamics, 132

Prior to 1990, 135
  Superconductivity, 199, 217
  TURBOMOLE software, 104
Transfer Matrix Method, 84
Tubules. *See* buckytubes,

van der Waals adducts, 13
  C_{60} with benzene, 14

Willmore Conjecture, 89

Ziman Equation, 209